Nonlinear Dynamics of Reservoir Mixtures

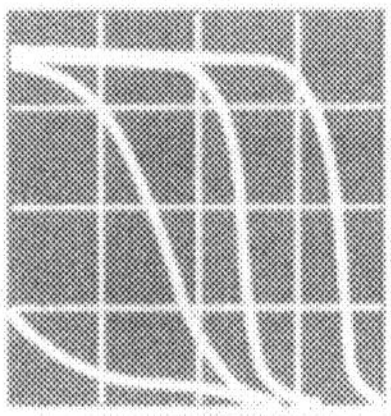

Vladimir S. Mitlin

CRC Press
Boca Raton Ann Arbor London Tokyo

Library of Congress Cataloging-in-Publication Data

Mitlin, Vladimir S. (Vladimir Solomon), 1959-
Nonlinear dynamics of reservoir mixtures / Vladimir S. Mitlin.
p. cm.
Includes bibliographical references and index.
ISBN 0-8493-4416-6 (alk. paper)
1. Oil reservoir engineering--Mathematical models. 2. Fluid dynamics--Mathematical models. 3. Two-phase flow--Mathematical models. I. Title.
TN871.M58 1993
622'.3382--dc20 92-34359
CIP

Direct all inquiries to CRC Press, Inc., 2000 Corporate Blvd., N.W., Boca Raton, Florida 33431.

International Standard Book Number 0-8493-4416-6

Library of Congress Card Number 92-34359

Printed in the United States of America 1 2 3 4 5 6 7 8 9 0

Printed on acid-free paper

To my parents,
Mina and Solomon Mitlin

TABLE OF CONTENTS

1

Introduction

Natural gas, condensate, and oil are complex hydrocarbon mixtures with additions (sometimes essential) of non-hydrocarbon components. In the course of the development of an oil-gas-condensate formation, the reservoir flow of such a system is accompanied by several phase transitions which consist of the evaporation of oil (condensate) into the gaseous phase, forming the liquid phase, and so on. The summary effect of all the processes on the actual conditions of natural (extraction) and artificial (injection of different kinds of active agents) action on a formation, determines the efficiency of hydrocarbon recovery.

The mathematical description of the processes enumerated is based on the theory of multicomponent reservoir flow (TMRF). The equations of TMRF remain nonlinear even for the simplest suppositions about the mechanism of reservoir flow. The nonlinearity of the process can lead to the formation of some typical spatially inhomogeneous flow structures (discontinuous structures, autowaves, etc.). Another peculiarity of the mathematical description is connected with the presence of two different space scales: first is the characteristic macroscopic flow scale and second is the characteristic pore size. To obtain a macroscopic hydrodynamic model, one can average the equations of fluid motion in pores at scales essentially exceeding the microscopic scale. One should expect that the microscopic scale is excluded from consideration by this procedure. However, when closing the macroscopic model one has to take into account the processes at the pore scale. At least, that is necessary when determining the phase permeabilities; in ''bad'' cases (collectors with low porosity) all the thermodynamic properties of reservoir fluid depend strongly on its interactions with the porous medium, whose influence can lead to qualitative changes in the phase state of the fluid.

The first goal of TMRF is the study of dynamic equations at the macrolevel, the second is obtaining the thermodynamic and rheological properties of reservoir systems at the microlevel, for the closure of macro-models. There are probably two main approaches to TMRF: (1) solving only those problems which can be solved with mathematical accuracy and (2) solving all the necessary problems and disregarding mathematical accuracy. Despite the efforts of researchers working in both directions, the hydrocarbon production is still non-zero.

The present monograph is devoted to the nonlinear problems of the theory of two-phase multicomponent reservoir flow and is based mainly on the results of studies carried out by the author during the last ten years. The outline of the book was formed as a result of preparation for the course of lectures in

"Mathematical Modeling of Multicomponent Reservoir Flow" given by the author to the graduate students and research workers of the Institute of the Oil and Gas Problems of Russian Academy of Sciences in 1989.

This monograph consists of ten chapters; references are given after each chapter. There has been no attempt to give a full review of the theoretical and experimental investigations concerned with multicomponent reservoir flow. To fill this gap, at least partially, a set of references from the last 40 years pertaining to TMRF is given in chronological order at the end of Chapter 1. The author extends apologies to researchers whose results have not been cited.

The complexity of the differential equations of TMRF, analytical study of which is possible in only a few cases, makes it necessary to use the technique of numerical modeling. We will discuss the questions of numerical modeling in conformity with the generally used large-scale approximation of TMRF. The full sequence of the stages of numerical modeling includes an accurate statement of the problem (Chapters 2 and 3), development of the algorithm and testing the computer program (Chapter 4), comparing the results of numerical modeling with the laboratory experiments (Chapter 5), and calculation of the reservoir flow processes for the real hydrocarbon formations (Chapter 6).

Chapter 7 discusses the theory of the autowave regimes of multicomponent reservoir flow. The fundamental point here is applying the ideas and methods of synergetics to TMRF. The general construction of synergetics — the loss of stability of the main system state at a certain critical value of external parameters and then forming a new stabilized state due to system nonlinearity — occurs in many areas of natural science. Therefore, using the "synergetic" approach to TMRF seems to be necessary. In Chapter 8 we consider another, quite new, class of synergetic phenomena — retardations in kinetically stable structures on the way to equilibrium, and discuss in detail some models where such spatially inhomogeneous structures appear (evolution of the discontinuous multicomponent reservoir flow structure by the inversion of the flow rate, retardation at the decomposition of the micro-structures by phase separation, and the dynamics of film rupture).

Chapter 9 is dedicated to the consideration of some surface phenomena using the gradient theory methods. The coalescence kinetics of embryos of the new phase, the thermodynamic properties of the thin liquid interlayer between two solids, changing the phase diagram of reservoir fluid due to the influence of porous media, and reservoir phase mobilities near the thermodynamic critical point are discussed. To analyze these nonlinear problems, the asymptotical technique is used.

The problems of TMRF requiring further development are enumerated in Chapter 10.

Where it is possible, consideration is not restricted to the framework of "petroleum" applications and an attempt is made to show the generality of the mathematical description to different natural objects [for example, the kinetics of phase transitions in fluids and in rock minerals (Sections 8.2, 9.1, and 9.2)].

ACKNOWLEDGMENTS

The book was written in 1990–1992 during my work, first, at the Laboratory AS of Theoretical Geophysics of the Institute of the Earth Physics of Russian Academy of Sciences, and then at the Departments of Petroleum Engineering and Chemical Engineering of the University of Texas at Austin. I would like to express my deep gratitude to I. Ya. Erukhimovich, N. A. Guzhov, B. V. Makeev, L. I. Manevich, V. N. Nikolaevskii, I. C. Sanchez, R. S. Schechter, and M. M. Sharma for numerous and useful discussions of the points considered in the book. Financial support of some investigations described in the book was provided by EOGRRP (Section 9.3) and AFOSR (Sections 8.4 and 9.4).

REFERENCES

1. **Muscat, M.,** *Physical Principles of Oil Production,* McGraw-Hill, New York, 1949.
2. **West, W. J., Garvin, W. W., and Sheldon, J. W.,** *Trans. Soc. Pet. Eng. AIME,* 201, 217, 1954.
3. **Bird, R. B., Steward, W. E., and Lightfood, E. N.,** *Transport Phenomena,* John Wiley & Sons, New York, 1960.
4. **Collins, R. E.,** *Flow of Fluids Through Porous Materials,* Van Nostrand Reinhold, New York, 1961.
5. **Welge, H. J., Jonson, E. F., Ewing, S. P., et al.** *J. Pet. Technol.,* 13(8), 787, 1961.
6. **Stone, H. L. and Garder, A. O.,** *Trans. Soc. Pet. Eng. AIME,* 222, 92, 1961.
7. **Zheltov, Yu. P. and Rozenberg, M. D.,** Filtration of multicomponent systems, in *Sci.-techn. Collected Articles in Oil Production (Nauchno-Tekhnicheskii Sbornik po Dobyche Nefti),* No. 18, Moscov, 1962, 9 (in Russian).
8. **Nikolaevskii, V. N.,** *Inzh. Zh.,* 12(3), 557, 1963, (in Russian).
9. **Kniazeff, V. J. and Naville, S. A.,** *Soc. Pet. Eng. J.,* 5(1), 37, 1965.
10. **Price, H. S. and Donohue, D. A. T.,** *Soc. Pet. Eng. J.,* 7(2), 205,1967.
11. **Nikolaevskii, V. N., Bondarev, E. A., Mirkin, M. I., et al.,** *Motion of Hydrocarbon Mixtures in a Porous Medium,* Nedra, Moscow, 1968 (in Russian).
12. **Rozenberg, M. D., Kundin, S. A., Kurbanov, A. K., et al.,** *Flow of a Gasified Liquid and Other Multicomponent Mixtures in Oil Reservoirs,* Nedra, Moscow, 1969 (in Russian).
13. **Culham, W. E., Faroug Ali S. M. and Stahl C. D.,** *Soc. Pet. Eng. J.,* 9(3), 323, 1969.
14. **Roebuck, J. F., Henderson, G. E., Douglas, J. J., et al.,** *Soc. Pet. Eng. J.,* 9(1), 115, 1969.
15. **Afanas'ev, E. F., Nikolaevskii, V. N. and Somov, B. E.,** Problem of the displacement of a multicomponent hydrocarbon mixture by gas injection, in *Theory and Practice of Petroleum Production,* Nedra, Moscow, 1971, 107 (in Russian).
16. **Van Quy, N., Simandoux, P. and Corteville, J.,** *Soc. Pet. Eng. J.,* 12(2), 171, 1972.
17. **Murkes, M. I., Rozhdestvenskii, V. A. and Shovkrinskii, G.Yu.,** *Zh. Vychisl. Mat. i Mat. Fiz.,* 17(3), 696, 1977; *USSR Comput. Math. Phys.,* 17(3), 133, 1977.
18. **Aziz, K. and Settari, A.,** *Petroleum Reservoir Simulation,* Applied Science Publishers, London, 1979.
19. **Coats, K. H.,** *Soc. Pet. Eng. J.,* 20(5), 363, 1980.
20. **Entov, V. M.,** *Usp. Mekhan.,* 4(3), 41, 1981 (in Russian).

21. **Bedrikovetskii, P. G.,** *Dokl. Akad. Nauk SSSR,* 262(1), 49, 1982; *Sov. Phys. Dokl.,* 27(1), 14, 1982.
22. **Mehra, R. K., Heidemann, R. A. and Aziz, K.,** *Soc. Pet. Eng. J.,* 22(1), 61, 1982.
23. **Panfilov, M. B.,** *Izv. Akad. Nauk SSSR, Mekh. Zhid. Gaza,* No. 4, 1985, 94; *Fluid Dyn. (USSR),* 20(4), 574, 1985.
24. **Levi, B. I. and Shakirov, H. G.,** *Izv. Akad. Nauk SSSR, Mekh. Zhid. Gaza,* No. 4, 1985, 101; *Fluid Dyn. (USSR),* 20(4), 581, 1985.
25. **Zazovskii, A. F.,** *Izv. Akad. Nauk SSSR, Mekh. Zhid. Gaza,* No. 5, 1985, 116; *Fluid Dyn. (USSR),* 20(3), 433, 1985.
26. **Lake, L. W.,** Enhanced Oil Recovery; Prentice-Hall, Englewood Cliffs, NJ, 1989.
27. **Mitlin, V. S.,** *J. Fluid. Mech.,* 222, 369, 1990.
28. **Barenblatt, G. I., Entov, V. M. and Ryzhik, V. M.,** *Theory of Fluid Flows Through Porous Media,* Klumer Academic Publ., Dordrecht, 1990.
29. **Nikolaevskii, V. N.,** *Mechaniks of Porous and Fractured Media,* Singapore World Science Publ., 1990.

2

Statement of the Problem of Two-Phase Multicomponent Reservoir Flow

The main equations of the equilibrium reservoir flow of multicomponent systems were proposed approximately 30 years ago in publications by Bird, Steward, and Lightfood,[1] Collins,[2] Zheltov and Rozenberg,[3] and Nikolaevskii.[4,5] The equations used by these authors differ in general by the choice of fundamental variables. For example, Zheltov and Rozenberg used $l - 2$ mass fractions of components in the liquid phase as such independent variables,[3] Collins used $l - 1$ overall mole fractions of the mixture components,[2] and Nikolaevskii preferred the ratio of the mass fraction of the intermediate component in the liquid phase to the sum of mass fractions of the intermediate and heavy components in the liquid phase (for three-component systems).[4] General thermodynamic analysis of the question about the number of independent variables in the multicomponent multiphase reservoir flow models, with and without taking into account the capillary forces, was carried out by Nikolaevskii based on the Gibbs approach.[5]

2.1. MATHEMATICAL MODELS CONSIDERED

A mathematical description of a multiphase multicomponent reservoir flow residues in the solution of the boundary problems for a system of nonlinear differential equations with partial derivatives, which is quite complicated even when using the simplifying assumptions about the mechanism of the process. For the most often used suppositions of the local thermodynamic phase equilibrium, correctness of the generalized Darcy law,

$$\mathbf{v}_{g,w} = -k\frac{f_{g,w}}{\eta_{g,w}}(\nabla P - \rho_{g,w}\mathbf{g}) \qquad (2.1.1)$$

and negligibly small influence of the capillary and diffusion forces, the isothermal reservoir flow of a *l*-component mixture is described by the following system of equations

$$\nabla(k\beta_i(\nabla P - \psi_i \mathbf{g})) = m\frac{\partial N z_i}{\partial t}$$

$$i = 1, \ldots, l$$

where

$$\beta_i = z_i\left(\frac{f_g\rho_g\mathcal{K}_i}{\eta_g M_g} + \frac{f_w\rho_w}{\eta_w M_w}\right)[1 + V_g(\mathcal{K}_i - 1)]^{-1}$$

$$\psi_i = \left(\frac{f_g\rho_g^2\mathcal{K}_i}{\eta_g M_g} + \frac{f_w\rho_w^2}{\eta_w M_w}\right)\left(\frac{f_g\rho_g\mathcal{K}_i}{\eta_g M_g} + \frac{f_w\rho_w}{\eta_w M_w}\right)^{-1}$$

$$N = S\frac{\rho_w}{M_w} + (1 - S)\frac{\rho_g}{M_g}$$

$$S = S_w = 1 - S_g \tag{2.1.2}$$

$\mathbf{g}$ is the gravity force acceleration; $\mathbf{v}_{g,w}$ are the reservoir flow rates of the gaseous and liquid phases; $P(\mathbf{r},t)$ is the pressure; $z_i(\mathbf{r},t)$ is the mole fraction of the i-th component in the mixture (overall fraction),

$$\sum_{i=1}^{l} z_i = 1$$

$k(\mathbf{r})$, $m(\mathbf{r})$ are the absolute permeability and porosity; $f_{g,w}$ are the relative phase permeabilities; $\rho_{g,w}$ and $\eta_{g,w}$ are the densities and viscosities of the phases; x_i, y_i are the mole fractions of the i-th component in the liquid and gaseous phases,

$$\sum_{i=1}^{l} y_i = \sum_{i=1}^{l} x_i = 1$$

$M_{g,w}$ are the mean molecular weights of the phases,

$$M_g = \sum_{i=1}^{l} y_i M_i$$

$$M_w = \sum_{i=1}^{l} x_i M_i$$

M_i is the molecular weight of the i-th component; $\mathcal{K}_i$ is the equilibrium constant of the i-th component; S is the liquid saturation of the porous medium; V_g is the mole fraction of the gaseous phase; $\mathbf{r} = (x,y,z)$ is the radius-vector of a point of medium; t is the time.

The quantities introduced above obey the following additional relations

$$y_i = \frac{z_i \mathcal{K}_i}{1 + V_g(\mathcal{K}_i - 1)} \tag{2.1.3}$$

$$x_i = \frac{z_i}{1 + V_g(\mathcal{K}_i - 1)} \tag{2.1.4}$$

$$S = \frac{(1 - V_g)\rho_g M_w}{(1 - V_g)\rho_g M_w + V_g \rho_w M_g} \tag{2.1.5}$$

$$\sum_{i=1}^{l} \frac{(\mathcal{K}_i - 1)z_i}{1 + V_g(\mathcal{K}_i - 1)} = 0 \tag{2.1.6}$$

The densities and viscosities of the phases are functions of the pressure and phase compositions, the equilibrium constants are derived as the functions of the pressure and the overall composition, and the relative phase permeabilities can depend both on the saturation and on other parameters of the mixture (say, composition).

The quantities to be determined in the model are the pressure P and the overall mole fractions of the components. Such a set of the basic variables has superiorities compared to others (for example, to the pressure and $l - 2$ fractions of the components in a phase), since in this case there is no necessity to outline specially by numerical modeling the flow regions which do not contain the phase corresponding to the basic variables. This allows us to apply the "through" finite difference methods to work out the numerical model.[6] The use of mole fractions is connected with the traditional character of representing the calculation results of the vapor-liquid equilibrium in the "mole" form.

Equations (2.1.2) are the balance relations for the amount of every component expressed in the differential form. Equations (2.1.3) and (2.1.4) express the connection between the component fractions in both phases and the overall component fractions. Equation (2.1.5) establishes the connection of the mole (V_g) and volume (S) phase fraction. The mole fraction of the gaseous phase is derived from solving Equation (2.1.6). If

$$\sum_{i=1}^{l} z_i \mathcal{K}_i \leq 1$$

then $V_g = 0$ (the liquid state of mixture); if

$$\sum_{i=1}^{l} \frac{z_i}{\mathcal{K}_i} \leq 1$$

then $V = 1$ (the gaseous state of mixture); if both of the conditions do not hold,

V_g is calculated directly by Equation (2.1.6) (in this case one can easily prove that only the root V_g exists within the interval (0, 1)).

Let us summarize the equations for the components and replace the last one by the obtained one. Then system (2.1.2) is equivalent to the following:

$$\nabla(k\beta_i(\nabla P - \psi_i \mathbf{g})) = m\frac{\partial N z_i}{\partial t}, \quad i = 1, \ldots, l - 1$$

$$\nabla(k\beta(\nabla P - \psi \mathbf{g})) = m\frac{\partial N}{\partial t} \tag{2.1.7}$$

where

$$\beta = \sum_{i=1}^{l} \beta_i, \quad \psi = \sum_{i=1}^{l} \beta_i \psi_i / \sum_{i=1}^{l} \beta_i$$

Equations (2.1.7) describe the balance of the mixture in the differential form.

Consider some consequences of system (2.1.7) which will be used further. In the case of a sufficiently thin horizontal seam, if the vertical motion of the fluid is neglected, Equations (2.1.7) transform to the following

$$\nabla(kh\beta_i \nabla P) = mh\frac{\partial N z_i}{\partial t}, \quad i = 1, \ldots, l - 1$$

$$\nabla(kh\beta \nabla P) = mh\frac{\partial N}{\partial t} \tag{2.1.8}$$

Here $h(\mathbf{r})$ is the thickness of the seam in the point $\mathbf{r} = (x,y)$, and all the quantities are understood in the sense of thickness-averaging.

In the case of a symmetric flow region (one space variable) Equations (2.1.7) are written in the form

$$\frac{1}{r^\alpha}\frac{\partial}{\partial r}\left(r^\alpha k\beta_i \frac{\partial P}{\partial r}\right) = m\frac{\partial N z_i}{\partial t}, \quad i = 1, \ldots, l - 1$$

$$\frac{1}{r^\alpha}\frac{\partial}{\partial r}\left(r^\alpha k\beta \frac{\partial P}{\partial r}\right) = m\frac{\partial N}{\partial t} \tag{2.1.9}$$

Here, r is the space coordinate; $\alpha = 0$ stands for the linear flow, $\alpha = 1$ for the plane-radial flow, $\alpha = 2$ for the spherically radial flow.

To construct a closed system for the multicomponent reservoir flow equations, it is necessary to define the phase densities, the phase viscosities, the equilibrium constants, and the relative phase permeabilities. Below, the functions $\rho_{g,w}$, $\eta_{g,w}$ will be used in the form

$$\rho_g = \rho_g^o\left(\frac{P}{P^o}\right)^{\lambda_1} \cdot \left(\frac{M_g}{M_g^o}\right)^{\lambda_2}, \quad \rho_w = \rho_w^o\left(\frac{P}{P^o}\right)^{\lambda_3} \cdot \left(\frac{M_w}{M_w^o}\right)^{\lambda_4}$$

$$\eta_g = \eta_g^o\left(\frac{P}{P^o}\right)^{\lambda_5} \cdot \left(\frac{M_g}{M_g^o}\right)^{\lambda_6}, \quad \eta_w = \eta_w^o\left(\frac{P}{P^o}\right)^{\lambda_7} \cdot \left(\frac{M_w}{M_w^o}\right)^{\lambda_8} \tag{2.1.10}$$

Here the index '0' relates to the a initial state of the reservoir system, λ_j are constant.

The equilibrium constants are represented by the polynomial dependences on the pressure and the composition parameter R defined in the form

$$R = \frac{z_{int}}{z_{int} + z_{he} + d_{tr}} \tag{2.1.11}$$

where z_{int} and z_{he} are the fractions of intermediate and heavy components correspondingly, d_{tr} is constant. Notice that the parameters in the form of Equation (2.1.11) were used earlier in describing the three-component reservoir flow.[5,7]

The phase permeabilities are derived in two forms. Firstly, the dependence of $f_{g,w}$ on the phase saturations only will be used. Secondly, it is known that for miscible displacement modeling it is important to take into account the dependence of $f_{g,w}$ on the reservoir mixture composition. We will consider the displacement of the reservoir gas-condensate mixture by the enriched gas. This process is characterized by a sharp decrease of surface tension. Therefore, the composition dependence of $f_{g,w}$ can be defined in terms of the surface tension σ taken in the form

$$\sigma^{1/4} = \sum_{i=1}^{l}(P_{ch})_i\left(x_i\frac{\rho_w}{M_w} - y_i\frac{\rho_g}{M_g}\right) \tag{2.1.12}$$

Here $(P_{ch})_i$ is the parachor of the i-th component.[5] A thorough discussion of the rearrangement of the phase permeabilities by miscible displacement and the concrete form of the dependences of $f_{g,w}$ on σ are given in Chapter 5; see also the consideration in Section 9.4.

Sometimes, by the calculation of the flow in the near-zone of the well, one has to take into account the dependence of the reservoir flow resistance coefficient on the flow rate. In the simplest form this effect is described by the quadratic flow law for both phases (in the case of a one-dimensional flow)

$$-\frac{\partial P}{\partial r} = \frac{\eta_{g,w}}{k\, f_{g,w}} v_{g,w} + \frac{\rho_{g,w}}{\mathcal{H}} v_{g,w}^2 \operatorname{sign} v_{g,w} \tag{2.1.13}$$

where $\mathcal{H}$ is the macro-roughness coefficient. Generalization of the numerical method, in the event of nonlinear Darcy law (2.1.13), will be considered in Chapter 4.

To obtain an unambiguous solution of the system in a region $\{\mathbf{r} \in \mathcal{G}, t \geq t_0\}$ it is necessary to impose the initial and boundary conditions. Analysis of the existing results showed the necessity of additional investigations to form a basis for the statement of the multidimensional boundary problems of the multicomponent reservoir flow. The corresponding methods for the correct imposition of the boundary conditions are necessary in order to work out the correct numerical method. The questions of imposing boundary conditions will be considered in Chapter 3, using the characteristics method. A description of the numerical method is presented in Chapter 4.

2.2. DISCUSSION OF THE STARTING HYPOTHESES

Let us discuss the physical suppositions, for which the above equations can be used.

Isothermal character of the process — A porous medium possesses a large specific surface. One can say that, regarding the formation fluid, porous media play the role of thermostat: the local thermal fluctuations in fluid have slight influence on the mean formation temperature of the reservoir system, which is defined by the porous medium itself. Therefore, in the study of the multicomponent reservoir flow without special thermal action on the seam, the heat transfer factor usually can be neglected.

Local thermodynamic equilibrium — Usually, the flow of a hydrocarbon mixture in a formation has a rate sufficiently slow for one to admit the correctness of the local equilibrium condition. This means that the characteristic time for the establishment of equilibrium in a elementary macrovolume of the porous medium is sufficiently smaller than the characteristic motion time of a mixture particle in the flow through the macrovolume.[5]

Neglecting the diffusion transfer — Omitting the molecular and convective diffusion terms is based on comparing the contributions of the "diffusion" and "Darcy" terms in the reservoir flow equations. The ratio of the contributions (the Pecle number) is of order of the ratio of the porous medium inner scale to the characteristic space flow scale, and is usually essentially less than 1.[8]

Neglecting the capillary forces — For the two-phase reservoir flow equations the contribution of the capillary forces in the flow process is defined by the so-called capillary number,[8] which usually is a small parameter in the order of 10^{-2} to 10^{-4}. Therefrom, one can make a conclusion about the applicability of the *large-scale approximation* in the reservoir flow theory: the corresponding equations allow acquisition of information about the structure of almost the entire flow (except in the narrow zones with the greatest values of the saturation gradient, corresponding to the discontinuity surfaces in the large-scale approximation).

It is important that the set of the hypotheses singled out as the large-scale approximation defines a minimal amount of the information which is necessary for the mathematical modeling and which could be obtained with sufficient precision from the preliminary experimental investigations.[8] Actually, even obtaining the diffusion coefficients of the complex hydrocarbon systems leads to the principal troubles in the statement and the carrying out of the corresponding physical experiments. The set of the interphase mass transfer coefficients, which appears by taking into account the nonequilibrium effects, also cannot be obtained experimentally for complex cases. Therefore, Equations (2.1.1) through (2.1.12), which contain the minimal set of input parameters for the description of a two-phase multicomponent reservoir flow, appear to be the most applicable for the purposes of numerical modeling.

REFERENCES

1. **Bird, R. B., Steward, W. E., and Lightfood, E. N.,** *Transport Phenomena,* John Wiley & Sons, New York, 1960.
2. **Collins, R. E.,** *Flow of Fluids Through Porous Materials,* Van Nostrand Reinhold, New York, 1961.
3. **Zheltov, Yu. P. and Rozenberg, M. D.,** Filtration of multicomponent systems, in *Sci.-Technical Collected Artickes in Oil Production (Nauchno-Tekhnicheskii Sbornik po Dobyche Nefti),* No. 18, Moscow, 1962, 9 (in Russian).
4. **Nikolaevskii, V. N.,** *Ingenernyi Zh.,* 12(3), 557, 1963 (in Russian).
5. **Nikolaevskii, V. N., Bondarev, E. A., Mirkin, M. I., et al.,** *Motion of Hydrocarbon Mixtures in Porous Media,* Nedra, Moscow, 1968 (in Russian).
6. **Aziz, K. and Settari, A.,** *Petroleum Reservoir Simulation,* Elsevier, London, 1980.
7. **Van Quy, N., Simandoux, P., and Corteville, J.,** *Soc. Pet. Eng. J.,* 12(2), 171, 1972.
8. **Barenblatt, G. I., Entov, V. M., and Ryzhik, V. M.,** *Motion of Fluids in Natural Rocks,* Klumer Academic Publ., Dordrecht, 1990.

3

A Study of the Statement of the Multidimensional Boundary Problems of Multicomponent Reservoir Flow by the Method of Characteristics

3.1. BASING THE STATEMENT OF TWO-DIMENSIONAL BOUNDARY PROBLEMS

In this section the analysis of the statement of the boundary problems for system (2.1.8) is carried out. For this purpose, the multidimensional method of characteristics will be used. Notice that the investigations concerned with basing the statement of some multiphase reservoir flow boundary problems were carried out by Lee and Fayers.[1] As will be clear from the following consideration, analysis of the question for system (2.1.8) becomes increasingly more complicated because of the multidimensional character of the problem.[2,3]

Let us represent Equation (2.1.8) in the form

$$\nabla(\beta_i B \nabla P) = A\frac{\partial N z_i}{\partial t}, \quad i = 1, \ldots, l-1$$

$$\nabla(\beta B \nabla P) = A\frac{\partial N}{\partial t}, \qquad (3.1.1)$$

and, continuing the transformations, in the form

$$B\sum_{j=1}^{l-1} a_{ij}\nabla z_j \nabla P + B b_i \nabla^2 P = AN\frac{\partial z_i}{\partial t} + B c_i(\nabla P)^2 - b_i \nabla B \cdot \nabla P, \quad i = 1, \ldots, l-1$$

$$B\sum_{j=1}^{l-1} \frac{\partial \beta}{\partial z_j}\nabla z_j \nabla P + B\beta\nabla^2 P = A\frac{\partial N}{\partial P}\frac{\partial P}{\partial t} +$$

$$+ A\sum_{j=1}^{l-1}\frac{\partial N}{\partial z_j}\frac{\partial z_j}{\partial t} - \beta\nabla B \cdot \nabla P - B\frac{\partial \beta}{\partial P}(\nabla P)^2 \qquad (3.1.2)$$

where

$$B(x,y) = kh,\ A(x,y) = mh,\ a_{ij} = \frac{\partial \beta_i}{\partial z_j} - z_i\frac{\partial \beta}{\partial z_j}\ ,\ b_i = \beta_i - z_i\beta\ ,$$

$$c_i = z_i\frac{\partial \beta}{\partial P} - \frac{\partial \beta_i}{\partial P}$$

And lastly, defining the new variables $\mathbf{w} = (w_x, w_y)$

$$w_x = \frac{\partial P}{\partial x}\ ,\ w_y = \frac{\partial P}{\partial y}$$

one can represent Equation (3.1.2) as a system of quasilinear first-order equations, which can be written in the matrix form

$$\boldsymbol{\Omega}_t\frac{\partial \mathbf{u}}{\partial t} + \boldsymbol{\Omega}_x\frac{\partial \mathbf{u}}{\partial x} + \boldsymbol{\Omega}_y\frac{\partial \mathbf{u}}{\partial y} = \mathbf{F} \tag{3.1.3}$$

where

$$\boldsymbol{\Omega}_t = \begin{pmatrix} -AN & \cdots & 0 & 0 & 0 & 0 \\ \vdots & & \vdots & \vdots & \vdots & \vdots \\ 0 & \cdots & -AN & 0 & 0 & 0 \\ -A\dfrac{\partial N}{\partial z_1} & \cdots & -A\dfrac{\partial N}{\partial z_{l-1}} & 0 & 0 & -A\dfrac{\partial N}{\partial P} \\ 0 & \cdots & 0 & 0 & 0 & 0 \\ 0 & \cdots & 0 & 0 & 0 & 0 \end{pmatrix}$$

$$\boldsymbol{\Omega}_x = \begin{pmatrix} w_x a_{11}B & \cdots & w_x a_{1,l-1}B & Bb_1 & 0 & 0 \\ \vdots & & \vdots & \vdots & \vdots & \vdots \\ w_x a_{l-1,1}B & \cdots & w_x a_{l-1,l-1}B & Bb_{l-1} & 0 & 0 \\ w_x\dfrac{\partial \beta}{\partial z_1}B & \cdots & w_x\dfrac{\partial \beta}{\partial z_{l-1}}B & B\beta & 0 & 0 \\ 0 & \cdots & 0 & 0 & -1 & 0 \\ 0 & \cdots & 0 & 0 & 0 & 0 \end{pmatrix}$$

$$\boldsymbol{\Omega}_y = \begin{pmatrix} w_y a_{11} B & \cdots & w_y a_{1,l-1} B & 0 & Bb_1 & 0 \\ \cdot & & \cdot & \cdot & \cdot & \cdot \\ \cdot & & \cdot & \cdot & \cdot & \cdot \\ \cdot & & \cdot & \cdot & \cdot & \cdot \\ w_y a_{l-1,1} B & \cdots & w_y a_{l-1,l-1} B & 0 & Bb_{l-1} & 0 \\ w_y \frac{\partial \beta}{\partial z_1} B & \cdots & w_y \frac{\partial \beta}{\partial z_{l-1}} B & 0 & B\beta & 0 \\ 0 & \cdots & 0 & 1 & 0 & 0 \\ 0 & \cdots & 0 & 0 & 0 & 1 \end{pmatrix}$$

$$\mathbf{F} = \begin{pmatrix} Bc_1(w_x^2 + w_y^2) - b_1 \nabla B \nabla P \\ \cdot \\ \cdot \\ \cdot \\ Bc_{l-1}(w_x^2 + w_y^2) - b_{l-1} \nabla B \nabla P \\ -B\frac{\partial \beta}{\partial P}(w_x^2 + w_y^2) - \beta \nabla B \nabla P \\ 0 \\ w_y \end{pmatrix}, \quad \mathbf{u} = \begin{pmatrix} z_1 \\ \cdot \\ \cdot \\ \cdot \\ z_{l-1} \\ w_x \\ w_y \\ P \end{pmatrix}$$

The next-to-last lines of these matrixes correspond to the equation

$$\frac{\partial w_x}{\partial y} - \frac{\partial w_y}{\partial x} = 0$$

Let it be necessary to find the solution of system (3.1.3) in the vicinity of a surface $\partial\mathcal{G}$ defined by the equation

$$\varphi_1(x, y, t) = 0$$

Following,[4] let us define the functions

$$\varphi_1 = \varphi_1(x, y, t) ,$$

$$\varphi_2 = \varphi_2(x, y, t) ,$$

$$\varphi_3 = \varphi_3(x, y, t) , \qquad (3.1.4)$$

so that in the vicinity of a given point (x,y,t) belonging to the surface $\partial\mathcal{G}$, the determinant of the matrix,

$$\begin{vmatrix} \frac{\partial\varphi_1}{\partial x} & \frac{\partial\varphi_1}{\partial y} & \frac{\partial\varphi_1}{\partial z} \\ \frac{\partial\varphi_2}{\partial x} & \frac{\partial\varphi_2}{\partial y} & \frac{\partial\varphi_2}{\partial z} \\ \frac{\partial\varphi_3}{\partial x} & \frac{\partial\varphi_3}{\partial y} & \frac{\partial\varphi_3}{\partial z} \end{vmatrix}$$

is non-zero.

The set of functions (3.1.4) can be considered as a new coordinate system, for which Equations (3.1.3) transform to

$$\left(\boldsymbol{\Omega}_t\frac{\partial\varphi_1}{\partial t} + \boldsymbol{\Omega}_x\frac{\partial\varphi_1}{\partial x} + \boldsymbol{\Omega}_y\frac{\partial\varphi_1}{dy}\right)\frac{\partial\mathbf{u}}{\partial\varphi_1} + \left(\boldsymbol{\Omega}_t\frac{\partial\varphi_2}{\partial t} + \boldsymbol{\Omega}_x\frac{\partial\varphi_2}{\partial x} + \right.$$

$$\left. + \boldsymbol{\Omega}_y\frac{\partial\varphi_2}{\partial y}\right)\frac{\partial\mathbf{u}}{\partial\varphi_2} + \left(\boldsymbol{\Omega}_t\frac{\partial\varphi_3}{\partial t} + \boldsymbol{\Omega}_x\frac{\partial\varphi_3}{\partial x} + \boldsymbol{\Omega}_y\frac{\partial\varphi_3}{\partial y}\right)\frac{\partial\mathbf{u}}{\partial\varphi_3} = \mathbf{F} \qquad (3.1.5)$$

For the solvability of the system with regard to $\partial\mathbf{u}/d\varphi_1$ the following condition has to be satisfied:

$$\det\boldsymbol{\Omega} \neq 0 \ , \ \boldsymbol{\Omega} = \boldsymbol{\Omega}_x\frac{\partial\varphi_1}{\partial x} + \boldsymbol{\Omega}_y\frac{\partial\varphi_1}{\partial y} + \boldsymbol{\Omega}_t\frac{\partial\varphi_1}{\partial t} \qquad (3.1.6)$$

The surface $\partial\mathcal{G}$, for which $\det\boldsymbol{\Omega} = 0$, is called the *characteristic surface*. For the existence of the solutions of system (3.1.3) near the characteristic surface, the additional compatibility condition has to be satisfied:

$$\text{rank}\boldsymbol{\Omega} = \text{rank}\boldsymbol{\Omega}^*$$

where $\boldsymbol{\Omega}^*$ is obtained from $\boldsymbol{\Omega}$ by adding the column

$$\mathbf{F}^* = \mathbf{F} - \left(\frac{\partial\varphi_2}{\partial t}\boldsymbol{\Omega}_t + \frac{\partial\varphi_2}{\partial x}\boldsymbol{\Omega}_x + \frac{\partial\varphi_2}{dy}\boldsymbol{\Omega}_y\right)\frac{\partial\mathbf{u}}{\partial\varphi_2} -$$

$$\left(\frac{\partial\varphi_3}{\partial t}\boldsymbol{\Omega}_t + \frac{\partial\varphi_3}{\partial x}\boldsymbol{\Omega}_x + \frac{\partial\varphi_3}{\partial y}\boldsymbol{\Omega}_y\right)\frac{\partial\mathbf{u}}{\partial\varphi_3}$$

From the point of view of basing the boundary conditions for system (2.1.8) or (3.1.3) one is interested in its solvability in a vicinity of the cylindrical surfaces defined by the relationships

$$\varphi_1(x, y) = 0 \ , \ \left(\frac{\partial\varphi_1}{\partial x}\right)^2 + \left(\frac{\partial\varphi_1}{\partial y}\right)^2 \neq 0$$

The boundary of the region of solution belongs to this class of surfaces.

Let us choose the local coordinate system in the form

$$\varphi_1 = \varphi_1(x, y) \ , \ \varphi_2 = \varphi_2(x, y) \ , \ \varphi_3 = t$$

where $\varphi_2(x,y) = 0$ is the curve, which is orthogonal to the curve $\varphi_1(x,y) = 0$ in a given point of boundary.

Since $\partial\varphi_1/\partial t = \partial\varphi_2/\partial t = 0$, the solvability condition (3.1.6) takes the form

$$B^l \frac{\partial\varphi_1}{\partial y}\left[\left(\frac{\partial\varphi_1}{\partial x}\right)^2 + \left(\frac{\partial\varphi_1}{\partial y}\right)^2\right]\left(\frac{\partial\varphi_1}{\partial x}w_x + \frac{\partial\varphi_1}{\partial y}w_y\right)^{l-1} \det\mathbf{\Omega}_1 \neq 0$$

where

$$\mathbf{\Omega}_1 = \begin{pmatrix} a_{11} & \cdots & a_{1,l-1} & b_1 \\ \vdots & & \vdots & \vdots \\ a_{l-1,1} & \cdots & a_{l-1,l-1} & b_{l-1} \\ \dfrac{\partial\beta}{\partial z_1} & \cdots & \dfrac{\partial\beta}{\partial z_{l-1}} & \beta \end{pmatrix}$$

Let us suppose $\det\mathbf{\Omega}_1 \neq 0$; then Equation (3.1.5) is solvable for $\partial\mathbf{u}/\partial\varphi_1$ in a vicinity of the boundary, if

$$\frac{\partial\varphi_1}{\partial x} w_x + \frac{\partial\varphi_1}{\partial y} w_y \neq 0$$

which is equivalent to $\partial P/\partial\mathbf{n} \neq 0$.

In this case, to obtain the solution of the problem near the sections of the boundary where the flow is directed inside the solution region, one has to give up $l - 1$ concentrations of components

$$\varphi_1(x, y) = 0 \ , \ \left.\frac{\partial P}{\partial\mathbf{n}}\right|_{\varphi_1=0} < 0 \ : \ z_i = z_i^!(x, y, t) \ , \ i = 1, \ldots, l-1 \qquad (3.1.7)$$

where $\mathbf{n}$ is the inner normal to the boundary.

The case of imposing the conditions

$$\varphi_1(x, y) = 0 \ : \ \frac{\partial P}{\partial\mathbf{n}} = 0 \qquad (3.1.8)$$

at the boundary is of special interest. In this situation the surface $\varphi_1(x,y) = 0$

is characteristic. The first $(l - 1)$ columns of matrix $\mathbf{\Omega}$ vanish, and rank $\mathbf{\Omega} = 3$.

For the existence of the solution to Equation (3.1.5), the compatibility conditions have to be satisfied

$$\text{rank} \left\| \mathbf{\Omega}, \mathbf{F} - \left(\frac{\partial \varphi_2}{\partial x} \mathbf{\Omega}_x + \frac{\partial \varphi_2}{\partial y} \mathbf{\Omega}_y \right) \frac{\partial \mathbf{u}}{\partial \varphi_2} - \mathbf{\Omega}_t \frac{\partial \mathbf{u}}{\partial t} \right\| = 3$$

This means that any determinant of the fourth-order matrixes collected by lines of the following matrix

$$\tilde{\mathbf{\Omega}} = \begin{pmatrix} Bb_1 \frac{\partial \varphi_1}{\partial x} & Bb_1 \frac{\partial \varphi_1}{\partial y} & 0 & F_1^* \\ \cdot & \cdot & \cdot & \cdot \\ \cdot & \cdot & \cdot & \cdot \\ \cdot & \cdot & \cdot & \cdot \\ Bb_{l-1} \frac{\partial \varphi_1}{\partial x} & Bb_{l-1} \frac{\partial \varphi_1}{\partial y} & 0 & F_{l-1}^* \\ B\beta \frac{\partial \varphi_1}{\partial x} & B\beta \frac{\partial \varphi_1}{\partial y} & 0 & F_l^* \\ \frac{\partial \varphi_1}{\partial y} & -\frac{\partial \varphi_1}{\partial x} & 0 & 0 \\ 0 & 0 & \frac{\partial \varphi_1}{\partial y} & w_y \end{pmatrix}$$

should vanish. Consequently, considering the existing possibilities, one can show that this condition is equivalent to the vanishing of determinants of the kind

$$\begin{vmatrix} b_i & F_i^* \\ b_j & F_j^* \end{vmatrix}, \quad \begin{vmatrix} b_i & F_i^* \\ \beta & F_l^* \end{vmatrix}$$

at all i and j. Therefore it follows that the vectors

$$\mathbf{B}^* = \begin{pmatrix} b_1 \\ \cdot \\ \cdot \\ \cdot \\ b_{l-1} \\ \beta \end{pmatrix}, \quad \mathbf{F}^* = \begin{pmatrix} F_1^* \\ \cdot \\ \cdot \\ \cdot \\ F_{l-1}^* \\ F_l^* \end{pmatrix}$$

are linearly dependent,

$$\mathbf{F}^* = \lambda \mathbf{B}^*$$

From the last line of the vector equality one can find the expression of λ. Substituting it into the other lines, we obtain

$$(w_x^2 + w_y^2)B\beta^2\frac{\partial}{\partial P}\left(\frac{\beta_i}{\beta}\right) + A\beta_i\frac{\partial N}{\partial t} - A\beta\frac{\partial Nz_i}{\partial t} +$$

$$+ B\left(\frac{\partial \varphi_2}{\partial x}w_x + \frac{\partial \varphi_2}{\partial y}w_y\right)\beta^2\sum_{j=1}^{l-1}\frac{\partial}{\partial z_j}\left(\frac{\beta_i}{\beta}\right)\frac{\partial z_j}{\partial \varphi_2} = 0 \,,$$

$$i = 1, \ldots, l-1 \qquad (3.1.9)$$

Choosing the coordinate function φ_2 so that $\partial/\partial\varphi_2$ gives the derivative along the boundary, and using the relationships

$$w_x^2 + w_y^2 = \left(\frac{\partial P}{\partial \varphi_2}\right)^2 \,, \quad w_x\frac{\partial \varphi_2}{\partial x} + w_y\frac{\partial \varphi_2}{\partial y} = \frac{\partial P}{\partial \varphi_2} \,,$$

$$\frac{\partial}{\partial \varphi_2} = \frac{\partial P}{\partial \varphi_2}\frac{\partial}{\partial P} + \sum_{j=1}^{l-1}\frac{\partial z_j}{\partial \varphi_2}\frac{\partial}{\partial z_j} \,,$$

let us represent Equation (3.1.9) in the form

$$B\beta\frac{\partial P}{\partial \varphi_2}\frac{\partial}{\partial \varphi_2}\left(\frac{\beta_i}{\beta}\right) = A\left(\frac{\partial Nz_i}{\partial t} - \frac{\beta_i}{\beta}\frac{\partial N}{\partial t}\right) \,, \quad i = 1, \ldots, l-1 \qquad (3.1.10)$$

In the particular case of a symmetric (plane-radial) flow $\partial P/\partial\varphi_2 = 0$, we obtain from Equation (3.1.10) the known relationships at the boundary of a circular, closed reservoir[5]

$$\frac{\partial (Nz_i)}{\partial P}\bigg/\frac{\partial N}{\partial P} = \frac{\beta_i}{\beta} \,, \quad i = 1, \ldots, l-1$$

In general, by satisfying the compatibility conditions, we obtain the equation

$$\begin{pmatrix} 0 \ldots 0 & Bb_1\frac{\partial \varphi_1}{\partial x} & Bb_1\frac{\partial \varphi_1}{\partial y} & 0 \\ \cdot & \cdot & \cdot & \cdot \\ \cdot & \cdot & \cdot & \cdot \\ \cdot & \cdot & \cdot & \cdot \\ 0 \ldots 0 & Bb_{l-1}\frac{\partial \varphi_1}{\partial x} & Bb_{l-1}\frac{\partial \varphi_1}{\partial y} & 0 \\ 0 \ldots 0 & B\beta\frac{\partial \varphi_1}{\partial x} & B\beta\frac{\partial \varphi_1}{\partial y} & 0 \\ 0 \ldots 0 & \frac{\partial \varphi_1}{\partial y} & -\frac{\partial \varphi_1}{\partial x} & 0 \\ 0 \ldots 0 & 0 & 0 & \frac{\partial \varphi_1}{\partial y} \end{pmatrix} \frac{\partial}{\partial \varphi_1} \begin{pmatrix} z_1 \\ \cdot \\ \cdot \\ \cdot \\ z_{l-1} \\ w_x \\ w_y \\ P \end{pmatrix} = \begin{pmatrix} F_1^* \\ \cdot \\ \cdot \\ \cdot \\ F_{l-1}^* \\ F_l^* \\ 0 \\ w_y \end{pmatrix}$$

from which it is possible to express the quantities $\partial w_x/\partial \varphi_1$, $\partial w_y/\partial \varphi_1$, as the functions of dependent variables and their derivatives along the boundary. The derivative $\partial P/\partial \varphi_1 = 0$ according to the starting premises. Since the coefficients by $\partial z_i/\partial \varphi_1$, $i = 1,.., l - 1$ vanish at the boundary, the values of the derivatives can be arbitrary and have no influence on the determination of the remaining quantities.

From the above analysis it follows that, in the case of a closed reservoir flow region, imposing the initial composition and pressure distributions is sufficient for finding the solution, and imposing boundary conditions for the composition is not required.

Thus, the two-phase multicomponent reservoir flow description reduces, in the two-dimensional case, to solving system (2.1.8) in the region

$$\{(x, y) \in \mathcal{G},\ t \geq t_o\}$$

where $\mathcal{G}$ is the arbitrary two-dimensional region with smooth boundary $\partial\mathcal{G}$, by imposing the initial conditions

$$t = 0,\ (x, y) \in \mathcal{G} : P = P^o(x, y),\ z_i = z_i^o(x, y),\ i = 1, \ldots\ l - 1$$

and the boundary conditions

$$(x, y) \in \partial\mathcal{G}_1 : P = P^I(x, y, t),\ z_i = z_i^I(x, y, t),\ i = 1, \ldots\ l - 1$$

$$(x, y) \in \partial\mathcal{G}_2 : P = P^{II}(x, y, t)$$

$$(x, y) \in \partial\mathcal{G}_3 : \frac{\partial P}{\partial \mathbf{n}} = 0$$

Here $\partial\mathcal{G}_1$ is the section of the boundary, where the reservoir flow is directed inside the region ($\partial P/\partial \mathbf{n} < 0$); $\partial\mathcal{G}_2$ is another section of the boundary where the flow is directed outside the region; $\partial\mathcal{G}_3$ is the impermeable section of the boundary.

The technique used above can be applied to the analysis of the mathematical statement of various boundary problems of reservoir flow.

3.2. BASING THE STATEMENT OF THREE-DIMENSIONAL BOUNDARY PROBLEMS

Let us consider the general case of a three-dimensional reservoir flow of a multicomponent mixture, described by system (2.1.7). To find its solution, one gives up the initial conditions as the values of the functions to be found, at a moment $t = t_o$

$$t = t_o : P = P^o(\mathbf{r})\ ,\ z_i = z_i^o(\mathbf{r})\ ,\ i = 1, \ldots, l - 1\ ,\ \mathbf{r} \in \mathcal{G}$$

For precision among the full set of boundary conditions let us study Equation (2.1.7), using the method of characteristics[4] and let us define, as in the previous section, the new variables[6]

$$w_x = \frac{\partial P}{\partial x},\ w_y = \frac{\partial P}{\partial y},\ w_z = \frac{\partial P}{\partial z},\ \mathbf{w} = (w_x, w_y, w_z)$$

and the additional relationships

$$\frac{\partial w_x}{\partial y} - \frac{\partial w_y}{\partial x} = 0,\ \frac{\partial w_y}{\partial z} - \frac{\partial w_z}{\partial y} = 0,\ \frac{\partial P}{\partial z} = w_z \tag{3.2.1}$$

Now Equations (2.1.7) and (3.2.1) can be represented together as a system of first-order quasilinear equations

$$\boldsymbol{\Omega}_t \frac{\partial \mathbf{u}}{\partial t} + \boldsymbol{\Omega}_x \frac{\partial \mathbf{u}}{\partial x} + \boldsymbol{\Omega}_y \frac{\partial \mathbf{u}}{\partial y} + \boldsymbol{\Omega}_z \frac{\partial \mathbf{u}}{\partial z} = \mathbf{F} \tag{3.2.2}$$

where

$$\mathbf{F} = \begin{pmatrix} kc_1(\nabla P)^2 - \mathbf{B}_1 \cdot \nabla k + kd_1\mathbf{g} \cdot \nabla P \\ \cdot \\ \cdot \\ \cdot \\ kc_{l-1}(\nabla P)^2 - \mathbf{B}_{l-1} \cdot \nabla k + kd_{l-1} \cdot \mathbf{g} \cdot \nabla P \\ -k\dfrac{\partial \beta}{\partial P}(\nabla P)^2 - \mathbf{B} \cdot \nabla k + k\dfrac{\partial(\beta\psi)}{\partial P} \cdot \mathbf{g} \cdot \nabla P \\ 0 \\ 0 \\ w_z \end{pmatrix},\ \mathbf{u} = \begin{pmatrix} z_1 \\ \cdot \\ \cdot \\ \cdot \\ z_{l-1} \\ w_x \\ w_y \\ w_z \\ P \end{pmatrix}$$

It is convenient to represent the square $(l + 3)$-matrixes Ω_ν as cellular:

$$\boldsymbol{\Omega}_\nu = \begin{pmatrix} \boldsymbol{\Omega}_\nu^{(1)} & \boldsymbol{\Omega}_\nu^{(2)} \\ \boldsymbol{\Omega}_\nu^{(3)} & \boldsymbol{\Omega}_\nu^{(4)} \end{pmatrix},\ \nu = t,x,y,z$$

where

$$\boldsymbol{\Omega}_t^{(1)} = \begin{pmatrix} -mN & \dots & 0 \\ \cdot & & \cdot \\ \cdot & & \cdot \\ \cdot & & \cdot \\ 0 & \dots & -mN \\ -m\dfrac{\partial N}{\partial z_i} & \dots & -m\dfrac{\partial N}{\partial z_{l-1}} \end{pmatrix},\ \boldsymbol{\Omega}_t^{(2)} = \begin{pmatrix} 0 & 0 & 0 & 0 \\ \cdot & \cdot & \cdot & \cdot \\ \cdot & \cdot & \cdot & \cdot \\ \cdot & \cdot & \cdot & \cdot \\ 0 & 0 & 0 & 0 \\ 0 & 0 & 0 & -m\dfrac{\partial N}{\partial P} \end{pmatrix}$$

$$\boldsymbol{\Omega}_x^{(1)} = \{kX_{ij}\}_{l,l-1},\ \boldsymbol{\Omega}_y^{(1)} = \{kY_{ij}\}_{l,l-1},\ \boldsymbol{\Omega}_z^{(1)} = \{kZ_{ij}\}_{l,l-1}$$

$$\boldsymbol{\Omega}_x^{(2)} = \begin{pmatrix} kb_1 & 0 & 0 & 0 \\ \cdot & \cdot & \cdot & \cdot \\ \cdot & \cdot & \cdot & \cdot \\ \cdot & \cdot & \cdot & \cdot \\ kb_{l-1} & 0 & 0 & 0 \\ k\beta & 0 & 0 & 0 \end{pmatrix},\ \boldsymbol{\Omega}_x^{(4)} = \begin{pmatrix} 0 & -1 & 0 & 0 \\ 0 & 0 & 0 & 0 \\ 0 & 0 & 0 & 0 \end{pmatrix}$$

$$\boldsymbol{\Omega}_y^{(2)} = \begin{pmatrix} 0 & kb_1 & 0 \\ \cdot & \cdot & \cdot \\ \cdot & \cdot & \cdot \\ \cdot & \cdot & \cdot \\ 0 & kb_{l-1} & 0 \\ 0 & k\beta & 0 \end{pmatrix},\ \boldsymbol{\Omega}_y^{(4)} = \begin{pmatrix} 1 & 0 & 0 & 0 \\ 0 & 0 & -1 & 0 \\ 0 & 0 & 0 & 0 \end{pmatrix}$$

$$\boldsymbol{\Omega}_z^{(2)} = \begin{pmatrix} 0 & 0 & kb_1 & 0 \\ \cdot & \cdot & \cdot & \cdot \\ \cdot & \cdot & \cdot & \cdot \\ \cdot & \cdot & \cdot & \cdot \\ 0 & 0 & kb_{l-1} & 0 \\ 0 & 0 & k\beta & 0 \end{pmatrix},\ \boldsymbol{\Omega}_z^{(4)} = \begin{pmatrix} 0 & 0 & 0 & 0 \\ 0 & 1 & 0 & 0 \\ 0 & 0 & 0 & 1 \end{pmatrix}$$

All the remaining cells $\boldsymbol{\Omega}_\nu^{(\,)}$ which were not mentioned in the above determinations consist of zero elements. In the expressions the following designations were used

$$\mathbf{B}_i = (\beta_i - Z_i\beta)\nabla P - (\beta_i\psi_i - Z_i\beta\psi)\mathbf{g},\ \mathbf{B} = \beta(\nabla P - \psi\mathbf{g});$$

$$c_i = z_i\frac{\partial\beta}{\partial P} - \frac{\partial\beta_i}{\partial P},\ d_i = \frac{\partial\beta_i\psi_i}{\partial P} - z_i\frac{\partial\beta\psi}{\partial P};$$

$$X_{ij} = a_{ij}w_x - e_{ij}g_x,\ Y_{ij} = a_{ij}w_y - e_{ij}g_y,\ Z_{ij} = a_{ij}w_z - e_{ij}g_z,\ i < l;$$

$$X_{lj} = \frac{\partial\beta}{\partial z_j}w_x - \frac{\partial\beta\psi}{\partial z_j}g_x,\ Y_{lj} = \frac{\partial\beta}{\partial z_j}w_y - \frac{\partial\beta\psi}{\partial z_j}g_y,$$

$$Z_{lj} = \frac{\partial\beta}{\partial z_j}w_z - \frac{\partial\beta\psi}{\partial z_j}g_z;\ b_i = \beta_i - z_i\beta$$

$$a_{ij} = \frac{\partial\beta_i}{\partial z_j} - z_i\frac{\partial\beta}{\partial z_j},\ e_{ij} = \frac{\partial\beta_i\psi_i}{\partial z_j} - z_i\frac{\partial\beta\psi}{\partial z_j}$$

Let us consider the new coordinate system

$$\varphi_\kappa = \varphi_\kappa\,(x,y,z,t), \qquad \kappa = 1,\ldots,4$$

which satisfies at a given point (x,y,z,t) the condition

$$\frac{D(\varphi_1,\varphi_2,\varphi_3,\varphi_4)}{D(x,y,z,t)} \neq 0$$

Using the new coordinates, system (3.2.2) takes the form

$$\sum_{\kappa=1}^{4}\left(\boldsymbol{\Omega}_t\frac{\partial\varphi_\kappa}{\partial t} + \boldsymbol{\Omega}_x\frac{\partial\varphi_\kappa}{\partial x} + \boldsymbol{\Omega}_y\frac{\partial\varphi_x}{\partial y} + \boldsymbol{\Omega}_z\frac{\partial\varphi_\kappa}{\partial z}\right)\frac{\partial\mathbf{u}}{\partial\varphi_\kappa} = \mathbf{F}$$

For its solvability, for example, regarding $\partial\mathbf{u}/\partial\varphi_1$ one has to satisfy the condition

$$\det\boldsymbol{\Omega} \neq 0,\ \boldsymbol{\Omega} = \boldsymbol{\Omega}_t\frac{\partial\varphi_1}{\partial t} + \boldsymbol{\Omega}_x\frac{\partial\varphi_1}{\partial x} + \boldsymbol{\Omega}_y\frac{\partial\varphi_1}{\partial y} + \boldsymbol{\Omega}_z\frac{\partial\varphi_1}{\partial z} \qquad (3.2.3)$$

For basing the boundary conditions for system (2.1.7) or (3.2.2) one is interested in the point of its solvability in a vicinity of $\partial\mathcal{G}$. Let the boundary of the region $\mathcal{G}$ consists of some sections

$$\partial\mathcal{G} = \sum_\nu \partial\mathcal{G}_\nu$$

Let us choose φ_κ, $\kappa = 1,.,4$, so that the surface

$$\varphi_1\,(x,y,z) = 0$$

coincides with $\partial\mathcal{G}_\nu$ and

$$\left(\frac{\partial\varphi_1}{\partial x}\right)^2 + \left(\frac{\partial\varphi_1}{\partial y}\right)^2 + \left(\frac{\partial\varphi_1}{\partial z}\right)^2 \neq 0$$

The other coordinate functions are defined by equalities

$$\varphi_2 = \varphi_2(x,y,z),\ \varphi_3 = \varphi_3(x,y,z),\ \varphi_4 = t$$

In this case, the solvability condition (3.2.3) takes the form

$$\frac{\partial\varphi_1}{\partial y}\cdot\frac{\partial\varphi_1}{\partial z}\left[\left(\frac{\partial\varphi_1}{\partial x}\right)^2 + \left(\frac{\partial\varphi_1}{\partial y}\right)^2 + \left(\frac{\partial\varphi_1}{\partial z}\right)^2\right]\det\boldsymbol{\Omega}_1 \neq 0$$

where

$$\boldsymbol{\Omega}_1 = \begin{pmatrix} \mathbf{A}_{11} \cdot \nabla\varphi_1 & \dots & \mathbf{A}_{1,l-1} \cdot \nabla\varphi_1 & b_1 \\ \cdot & & \cdot & \cdot \\ \cdot & & \cdot & \cdot \\ \cdot & & \cdot & \cdot \\ \mathbf{A}_{l-1,1} \cdot \nabla\varphi_1 & \dots & \mathbf{A}_{l-1,l-1} \cdot \nabla\varphi_1 & b_{l-1} \\ \mathbf{A}_1^* \cdot \nabla\varphi_1 & \dots & \mathbf{A}_{l-1}^* \cdot \nabla\varphi_1 & \beta \end{pmatrix}$$

$$\mathbf{A}_{ij} = a_{ij}\nabla P - e_{ij} \cdot \mathbf{g},\ \mathbf{A}_j^* = \frac{\partial\beta}{\partial z_j}\nabla P - \frac{\partial(\beta\psi)}{\partial z_j}\mathbf{g}$$

Vanishing det $\boldsymbol{\Omega}_1$ should correspond to the linear dependence between lines of $\boldsymbol{\Omega}_1$; therefrom it should follow that the quantities $\nabla P \cdot \nabla\varphi_1$ and $\mathbf{g} \cdot \nabla\varphi_1$ synchronously should have to satisfy the relationship

$$\nabla P \cdot \nabla\varphi_1 \left(\lambda_l \frac{\partial\beta}{\partial z_j} + \sum_{\kappa=1}^{l-1} \lambda_\kappa a_{\kappa j}\right) - \mathbf{g} \cdot \nabla\varphi_1 \left(\lambda_l \frac{\partial(\beta\psi)}{\partial z_j} + \sum_{\kappa=1}^{l-1} \lambda_\kappa e_{\kappa j}\right) = 0 \qquad (3.2.4)$$
$$j = 1, \dots, l-1$$

or

$$\nabla P \cdot \nabla\varphi_1 \cdot \sum_{\kappa-1}^{l-1} \lambda^{(\kappa)} a_{i\kappa} - \mathbf{g} \cdot \nabla\varphi_1 \sum_{\kappa=1}^{l-1} \lambda^{(\kappa)} e_{i\kappa} = \lambda^{(l)} b_\kappa \qquad (3.2.5)$$
$$i = 1, \dots, l-1$$

for a certain set of constants λ_κ, $\lambda^{(\kappa)}$. Equalities (3.2.4) and (3.2.5) are equivalent to the possibility to express the quantity

$$\nabla P \cdot \nabla\varphi_1 = \frac{\partial P}{\partial \mathbf{n}}$$

through a_{ij}, e_{ij}, $\nabla\varphi_1 \cdot \mathbf{g}$ by several ways. This should mean that system (3.2.4) (or (3.2.5)) is incompatible. This contradiction shows that in the general case of arbitrary $f_{g,w}$, when we impose no limitations on the quantities β_i, β, and $\nabla\varphi_1 \cdot \mathbf{g}$, quantity det $\boldsymbol{\Omega}_1$ cannot vanish at any section of boundary $\partial\mathcal{G}_\nu$.

Notice that, in the multicomponent reservoir flow model under consideration, the two-phase system with both mobile phases cannot exist near an impermeable section of the boundary. Actually, in general, the requirements of the absence of flow of all the components through the impermeable section

$$(\mathbf{n},\mathbf{J}_i) = 0,\ \mathbf{J}_i = -k\beta_i(\nabla P - \psi_i\mathbf{g}),\ i = 1, \dots, l \qquad (3.2.6)$$

are incompatible: by allowing the compatibility of equalities (3.2.6) one deduces that the quantity $\nabla P \cdot \mathbf{n}$ can be expressed through $\psi_1 \cdot \mathbf{g} \cdot \mathbf{n}$ by several ways, and that is impossible. The only geometrical exception corresponds to the exactly vertical sections of the boundary, when $\mathbf{g} \cdot \mathbf{n} = 0$ and the equalities (3.2.6) are compatible. But this case is not general, i.e., it can be removed by a small turn of the boundary; one can easily see that it is then reduced to the case considered in the previous section, and we will not discuss it here.

The above arguments exactly repeat the consideration of the conditions of vanishing det $\boldsymbol{\Omega}_1$. Therefore, the possibility of the solvability of the Cauchy problem for system (3.2.2) near $\partial\mathcal{G}_\nu$ is directly connected with the impossibility of the existence of a two-phase system near an impermeable boundary in the case of both mobile phases. However, in the case of the one-phase system and the two-phase system with at least one immobile phase, conditions (3.2.6) can already be compatible, and correspondingly, det $\boldsymbol{\Omega}_1$ can vanish at a certain section of the boundary.

Let us consider the possible cases from the point of view of the solvability of system (3.2.2) at $\partial\mathcal{G}_\nu$, in regard to $\partial\mathbf{u}/\partial\varphi_1$.

One-Phase System

In this case the matrix $\boldsymbol{\Omega}_1$ is triangular:

$$\boldsymbol{\Omega}_1 = \begin{pmatrix} \beta(\nabla\varphi_1, \mathbf{w} - \rho\mathbf{g}) & 0 & \cdots & 0 & 0 \\ 0 & \beta(\nabla\varphi_1, \mathbf{w} - \rho\mathbf{g}) & & \cdot & \\ \cdot & & & \cdot & \\ \cdot & & & \cdot & \\ \cdot & & & & \\ 0 & \cdot & \cdot & \beta(\nabla\varphi_1, \mathbf{w} - \rho\mathbf{g}) & 0 \\ \left(\nabla\varphi_1, \mathbf{w}\dfrac{\partial\beta}{\partial z_1} - \mathbf{g}\dfrac{\partial\beta\psi}{\partial z_1}\right) & \cdots & & \left(\nabla\varphi_1, \mathbf{w}\dfrac{\partial\beta}{\partial z_{l-1}} - \mathbf{g}\dfrac{\partial\beta\psi}{\partial z_{l-1}}\right) & 0 \end{pmatrix}$$

The solvability condition takes the form

$$\det \boldsymbol{\Omega}_1 = (\nabla P - \rho\mathbf{g}, \nabla\varphi_1)^{l-1} \cdot \beta^l \neq 0$$

or

$$\frac{\partial P}{\partial \mathbf{n}} - \rho\,(\mathbf{g}, \mathbf{n}) \neq 0 \tag{3.2.7}$$

where ρ is the density of the one-phase system under consideration.

Relation (3.2.7) is the permeability condition of the surface $\mathcal{G}_\nu$. The surface, where Equation (3.2.7) is not satisfied, is characteristic.

Two-Phase System With One Immobile Phase

Allow, for purposes of this discussion, that the liquid phase is immobile. Then, after simple transformations of the lines and columns of matrix $\mathbf{\Omega}_1$, we obtain the solvability condition in the form

$$\det \mathbf{\Omega}_1 = (\nabla P - \rho_g \mathbf{g}, \nabla \varphi_1)^{l-1} \cdot$$

$$\det \cdot \begin{pmatrix} \frac{\partial \beta_1}{\partial z_1} & \cdots & \frac{\partial \beta_1}{\partial z_{l-1}} \\ \vdots & & \\ \frac{\partial \beta_{l-1}}{\partial z_1} & \cdots & \frac{\partial \beta_{l-1}}{\partial z_{l-1}} \\ \frac{\partial \beta}{\partial z_1} & \cdots & \frac{\partial \beta}{\partial z_{l-1}} \end{pmatrix} \neq 0$$

or

$$\frac{\partial P}{\partial \mathbf{n}} - \rho_g (\mathbf{g}, \mathbf{n}) \neq 0 \tag{3.2.8}$$

When condition (3.2.8) does not hold, the impermeability conditions are satisfied for all components, as in the one-phase case, i.e., the system of conditions (3.2.6) is compatible.

System With Both Mobile Phases

As was pointed out above, the solvability condition always holds in this case. Thus, at the permeable sections, in accordance with the kind of equations considered, the boundary conditions for the pressure and for the concentrations of the components whose flows are directed inside the boundary, have to be imposed.

Summarizing the results of the consideration, one can conclude that the reservoir system at the impermeable boundary can only be in a one-phase state or in a two-phase state with one immobile phase.

For the existence of solution in a vicinity of the characteristic surface $\partial\mathcal{G}_\nu$, an additional compatibility condition must be satisfied, which in the case of a three-dimensional reservoir flow has the form

$$\text{rank } \mathbf{\Omega} = \text{rank } \mathbf{\Omega}^* = 4 \tag{3.2.9}$$

where the matrix $\boldsymbol{\Omega}^*$ is obtained by adding the column

$$\mathbf{F}^* = \mathbf{F} - \sum_{\kappa=2}^{4} \left(\boldsymbol{\Omega}_t \frac{\partial \varphi_\kappa}{\partial t} + \boldsymbol{\Omega}_x \frac{\partial \varphi_\kappa}{\partial x} + \boldsymbol{\Omega}_y \frac{\partial \varphi_\kappa}{\partial y} + \boldsymbol{\Omega}_z \frac{\partial \varphi_\kappa}{\partial z} \right) \frac{\partial \mathbf{u}}{\partial \varphi_\kappa}$$

to the matrix Ω. As in the previous section, one can show that this condition is equivalent to the following equalities

$$\beta F_i^* = b_i F_l^*, \qquad i = 1, \ldots, l - 1$$

In the case of a one-phase system ($b_i = 0$) these equalities are

$$F_i^* = 0, \qquad i = 1, \ldots, l = 1$$

Owing to the arbitrariness of the local coordinates chosen, let us define the function φ_2 so that $\varphi_2 = \varphi_2\,(x,y,z)$ is the natural parametrization of the flow line, which issues from a given point of space in a given moment of time and is determined by vector (w_x, w_y, w_z). In addition, taking the family of curves φ_3 orthogonal to the family of curves φ_2 and using the relationships

$$w_x^2 + w_y^2 + w_z^2 = \left(\frac{\partial P}{\partial \varphi_2}\right)^2$$

$$w_x \frac{\partial \varphi_2}{\partial x} + w_y \frac{\partial \varphi_2}{\partial y} + w \frac{\partial \varphi_2}{\partial z} = \frac{\partial P}{\partial \varphi_2}$$

$$w_x \frac{\partial \varphi_3}{\partial x} + w_y \frac{\partial \varphi_3}{\partial y} + w_z \frac{\partial \varphi_3}{\partial z} = 0$$

one can represent condition (3.2.9) for the case of a two-phase system with one immobile phase in the form

$$m\left(\frac{\partial N z_i}{\partial t} - y_i \frac{\partial N}{\partial t}\right) = k\beta\left(-g_v \rho_g + \frac{\partial P}{\partial \varphi_2}\right) \frac{\partial y_i}{\partial \varphi_2}, \qquad i = 1, \ldots, l - 1 \tag{3.2.10}$$

and for the case of a one-phase system in the form

$$mN \frac{\partial z_i}{\partial t} = k\beta\left(-g_v \rho + \frac{\partial P}{\partial \varphi_2}\right) \frac{\partial z_1}{\partial \varphi_2} \qquad i = 1, \ldots, l - 1 \tag{3.2.11}$$

and in addition to the impermeability condition one has not to impose any boundary conditions on the composition. The quantity g_v in Equations (3.2.10) and (3.2.11) is the projection of the vector $\mathbf{g}$ on the flow line,

$$g_v = (\mathbf{g}, \nabla P/|\nabla P|)$$

and the symbol $\partial/\partial\varphi_2$ designates the derivative along the flow line.

Going over in expressions (3.2.10) and (3.2.11) from the derivatives along the flow line to the more traditional form of representation we obtain

$$(\nabla P)^2 m\left(\frac{\partial N z_i}{\partial t} - y_i\frac{\partial N}{\partial t}\right) = k\beta(\nabla P - \rho_g\mathbf{g}, \nabla P)(\nabla y_i, \nabla P) \qquad (3.2.12)$$
$$i = 1, \ldots, l - 1$$

in the case of a two-phase system with one immobile phase, and

$$(\nabla P)^2 mN\frac{\partial z_i}{\partial t} = k\beta(\nabla P - \rho\mathbf{g}, \nabla P)(\nabla z_i, \nabla P) \qquad i = 1, \ldots, l - 1 \qquad (3.2.13)$$

In the case of a thin seam the conditions of compatibility, (3.2.10) through (3.2.13), hold at the impermeable sections of the boundary line and the family of functions $y_2 = y_2(x,y,z)$ corresponds to its natural parametrization.

Notice that the impossibility of the coexistence of two mobile phases at the impermeable sections of $\partial\mathcal{G}$ is connected when taking into account the gravitational forces in the problem statement considered. Neglecting the gravitational forces the solvability condition is $\partial P/\partial\mathbf{n} \neq 0$; if it does not hold, the impermeability conditions $(\mathbf{n},\mathbf{J}_i) = 0$ are satisfied for all components at the corresponding sections of the boundary. Fulfilling the calculations, one can obtain the following system of compatibility conditions for the general case of the coexistence of two phases at the impermeable boundary in a model without gravitation

$$m\left(\frac{\partial N z_i}{\partial t} - \frac{\beta_i}{\beta}\frac{\partial N}{\partial t}\right) = k\beta\frac{\partial P}{\partial\varphi_2}\frac{\partial}{\partial\varphi_2}\left(\frac{\beta_i}{\beta}\right) \qquad i = 1, \ldots, l - 1 \qquad (3.2.14)$$

for the arbitrary phase mobilities. For the limiting case of a thin and horizontal seam, conditions (3.2.14) go over into system (3.1.10) in the previous section.

Comparison of the results obtained in this and in the previous sections shows that taking into account the gravitational forces brings some important features into the multicomponent reservoir flow problem: namely, the impossibility of coexistence of two phases with non-zero mobilities at the impermeable boundary, and the existence of the individual flow directions of single components, which can differ from the general direction of the mixture flow through the permeable boundary. The principal point of the above analysis is the conclusion that the compatibility conditions obtained are differential consequences of the flow equations and, thus, the "compositional" boundary conditions at the impermeable boundary sections have not been imposed. This situation seems to be typical for the mathematical statements of heterogeneous reservoir flow problems by neglecting capillary and diffusion forces.

REFERENCES

1. **Lee, E. H. and Fayers, F. J.,** *Trans. Soc. Pet. Eng. AIME,* 216, 284, 1959.
2. **Mitlin, V. S. and Tsybulskii, G. P.,** Mathematical model of two-phase multicomponent reservoir flow of a gas-condensate mixture in *Scientific and Technical Problems of Planning the Exploitation of Gas, Gas-Condensate and Gas-Oil Deposits,* VNIIGAS, Moscow, 1983, 97 (in Russian).
3. **Mitlin, V. S.,** New Methods of Calculating the Enriched Gas Action Upon a Gas-Condensate Seam., Ph.D. thesis, VNIIGAS, Moscow, 1987 (in Russian).
4. **Godunov, S. K.,** *Equations of Mathematical Physics,* Nauka, Moscow, 1971 (in Russian).
5. **Rozenberg, M. D. and Kundin, S. A.,** *Multiphase Multicomponent Reservoir Flow in Oil and Gas Recovery,* Nedra, Moscow, 1976 (in Russian).
6. **Mitlin, V. S. and Tsybulskii, G.P.,** Statement of three-dimensional problems of two-phase reservoir flow of multicomponent systems, in *Dynamics of Multiphase Media,* ITPM SO AN SSSR, Novosibirsk, 1985, 133 (in Russian).

4

Numerical Method of Solving Multicomponent Reservoir Flow Problems

This chapter is concerned with the description of a finite-difference method for multicomponent reservoir flow problems. The method was applied in the reservoir flow calculations in cases of one and two space variables, and has demonstrated high efficiency in computers with moderate computational capabilities.

Since it is possible to find analytical solutions of the problems of TMRF only in some simplest cases, working out effective algorithms for computer solving is of principal importance. Owing to the fact, that numerical solving of the system of l equations (2.1.8) in a two-dimensional region takes an extreme amount of processor time, even for small l, the main condition for improving the algorithm efficiency appears to be a possible decrease in the number of iteration procedures. The method of calculation of the pressure and composition fields without iterations is described below.

Suggestions on how to work out the methods without iterations for the multicomponent reservoir flow problems were discussed earlier by Murkes et al.[1] However, the applicability of the method is limited by reservoir systems with practically invariant density. In many real situations it is impossible to neglect the dependence of the formation density on pressure. The method described below takes this into account.

Another specific feature of the algorithm below is in the approach to the phase equilibrium calculation. Since the system of phase equilibrium equations contains l nonlinear algebraic relationships, the calculation of its solution for each site of the finite-difference grid and for each time layer, as is required for numerically solving Equations (2.1.8) or (2.1.9), would hardly be justified because of the processor time expenses. A specific of the phase equilibrium calculations shown in the following method is to consider the equilibrium constants $\mathcal{K}_i$ to depend on the pressure, temperature, and parameters of overall composition. The dependences are represented by the interpolation formulas which are obtained using standard program complexes of phase equilibria computation. First, the phase equilibrium is calculated and the values of $\mathcal{K}_i$, $\rho_{g,w}$, and $\eta_{g,w}$ are obtained for a finite number of pressure and composition points using a program complex. Then the values are used in the calculations of the interpolation polynomial coefficients for $\mathcal{K}_i$, $\rho_{g,w}$, and $\eta_{g,w}$ and then the polynomials are applied in a multicomponent reservoir flow simulation.

4.1. FINITE-DIFFERENCE EQUATIONS AND ALGORITHMS FOR SOLVING

To simulate the reservoir flow in a formation drained by a system of extracting and injecting wells, one should rewrite Equation (2.1.8) taking into account the sink and source terms, which replace the boundary conditions at the wells. Let us adduce the form of the term Q_κ for the κ -th component in three cases of the flow rate Q definition: (1) in units of the mass extracted (injected) per a time unit, (2) in units of the mole amount extracted (injected) per a time unit and (3) in units of the fluid volume extracted (injected) per a time unit and calculated by normal (P_a and T_a) conditions.

In the case (1) term Q_κ in the equation for the κ -th component has the following form: for the injecting wells, Q_κ is a known quantity, for the extracting wells

$$Q_\kappa = \frac{Q\epsilon_\kappa}{\sum\limits_{\kappa=1}^{l} M_\kappa \epsilon_\kappa}, \qquad \epsilon_\kappa = \frac{\beta_\kappa}{\beta} \tag{4.1.1}$$

In the case (2) term Q_κ has the form: for the injecting wells Q_κ is a known quantity, for the extracting wells

$$Q_\kappa = Q \cdot \epsilon_\kappa \tag{4.1.2}$$

In the case (3) term Q_κ has the form: for the injecting wells Q_κ is a known quantity, for the extracting wells

$$Q_\kappa = Q \cdot N(\epsilon_1, \ldots, \epsilon_{l-1}, P_a, T_a)\epsilon_\kappa \tag{4.1.3}$$

where ϵ_κ was defined in Equation (4.1.1).

Using Equations (4.1.1) to (4.1.3), system (2.1.8) is rewritten in the form

$$\begin{aligned} &\mathrm{div}(kh\beta_\kappa \,\mathrm{grad}\, P) + \sum_{j=1}^{W_{ex}} Q_{j\kappa}\delta(\mathbf{r} - \mathbf{r}_j^{(ex)}) \quad + \\ &+ \sum_{j=1}^{W_{in}} Q_{j\kappa}\delta(\mathbf{r} - \mathbf{r}_j^{(in)}) = \frac{\partial(Nz_\kappa)}{\partial t} mh, \qquad \kappa = 1, \ldots, l \end{aligned} \tag{4.1.4}$$

$$\operatorname{div}(kh\beta \operatorname{grad} P) + \sum_{j=1}^{W_{ex}} Q_{j\kappa}\delta(\mathbf{r} - \mathbf{r}_j^{(ex)}) + \\ + \sum_{j=1}^{W_{in}} Q_{j\kappa}\delta(\mathbf{r} - \mathbf{r}_j^{(in)}) = \frac{\partial N}{\partial t} mh \tag{4.1.5}$$

where W_{ex} and W_{in} are the numbers of the extracting and injecting wells; $\mathbf{r}_j^{(ex)}$, $\mathbf{r}_j^{(in)}$ are the coordinates of the extracting and injecting wells; $\delta(x,y)$ is the Dirak delta-function. Systems (4.1.4) and (4.1.5), unlike Equation (2.3.8), contains $l + 1$ equalities. Equation (4.1.5) is obtained by taking into account the condition $\sum_{\kappa=1}^{l} z_\kappa = 1$ and will be used in the algorithm instead of it.

Approximating systems (4.1.4) and (4.1.5) in the sites of a rectangular grid by the system of finite-difference equations, we obtain (for simplification, the grid, which is uniform by each coordinate, is considered)[2]

$$\begin{aligned} & k_{i+1/2,j} \cdot h_{i+1/2,j} \cdot \beta^{(n)}_{i+1/2,j,\kappa} \cdot (P^{(n+1)}_{i+1,j} - P^{(n+1)}_{i,j})/(\Delta x)^2 + \\ & k_{i-1/2,j} \cdot h_{i-1/2,j} \cdot \beta^{(n)}_{i-1/2,j,\kappa} \cdot (P^{(n+1)}_{i-1,j} - P^{(n+1)}_{i,j})/(\Delta x)^2 + \\ & + k_{i,j+1/2} \cdot h_{i,j+1/2} \cdot \beta^{(n)}_{i,j+1/2,\kappa} \cdot (P^{(n+1)}_{i,j+1} - P^{(n+1)}_{i,j})/(\Delta y)^2 + \\ & + k_{i,j-1/2} \cdot h_{i,j-1/2} \cdot \beta^{(n)}_{i,j-1/2,\kappa} \cdot (P^{(n+1)}_{i,j-1} - P^{(n+1)}_{i,j})/(\Delta y)^2 + Q^{(n)}_{i,j,\kappa} = \\ & = \frac{m_{i,j} \cdot h_{i,j}}{\Delta t} \left[\left(\frac{N}{P}\right)^{(n+\alpha_3)}_{i,j} \cdot P^{(n+1)}_{i,j} \cdot z^{(n+1)}_{i,j,\kappa} - \left(\frac{N}{P}\right)^{(n+\alpha_1)}_{i,j} \cdot P^{(n)}_{i,j} \cdot z^{(n)}_{i,j,\kappa} \right], \\ & \kappa = 1, \ldots, l \end{aligned} \tag{4.1.6}$$

$$\begin{aligned} & k_{i+1/2,j} \cdot h_{i+1/2,j} \cdot \beta^{(n)}_{i+1/2,j} \cdot (P^{(n+1/2)}_{i+1,j} - P^{(n+1/2)}_{i,j})/(\Delta x)^2 + \\ & + k_{i-1/2,j} \cdot h_{i-1/2,j} \cdot \beta^{(n)}_{i-1/2,j} \cdot (P^{(n+1/2)}_{i-1,j} - P^{(n+1/2)}_{i,j})/(\Delta x)^2 + \\ & + k_{i,j+1/2} \cdot h_{i,j+1/2} \cdot \beta^{(n)}_{i,j+1/2} \cdot (P^{(n)}_{i,j+1} - P^{(n)}_{i,j})/(\Delta y)^2 + \\ & + k_{i,j-1/2} \cdot h_{i,j-1/2} \cdot \beta^{(n)}_{i,j-1/2} \cdot (P^{(n)}_{i,j-1} - P^{(n)}_{i,j})/(\Delta y)^2 + Q^{(n)}_{i,j} = \\ & = \frac{2 \cdot m_{i,j} \cdot h_{i,j}}{\Delta t} \cdot \left[\left(\frac{N}{P}\right)^{(n+\alpha_2)}_{i,j} \cdot P^{(n+1/2)}_{i,j} - \left(\frac{N}{P}\right)^{(n+\alpha_1)}_{i,j} \cdot P^{(n)}_{i,j} \right] \end{aligned} \tag{4.1.7}$$

$$\begin{aligned} & k_{i+1/2,j} \cdot h_{i+1/2,j} \cdot \beta^{(n)}_{i+1/2,j} \cdot (P^{(n+1/2)}_{i+1,j} - P^{(n+1/2)}_{i,j})/(\Delta x)^2 + \\ & + k_{i-1/2,j} \cdot h_{i-1/2,j} \cdot \beta^{(n)}_{i-1/2,j} \cdot (P^{(n+1/2)}_{i-1,j} - P^{(n+1/2)}_{i,j})/(\Delta x)^2 + \\ & + k_{i,j+1/2} \cdot h_{i,j+1/2} \cdot \beta^{(n)}_{i,j+1/2} \cdot (P^{(n+1)}_{i,j+1} - P^{(n+1)}_{i,j})/(\Delta y)^2 + \\ & + k_{i,j-1/2} \cdot h_{i,j-1/2} \cdot \beta^{(n)}_{i,j-1/2} \cdot (P^{(n+1)}_{i,j-1} - P^{(n+1)}_{i,j})/(\Delta y)^2 + Q^{(n)}_{i,j} = \\ & = \frac{2 \cdot m_{i,j} \cdot h_{i,j}}{\Delta t} \left[\left(\frac{N}{P}\right)^{(n+\alpha_3)}_{i,j} \cdot P^{(n+1)}_{i,j} - \left(\frac{N}{P}\right)^{(n+\alpha_2)}_{i,j} \cdot P^{(n+1/2)}_{i,j} \right] \end{aligned} \tag{4.1.8}$$

Here i,j is the numeration of the sites by coordinates x,y; κ is the number of a component; n is the numeration of the time steps; $\Delta x, \Delta y, \Delta t$ are the space and time steps; $Q^{(n)}_{i,j,\kappa}$ is the injection (extraction) rate of the κ -th component in a given site; $Q^{(n)}_{i,j}$ is the total injection (extraction) rate in a given site; $\alpha_2 - \alpha_1 = \alpha_3 - \alpha_2 = 1/2$.

The algorithm for solving systems (4.1.6) to (4.1.8) consists of the following. The pressure field is found from Equations (4.1.7) and (4.1.8) by the method of changing variables.[3] To solve the appearing systems of linear equations with three-diagonal matrixes the run-through method is used. Then the composition fields are found explicitly from Equation (4.1.6). To determine the mole fraction of the gaseous phase V_g at each site, Equation (2.1.6) is solved using the Newton method. The conductivity coefficients in the "halved" sites ($\beta_{i \pm 1/2,j,\kappa}$) are calculated at the n -th time layer using the rule "against flow".[4]

One should mention the peculiarities of the approximation of the time derivative in Equations (4.1.6) to (4.1.8). This approximation has, as one will easily find, first-order precision in regard to Δt. The quantity α_3 can be most naturally determined as 0 or 1/2; in the computer program worked out the second variant was accepted to decrease approximation error. The quantity $(N/P)^{(n+1/2)}$ was determined by the extrapolation

$$\left(\frac{N}{P}\right)^{(n+1/2)} = 1.5\left(\frac{N}{P}\right)^{(n)} - 0.5\left(\frac{N}{P}\right)^{(n-1)}$$

The physical validity of the approximation suggested is connected with the fact that when expanding the mole phase densities by P the first term corresponds to ideal gas behavior. Exact execution of the ideal gas law leads to the constant value of N/P. The numerous numerical experiments show good validity of the approximation of the time derivative, even when the formation system modeled is very far from the ideal gas law behavior. The time derivatives in the differential equations for the composition are approximated issuing from the local conservation condition $\sum_{\kappa=1}^{l} z_\kappa = 1$ which has to hold at each site of the finite-difference grid.

The finite-difference scheme is three-layered; consequently, to provide the calculations at the first time step it is necessary either to use an iteration algorithm, which should be different from the given one, or (as it has been seen in the program) to carry out calculations in the very beginning, consequently increasing either Δt or the flow rate values at the wells. In the last case, the time derivatives at the first time step are approximated by the following expressions of zero-order precision

$$\frac{\partial N}{\partial t} = \left(\frac{N}{P}\right)^{(o)} \cdot \frac{P^{(1)} - P^{(o)}}{\Delta t} + O(1)$$

$$\frac{\partial N z_\kappa}{\partial t} = \left(\frac{N}{P}\right)^{(o)} \cdot \frac{P^{(1)} z_\kappa^{(1)} - P^{(o)} z_\kappa^{(o)}}{\Delta t} + O(1)$$

and the approximation error is small to the extent that the perturbations of the initial P and z_κ distributions are small. The numerical experiments showed that the artificial flow rate acceleration was more preferable than the consequent Δt increase.

The proposed three-layered method of the first-order precision is implicit in regard to P and explicit in regard to composition; owing to this fact it is conditionally stable in regard to z_κ, which imposes a certain limitation of the time step and the characteristic pressure gradient value (see Section 4.2). The control of the calculation correctness is provided by testing of the conditions: (1) total and component integral balance, and (2) $\sum_{\kappa=1}^{l} z_\kappa = 1$ at any site.

Besides describing the reservoir dynamics in general, it is often necessary to calculate the one-dimensional reservoir flow (near a single well, between two linear galleries, and so on). For such purposes an algorithm for solving the one-dimensional multicomponent reservoir flow problems, which is a simplification of the corresponding two-dimensional algorithm, was worked out. The finite-difference scheme has the form (for a uniform grid)

$$\left(\frac{r_{i+1/2}}{r_i}\right)^{\alpha} k_{i+1/2}\beta^{(n)}_{i+1/2,\kappa}\frac{P^{(n+1)}_{i+1} - P^{(n+1)}_{i}}{(\Delta r)^2} + \left(\frac{r_{i-1/2}}{r_i}\right)^{\alpha} k_{i-1/2}\beta^{(n)}_{i-1/2,\kappa}\frac{P^{(n+1)}_{i-1} - P^{(n+1)}_{i}}{(\Delta r)^2} = \frac{m_i}{\Delta t}\left[\left(\frac{N}{P}\right)^{(n+\alpha_2)}_i \cdot P^{(n+1)}_i \cdot z^{(n+1)}_{i,\kappa} - \left(\frac{N}{P}\right)^{(n+\alpha_1)}_i \cdot P^{(n)}_i \cdot z^{(n)}_{i,\kappa}\right], \quad \kappa = 1, \ldots, l \tag{4.1.9}$$

$$\left(\frac{r_{i+1/2}}{r_i}\right)^{\alpha} k_{i+1/2}\cdot\beta^{(n)}_{i+1/2}\cdot\frac{P^{(n+1)}_{i+1} - P^{(n+1)}_{i}}{(\Delta r)^2} + \left(\frac{r_{i-1/2}}{r_i}\right)^{\alpha} k_{i-1/2}\cdot\beta^{(n)}_{i-1/2}\frac{P^{(n+1)}_{i-1} - P^{(n+1)}_{i}}{(\Delta r)^2} = \frac{m_i}{\Delta t}\left[\left(\frac{N}{P}\right)^{(n+\alpha_2)}_i \cdot P^{(n+1)}_i - \left(\frac{N}{P}\right)^{(n+\alpha_1)}_i \cdot P^{(n)}_i\right] \tag{4.1.10}$$

Here i is the site numeration by space coordinate; $\alpha_2 = 1/2$; r_i is the coordinate of the i -th site, $r_{i\pm 1/2} = (r_i + r_{i\pm 1})/2$; $\alpha_2 - \alpha_1 = 1$; the quantity α was defined in Equation (2.3.9); the remaining denotations are the same as in Equations (4.1.6) to (4.1.8).

The algorithm for solving equations (4.1.9) and (4.1.10) consists of the following. First, the pressure field is found from Equation (4.1.10) by the "run-through" method, the first and last lines of the corresponding three-diagonal matrix being formed in accordance with the type of boundary conditions imposed. Then the composition field z_κ is found from Equation (4.1.9). Before the calculations of P and z_κ values at the (n + 1)-th time layer, the phase equilibrium is calculated at each site, and as a result, the quantities β_κ and N are obtained at the n -th time layer.

Recognition of the quadratic Darcy law for the both phases can be provided in the algorithm in the following way. Let us express $v_{g,w}$ through $\partial P/\partial r$ using relation (2.1.13):

$$v_{g,w} = -\frac{kf_{g,w}}{\eta_{g,w}}\frac{\partial P}{\partial r}\left[\frac{1}{2} + \left(\frac{1}{4} + \left|\frac{\partial P}{\partial r}\right|\frac{\rho_{g,w}\cdot k^2\cdot f_{g,w}}{\mathcal{H}\cdot\eta_{g,w}^2}\right)^{1/2}\right]^{-1}$$

then the quantities β_κ and β in Equation (2.1.9) will be presented in the form

$$\beta_\kappa = z_\kappa\left(\frac{f_g\rho_g\mathcal{K}_\kappa}{\eta_g M_g}D_g + \frac{f_w\rho_w}{\eta_w M_w}D_w\right)\left[1 + V_g(\mathcal{K}_\kappa - 1)\right]^{-1}, \qquad (4.1.11)$$

$$\beta = \frac{f_g\rho_g D_g}{\eta_g M_g} + \frac{f_w\rho_w D_w}{\eta_w M_w}$$

where

$$D_{g,w}^{-1} = \frac{1}{2} + \left(\frac{1}{4} + \left|\frac{\partial P}{\partial r}\right|\frac{\rho_{g,w}k^2f_{g,w}^2}{\mathcal{H}\cdot\eta_{g,w}^2}\right)^{1/2}$$

Now by solving systems (4.1.9) and (4.1.10) the quantities β_κ and β are determined from Equation (4.1.11); thus, the numerical method without iterations also can be presented in this case.

In concluding the section, let us discuss the matter of determining the flow parameters in the near-zone of a well using data obtained for a finite-difference cell containing the well. This problem is important for large-scale reservoir simulation with the characteristic size of a finite-difference cell in the order of hundreds of meters to kilometers. In this case the flow parameters in the near-zone can differ strongly from the averaged cell parameters presented by large-scale calculations. To solve the problem, the direct method possesses the most precision. Using the method, one carries out the main large-scale calculations together with parallel calculation of the small-scale boundary problems for the finite-difference cells containing the wells, the boundary conditions for the small-scale problems being determined from the large-scale solution. In order to use computers with moderate efficiency, another method based on the quasi-stationary flow behavior in the near-zone, can be applied.[5] Let us suppose that the small-scale flow at the well follows large-scale changes of the pressure and composition, that is, the ratio of characteristic times of flow changing inside a cell and for the formation scale is a small value. Using the formula of stationary inflow to a well, one can write

$$Q = \frac{2\pi kh}{\ln(R_c/r_b)} [H_f(P(R_c)) - H_f(P(r_b))] = \frac{2\pi kh}{\ln(R_c/r_b)} \cdot \int_{P(r_b)}^{P(R_c)} \beta dP \approx \tag{4.1.12}$$

$$\approx \beta(P(R_c)) \cdot \frac{2\pi kh}{\ln(R_c/r_b)} [P(R_c) - P(r_b)]$$

Here H_f is the flow potential; r_b is the well radius; R_c is the radius of a circular contour whose area would be equivalent to the area of finite-difference cell,

$$R_c = \left(\frac{\Delta x \cdot \Delta y}{\pi}\right)^{1/2}$$

The approximate replacing the integral in Equation (4.1.12) assumes that the difference between the pressure at the well and in the cell is essentially small compared to the pressure in the cell itself. Owing to the supposition of the method, $P(R_c)$ coincides with the pressure in the cell, therefore the pressure at the well $P(r_b)$ can be determined from Equation (4.1.12). The flow composition at the well, owing to the quasi-stationary flow behavior inside the cell, can be determined through the large-scale conductivity coefficients in the cell, as $\epsilon_\kappa = \beta_\kappa/\beta$. The numerical calculations show that the precision of the approximate method is quite satisfactory for practical purposes.

4.2. STABILITY OF NUMERICAL METHOD

Let us consider the case of a one-dimensional linear flow; the stability analysis in a two-dimensional case would be carried out in the same manner.

As we know, two stability limitations for computing the two-phase reservoir flow equations by using the IMPES method (implicit by the pressure, explicit by the saturation), have been outlined.[4] The first of them is concerned with explicitly taking into account the capillary forces in the scheme, the second is connected with the explicit determination of the conductivity coefficients. Since the capillary forces are not considered in the model presented, possible appearance of the numerical instability can be connected with the compositional dependence of β_κ. Let k and m are constant. Denote

$$\gamma_\kappa = \frac{k\beta_\kappa}{m \cdot z_\kappa}, \qquad \gamma = \frac{k\beta}{m}$$

and designate the z_κ and P errors as e_κ and e_p. The quantities e_κ and e_p correspond to the computational errors of the difference equations (4.1.9) and (4.1.10) by neglecting the approximation errors. Let the flow rate field have no singularities; this means that, for a given i -th site, the signs of $P_{i+1} - P_i$ and $P_i - P_{i-1}$ coincide. For definiteness, let the flow is directed toward decrease the space coordinate. Let us neglect the difference of the quantity γ_κ in the (i + 1) -th, i -th, and (i − 1) -th sites; this supposition is equivalent to neglecting the second term in the right-hand side of the expression

$$\frac{\partial \gamma_\kappa z_\kappa}{\partial r} = \gamma_\kappa \frac{\partial z_\kappa}{\partial r} + z_\kappa \sum_{j=1}^{l} \frac{\partial \gamma_\kappa}{\partial z_j} \frac{\partial z_j}{\partial r}$$

Then, make the same supposition in regard to γ. Taking into account the "against flow" rule for choosing $\beta_{\kappa,i\pm 1/2}$, we obtain the equations for the errors in the i -th site

$$\left(\frac{N}{P}\right)_i^{(n+\alpha_2)} \cdot (e_{p,i}^{(n+1)} \cdot z_{\kappa,i}^{(n+1)} + e_{\kappa,i}^{(n+1)} \cdot P_i^{(n+1)}) -$$

$$\left(\frac{N}{P}\right)_i^{(n+\alpha_1)} \cdot (e_{p,i}^{(n)} \cdot z_{\kappa,i}^{(n)} + e_{\kappa,i}^{(n)} \cdot P_i^{(n)}) =$$

$$= \frac{(\Delta r)^2}{\Delta t} \gamma_\kappa \, [z_{\kappa,i}^{(n)} \cdot \Delta_i^2 e_p^{(n+1)} + (e_{\kappa,i+1}^{(n)} - e_{\kappa,i}^{(n)}) \cdot (P_{\kappa,i+1}^{(n)} - P_{\kappa,i}^{(n)})] \, ,$$

$$\kappa = 1, \ldots, l$$

$$\left(\frac{N}{P}\right)^{(n+\alpha_2} \cdot e_{p,i}^{(n+1)} - \left(\frac{N}{P}\right)_i^{(n+\alpha_1)} \cdot e_{p,i}^{(n)} = \frac{(\Delta r)^2}{\Delta t} \gamma \cdot \Delta_i^2 e_p^{(n+1)} \tag{4.2.1}$$

Here Δ_i^2 is the second difference in the i -th site.

Representing e_κ and e_p in the form

$$e_{\kappa,i}^{(n)} = \sum_q \xi_\kappa^n \exp(jqi\Delta r), \; e_{p,i}^{(n)} = \sum_q \xi_p^n \exp(jqi\Delta r), \; j^2 = -1 \tag{4.2.2}$$

substituting Equation (4.2.2) into Equation (4.2.1) and combining the coefficients before the exponents with equal q, after some transformations we obtain

$$\left(\frac{N}{P}\right)_i^{(n+\alpha_2)} \cdot \xi_p - \left(\frac{N}{P}\right)_i^{(n+\alpha_1)} = -4 \frac{\Delta t}{(\Delta r)^2} \cdot \gamma \xi_p \sin^2 \frac{q\Delta r}{2} \, , \tag{4.2.3}$$

$$\left(\frac{N}{P}\right)_i^{(n+\alpha_2)} \cdot P_i^{(n+1)} \cdot \xi_\kappa^{n+1} = \left[\left(\frac{N}{P}\right)_i^{(n+\alpha_1)} \cdot P_i^{(n)} + \right.$$

$$\left. \frac{\Delta t}{(\Delta r)^2} \gamma_\kappa \, (P_{i+1}^{(n+1)} - P_i^{(n+1)}) \cdot (\exp(j\Delta r q) - 1) \right] \xi -$$

$$\xi_p^n \left[\left(\frac{N}{P}\right)_i^{(n+\alpha_2)} \cdot \xi_p + 4 \frac{\Delta t}{(\Delta r)^2} \gamma \sin^2 \frac{q\Delta r}{2} - \left(\frac{N}{P}\right)_i^{(n+\alpha_1)}\right] z_{\kappa,i}^{(n)} +$$

$$+ \xi_p^n \left[4 \frac{\Delta t}{(\Delta r)^2} (\gamma - \gamma_\kappa) \, z_{\kappa,i}^{(n)} \cdot \sin^2 \left(\frac{q\Delta r}{2}\right) \cdot \xi_p - \right.$$

$$\left. \left(\frac{N}{P}\right)_i^{(n+\alpha_1)} \cdot (z_{\kappa,i}^{(n+1)} - z_{\kappa,i}^{(n)})\right] , \qquad \kappa = 1, \ldots, l \tag{4.24}$$

Equations (4.2.3) and (4.2.4) are rewritten in the form

$$\begin{pmatrix} \xi_\kappa^{n+1} \\ \xi_p^{n+1} \end{pmatrix} = \begin{pmatrix} a_{11} & a_{12} \\ a_{21} & a_{22} \end{pmatrix} \begin{pmatrix} \xi_\kappa^{n} \\ \xi_p^{n} \end{pmatrix} ,$$

$$a_{11} = \left[\left(\frac{N}{P}\right)_i^{(n+\alpha_1)} \cdot P^{(n)} + \frac{\Delta t}{(\Delta r)^2} \cdot \gamma_\kappa \cdot (P_{i+1}^{(n+1)} - P_i^{(n+1)}) \right.$$

$$\left. (\exp(j\Delta r q) - 1) \right] \Big/ \left[\left(\frac{N}{P}\right)_i^{(n+\alpha_2)} \cdot P_i^{(n+1)} \right] , \; a_{21} = 0 ,$$

$$a_{22} = \left(\frac{N}{P}\right)^{(n+\alpha_1)} \Big/ \left[\left(\frac{N}{P}\right)^{(n+\alpha_2)} + 4 \frac{\Delta t}{(\Delta r)^2} \gamma \sin^2 \frac{q\Delta r}{2} \right]$$

and the stability conditions have the form

$$|a_{11}| < 1 , \qquad |a_{22}| < 1 \tag{4.2.5}$$

Conditions (4.2.5) can be transformed as

$$\frac{\Delta t}{\Delta r} \gamma_\kappa \frac{\partial P}{\partial r} < 2 \left(\frac{N}{P}\right)^{(n+\alpha_2)} \cdot P^{(n+1)} - \left(\frac{N}{P}\right)^{(n+\alpha_1)} \cdot P^{(n)} \tag{4.2.6}$$

$$\kappa = 1, \ldots , l$$

$$\frac{\left(\frac{N}{P}\right)^{(n+\alpha_1)}}{\left(\frac{N}{P}\right)^{(n+\alpha_2)} + 4 \gamma \frac{\Delta t}{(\Delta r)^2} \sin^2 \frac{q\Delta r}{2}} < 1 \tag{4.2.7}$$

Condition (4.2.7) always holds in the case of a sufficiently small changing of N and P during one time-step, and condition (4.2.6) goes over into the following:

$$\frac{\Delta t}{\Delta r} \frac{k}{m} \max_i \left(\frac{\frac{\partial P}{\partial r} \beta_\kappa}{N z_\kappa} \right) < 1 , \qquad \kappa = 1, \ldots , l \tag{4.2.8}$$

It follows from Equation (4.2.8) that the stability of the method becomes worse with an increase in the pressure gradient of the formation, with a decrease in the mole density of a given component, and with an increase in β_κ which characterizes mobility of a given component. Adding equations (4.2.8), we obtain one more stability condition

$$\frac{\Delta t}{\Delta r}\frac{k}{m}\max_{i}\left(\frac{\frac{\partial P}{\partial r}\beta}{N}\right) < 1$$

Defining the rate of the κ -th component v_κ as its mole flow rate $k\beta_\kappa\ \partial P/\partial r$ divided by the number of moles of the κ -th component per unit of volume mNz_κ, one can rewrite Equation (4.2.8) in the form

$$\frac{\Delta t}{\Delta r} v_\kappa < 1\ , \qquad \kappa = 1, \ldots, l \tag{4.2.9}$$

i.e., to provide calculation stability, the distance overcome by the particles of each component during one temporal step must not exceed the spatial step of the difference grid. Conditions (4.2.9) present the "multicomponent" generalization of the well-known stability condition of finite-difference schemes for two-phase reservoir flow models.

The above results can be generalized for the case of two space variables (the considerations for the two-phase reservoir flow of immiscible liquids are taken from Aziz and Settari[4]). The system for the stability conditions for schemes (4.1.6) to (4.1.8) has the form

$$\Delta t\left(\frac{v_{x,\kappa}}{\Delta x} + \frac{v_{y,\kappa}}{\Delta y}\right) < 1\ , \qquad \kappa = 1, \ldots, l \tag{4.2.10}$$

where $v_{x,k}$ and $v_{y,k}$ are the flow rates of the κ -th component in the directions x and y. The stability limitation in the two-dimensional case is more essential than in the one-dimensional one. Actually, let us compare conditions (4.2.9) and (4.2.10) for the reservoir flow problem with a single well.[4] One can write $v_{x,\kappa} = v_{y,\kappa} = v_\kappa$, $\Delta x = \Delta y = \Delta r$, and instead of Equation (4.2.10) we obtain

$$\Delta t \cdot 2v_\kappa < \Delta r\ , \qquad \kappa = 1, \ldots, l$$

i.e., for the two-dimensional scheme the upper limit of Δt is two times less than the corresponding value for the one-dimensional scheme.

Notice that representation of the divergent terms in the form $\nabla(\gamma_\kappa z_\kappa \nabla P)$ was used in many papers concerned with the numerical study of multicomponent reservoir flow.[1,6] The stability analysis given above became possible for the general case of the *l*-component reservoir flow by using the supposition of a weak γ_κ dependence on composition. Taking the dependence into account leads to the necessity to use the matrix, which is inverse to the Jacobi matrix $\{\partial \gamma_\kappa / \partial z_i\}$, and the consideration becomes absolutely nonconstructive. Nevertheless, since γ_κ can depend quite strongly on the composition, condition (4.2.9) should be considered as a certain stability estimation only.

We have omitted the point about the approximation error of the differential equations by finite-difference ones in the above analysis. It should be noted that, though the "against flow" rule for the conductivities is necessary to provide stability of the calculations, sometimes (namely, in the points of changing the flow orientation) such approximation is of zero-order precision. Actually, let $P_{i\pm 1} > P_i$ for the i -th site; the determination of conductivities "against flow" leads to the following approximation of expression $\partial/\partial r\ (\beta\ \partial P/\partial r)$:

$$\frac{\partial}{\partial r}\left(\beta \frac{\partial P}{\partial r}\right) = \frac{\beta_{i-1}}{(\Delta r)^2}(P_{i-1} - P_i) + \frac{\beta_{i+1}}{(\Delta r)^2}(P_{i+1} - P_i) \qquad (4.2.11)$$

On the other side, the approximation

$$\frac{\partial}{\partial r}\left(\beta \frac{\partial P}{\partial r}\right) = \frac{\beta_i\,(P_{i+1} + P_{i-1} - 2P_i)}{(\Delta r)^2} + \frac{(\beta_{i+1} - \beta_i)(P_{i+1} - P_i)}{(\Delta r)^2} \qquad (4.2.12)$$

has first-order precision in regard to Δr. Subtracting Equation (4.2.12) from Equation (4.2.11), one can see that the approximation error for the expression in Equation (4.2.11) is not less than

$$\frac{(\beta_i - \beta_{i-1})}{\Delta r} \cdot \frac{(P_i - P_{i-1}}{\Delta r},$$

i.e., this is the zero-order approximation. Thus, the scheme diffusion, which is connected with approximation errors, will especially strongly affect the precision of the solution by carrying out the calculations near the discontinuities, sources, and sinks. Note that a special numerical method for equations of TMRF, that has high precision due to "adjusting" the difference approximation near the discontinuities, has been developed by Levi and Shakirov.[7]

4.3. AN EXACT SOLUTION OF THE PLANE-RADIAL DISPLACEMENT PROBLEM

An essential stage of mathematical modeling is comparing the results of solving finite-difference equations and an exact solution. However, the multicomponent reservoir flow model rarely allows for an exact solution. The necessity for testing the computer program in the case of the mixture flow with a sufficiently large number of components requires a special construction of the exact solution. Let us show that the solution for an arbitrary l can be constructed by choosing a special form of the phase densities, viscosities, and permeabilities.[2]

Let the ideal gas law hold for the densities

$$\rho_{g,w} = \rho_o \frac{PM_{g,w}}{P_oM_o} \tag{4.3.1}$$

It follows from Equations (4.3.1) and (2.1.5) that $v_g = 1 - S$, i.e., the molar phase fractions are equal in this case to the volume fractions. The phase viscosities can be presumed to be determined as

$$\eta_{g,w} = \eta_o \frac{P}{P_o} \tag{4.3.2}$$

and the phase permeabilities are taken in the form

$$f_{g,w} = S_{g,w} \tag{4.3.3}$$

Suppose that no limitations are imposed on the equilibrium constants $\mathcal{K}_i$ behavior. Taking into account Equations (4.3.1) to (4.3.3), one rewrites Equation (2.1.2) in the form (neglecting gravitation)

$$\nabla(kz_i\nabla P) = \frac{\eta_o}{P_o}\frac{\partial}{\partial t}(mPz_i), \qquad i = 1, \ldots, l \tag{4.3.4}$$

In case of one-dimensional plane-radial flow and the constant k and m we have

$$\frac{1}{r}\frac{\partial}{\partial r}\left(rz_i\frac{\partial P}{\partial r}\right) = B_o\frac{\partial Pz_i}{\partial t}, \qquad i = 1, \ldots, l \tag{4.3.5}$$

where $B_o = m\eta_o/kP_o$. Adding Equation (4.3.5), we obtain

$$\frac{1}{r}\frac{\partial}{\partial r}\left(r\frac{\partial P}{\partial r}\right) = B_o\frac{\partial P}{\partial r} \tag{4.3.6}$$

Taking Equation (4.3.6) into account, one can rewrite Equation (4.3.5) in the form

$$\frac{\partial z_i}{\partial r}\cdot\frac{\partial P}{\partial r} = B_oP\frac{\partial z_i}{\partial t}, \qquad i = 1, \ldots, l \tag{4.3.7}$$

Equations (4.3.6) and (4.3.7) possess a self-similar solution by imposing the following boundary conditions: for the pressure

$$Q = 2\pi r h \frac{k\rho_o}{M_o \eta_o} \frac{\partial P(r,t)}{\partial r}\bigg|_{r\to 0},$$

$$P(r,0) = P(\infty,t) = P_1$$

where the first condition corresponds to giving up the constant molar flow rate and is obtained using relationships (4.3.1) to (4.3.3); the compositional boundary conditions are

$$z_i(r,0) = z_i(\infty,t) = z_i^I, \qquad i = 1, \ldots, l$$

and in case of negative Q (injection) the additional condition

$$z_i(0,t) = z_i^{II}, \qquad i = 1, \ldots, l$$

appears. Going over to the self-similar variable $\theta = r/\sqrt{t}$, we obtain instead of Equations (4.3.6) and (4.3.7)

$$\begin{aligned} &\frac{1}{\theta}\frac{\partial}{\partial \theta}\left(\theta \frac{\partial P}{\partial \theta}\right) = -B_o \frac{\theta}{2}\frac{\partial P}{\partial \theta}, \\ &\frac{\partial z_i}{\partial \theta}\left(\frac{\partial P}{\partial \theta} + P \frac{B_o}{2}\theta\right) = 0, \qquad i = 1, \ldots, l \end{aligned} \tag{4.3.8}$$

and the boundary conditions for Equation (4.3.8) have the form

$$Q = 2\pi h \frac{k_{\rho o}}{M_o \mu_o} \theta \frac{\partial P}{\partial \theta}\bigg|_{\theta\to 0},$$

$P = P_1$ at $\theta \to \infty$,

$z_i = z_i^I$ at $\theta \to \infty$,

$z_i = z_i^{II}$ at $\theta \to 0$ (the case of injection)

From the first Equation (4.3.8) we obtain

$$P = -\frac{QM_o\eta_o}{4\pi h k \rho_o} \mathrm{Ei}\left(-\frac{B_o\theta^2}{4}\right) + P_1, \quad \mathrm{Ei}(-x) = \int_x^\infty \frac{e^{-u}}{u} du \tag{4.3.9}$$

From the last l Equations (4.3.8) one can see that in the case of injection the solution for the composition has the form

$$\begin{aligned} z_i(\theta) &= z_i^I, \ \theta > \theta_o \\ z_i(\theta) &= z_i^{II}, \ \theta < \theta_o \end{aligned} \tag{4.3.10}$$

where the saltus coordinate θ_o can be obtained, by using Equation (4.3.9), from the following condition

$$P_1 + \frac{QM_o\eta_o}{4\pi hk\rho_o}\left[-\operatorname{Ei}\left(-\frac{B_o\theta_o^2}{4}\right) + \frac{4\exp\left(-\frac{B_o\theta_o^2}{4}\right)}{B_o\eta_o^2}\right] = 0 \qquad (4.3.11)$$

In the case of inflow to the well (Q > 0) the z_i discontinuities are impossible, and z_i are constant at all θ. The mole fraction of the gaseous phase is obtained from Equation (2.1.6)

$$\begin{aligned} V_g &= V_g(z_i^I, \ldots, z_{l-1}^I, P) \qquad \text{at } \theta > \theta_o \\ V_g &= V_g(z_i^{II}, \ldots, z_{l-1}^{II}, P) \qquad \text{at } \theta < \theta_o \end{aligned} \qquad (4.3.12)$$

One can see that the phase saturations can change with P even by constant z_i. The possibility of a more complicated structure of the composition discontinuities could be connected with the presence of nonlinear dependences $f_{g,w}$ (*S*).[8] However, the saturation profile also can be non-trivial in the case considered. As an example, one could describe the saturation distributions in the problem of injecting a "dry" gas into a seam.

The displacement front coordinate θ_o is determined unambiguously by the seam parameters, pressure P_1, and the molar flow rate Q from Equation (4.3.11). The pressure at the saltus P_2 (θ_o) is determined by the same quantities from Equations (4.3.9).

In Figures 1(a), 2(a), and 3(a) we show all the possible cases of the P_1 and P_2 mutual disposition at the phase diagram of the reservoir system. The curves of equal saturations are shown by dotted lines. Figure 1(a) corresponds to the case, when both values of P_1 and P_2 are disposed in the direct evaporation region; in Figure 2(a) the value of P_2 is disposed in the retrograde condensation region; in Figure 3(a) both values of P_1 and P_2 are disposed in the retrograde condensation region. The corresponding dependences *S* (θ) are shown in Figures 1(b), 2(b), and 3(b). In the first case, saturation decreases monotonically at $\theta > \theta_o$, in the second case the "bank" of *S* is formed, and in the third case *S* increases monotonically. The quantity S_{max} in Figure 2 is the saturation corresponding to the maximum condensation pressure at a given reservoir temperature.

As well as the solution presented by Equations (4.3.9) to (4.3.12), the exact solutions of the multicomponent reservoir flow problems can be written for the first-order boundary conditions at $\theta = 0$ and for the cases of linear and sphere-radial flow. Such solutions can be used in testing the computer programs for the multicomponent reservoir flow problems in one-, two-, and three-dimensional cases. It is important that the pressure and the composition are "separated" in some sense for such solutions, and that the quality of the phase equilibrium subroutine and the hydrodynamic calculation subroutine can be estimated independently.

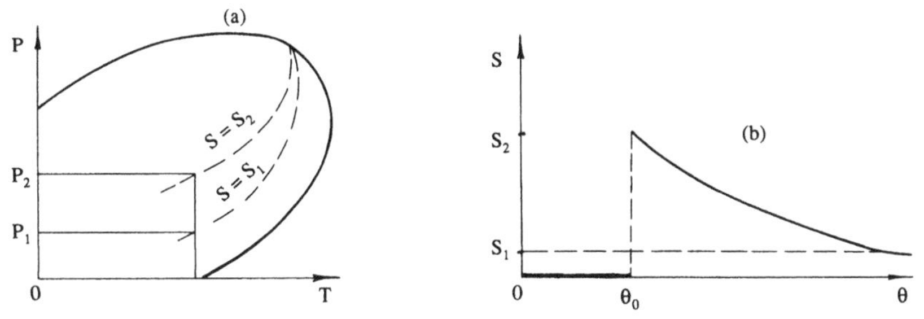

FIGURE 1. (a) P-T phase diagram where the pressures at the saltus and at infinity are shown for a given reservoir temperature; (b) dependence S (θ) decreasing at $\theta > \theta_o$.

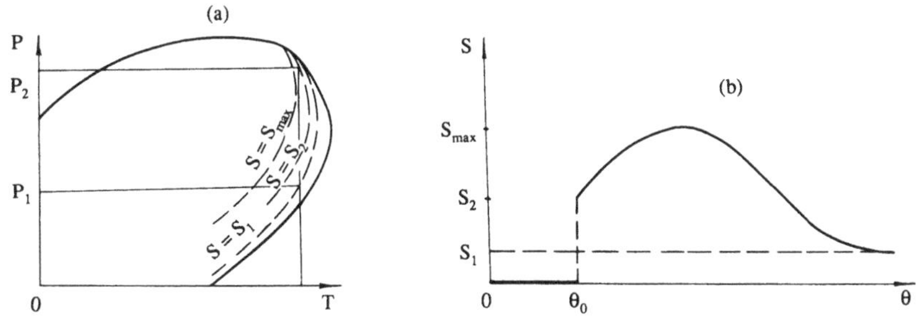

FIGURE 2. The same as that in Figure 1. Another disposition of P_1 and P_2 leads to nonmonotone behavior of S (θ) at $\theta > \theta_o$.

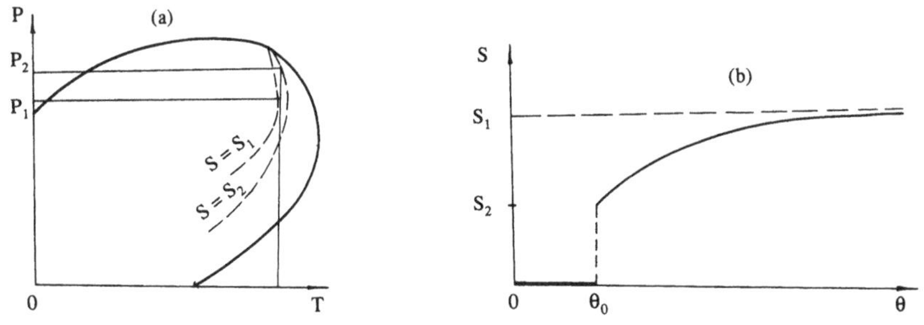

FIGURE 3. The same as that in Figure 1. Here S (θ) increases monotonically.

4.4. COMPARISON OF NUMERICAL AND EXACT SOLUTIONS

Below we describe the results of testing the one- and two-dimensional multicomponent reservoir flow computation programs, using the exact solution (4.3.9) to (4.3.12). The model reservoir composition consisted of ten components: N_2, CH_4, C_2H_6, C_3H_8, C_4H_{10}, C_5H_{12}, C_6H_{14}, C_7H_{16}, C_8H_{18}, and a conditional heavy component with a molecular weight of 0.2125 kg/mol. The porosity, thickness, and permeability were chosen constants: m = 0.222, h = 7 m, k = 10^{-14} m^2.

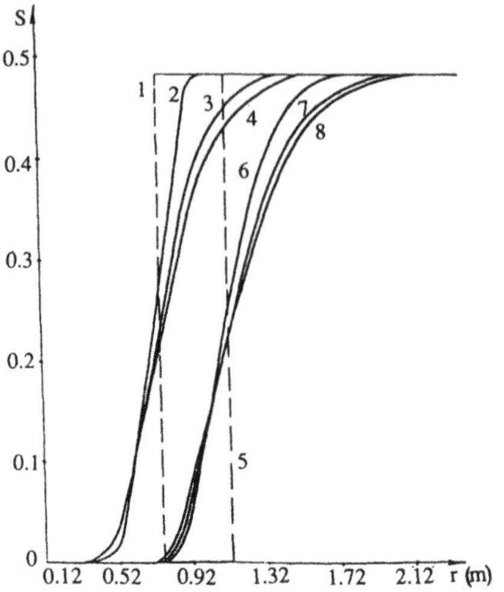

FIGURE 4. Comparison of exact and numerical solutions—one-dimensional model.

The initial composition of the reservoir system was: 0.002 (N_2), 0.59 (CH_4), 0.08 (C_2H_6), 0.05 (C_3H_8), 0.032 (C_4H_{10}), 0.022 (C_5H_{12}), 0.03 (C_6H_{14}), 0.072 (C_7H_{16}), 0.055 (C_8H_{18}), 0.057 (C_{10+}); the phase viscosities, densities, and permeabilities corresponded to Equations (4.3.1) to (4.3.3). The initial pressure was 10.5 MPa, $S = 1 - V = 0.48$. The problem of injecting a mixture with composition 0.50 (CH_4), 0.24 (C_2H_6), 0.17 (C_3H_8), and 0.09 (C_4H_{10}) by constant injection rate $Q = -0.54$ mol/sec in a seam was solved.

For testing the one-dimensional computational method, the circular region with the contour radius $R_c = 4.12$ m and well radius $r_b = 0.12$ m was considered. The scheme spatial step was $\Delta r = 0.2$ m, and the initial pressure was a supported invariant at the contour. The number of sites of the difference grid was 21. Since the exact solution was proposed for a infinite region, the calculation and comparison were made until the composition perturbations at the contour became essential. The quantity θ_o was determined from the solution of the transcendent equation (4.3.11) and equals $0.003435 \text{ m} \cdot \text{sec}^{-1/2}$.

Figure 4 shows the form of S profiles for the exact solution and for the numerical solution. Curves 1 to 4 correspond to the moment $t = 0.555$ day, curves 5 to 8 correspond to the moment $t_2 = 1.11$ day. The profiles pointed by the numbers 1 and 5 correspond to the exact solution (in the calculations we used $\mathcal{K}_i$, which were independent on P, i.e., the S values before and behind saltus should be independent on η); the profiles pointed by the numbers 2 and 6, 3 and 7, and 4 and 8 were obtained in the calculations with a time step of 0.0925 day, 0.04625 day, and 0.01542 day, respectively. Clearly, in the case of maximum Δt still permitting numerical stability, saltus smoothing occurs within a minimum space region. This is connected with the fact that the leading term in the expression of the approximation error, which defines artificial diffusion, vanishes at Δt of the order of the scheme stability limit.[4] In the computation process, one observes that the saltus smooths more and more strongly with time (compare curves 2 to 4 and 6 to 8).

The characteristics of the numerical method obtained by solving the problem at $t_2 = 1.11$ day are given in Table 1. The second column contains the relative error in the mixture integral balance calculation (in regard to the initial resources), the third contains the relative error in the methane integral balance calculation, the fourth contains the relative error in the heavy component (C_{10+}) integral balance calculation, and the fifth contains the maximum deviation of $\sum_{\kappa=1} z_\kappa$ from 1 for all the difference sites. The numbers of calculations are given in the first column, increasing with a Δt decrease.

The computation required 60 K of computer memory. One time step was calculated in 1.5 sec of the processor time on computer ES-1033.

To study the influence of the number of steps of the artificial flow rate acceleration (see Section 4.1) on the calculation stability, another plane-radial displacement problem was solved. In the above problem, according to Equation (4.3.1), the quantity N/P was constant, and the $\partial N/\partial t$ approximation, even at the first time step, had first-order precision. Therefore, in the given calculation we especially used the cubic dependences of the phase densities on the dimensionless pressure, instead of the linear form (4.3.1). The initial data were the same as that above, but $k = 0.5 \cdot 10^{-15}\ m^2$. Figure 5 shows the pressure difference ΔP between the boundaries of the solution region at 21 grid sites and Δt = 0.04625 day in the case of absence of AFRA (curve 1), two steps of AFRA (curve 2), and four steps of AFRA (curve 3) (during n steps the flow rate at the well was increased by the quantity Q/n). One can see that in all the cases ΔP approaches the same value, and the greater the number of AFRA steps, the faster pressure establishes after the AFRA end. One observed the oscillations without AFRA (curve 1). Curve 4 in Figure 5 corresponds the calculation with the same Δt and with 11 grid sites. The quantity ΔP approaches the value, which is a little less than for smaller Δr. Figure 6 shows the saturation profiles for this calculation at 0.555 day (curve 1) and 1.11 day (curve 3); curves 2 and 4 correspond to the calculation with 21 grid sites. Clearly, for the same Δt, saltus smoothing is stronger if Δr is higher. This is because the difference scheme is farther from the stability limit and the numerical diffusion affects it stronger, if Δr increases.

For testing the two-dimensional multicomponent flow program, one solved numerically the problem of gas injection through the well disposed in the center of a square seam. The space steps were $\Delta x = \Delta y = 0.2$ m, the initial pressure was fixed at the boundary. The number of sites was 23 × 23. All the initial conditions were the same as those in the one-dimensional (first calculation in this section) case; the densities, viscosities, and permeabilities were defined by Equations (4.3.1) to (4.3.3).

Tables 2 and 3 show the matrix of fraction of a component (C_3H_8 is presented) in the sites of the difference grid at 0.555 day (Table 2) and 1.11 day (Table 3) in the calculation with $\Delta t = 0.0077$ day. One can see that the lines of equal fraction form the concentric circles. The same thing also was observed for the

TABLE 1

Number of Calculation	Relative Error in Mixture Integral Balance (%)	Relative Error in CH_4 Integral Balance (%)	Relative Error in C_{10+} Component Integral Balance (%)	Maximum Deviation of $\sum_{\kappa=1}^{l} z_\kappa$ from 1 Over All Sites
1	0.15	0.77	1.03	0.015
2	0.15	0.86	0.71	0.003
3	0.15	0.86	0.8	0.0004

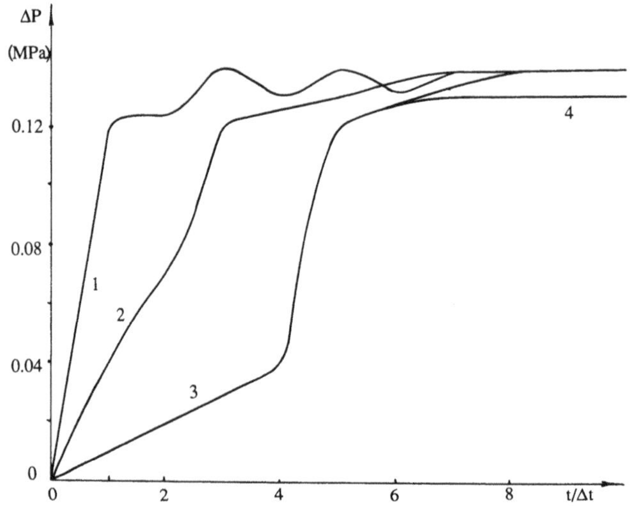

FIGURE 5. Temporal dependences of the pressure difference between the boundaries at different numbers of steps of an artificial flow rate acceleration.

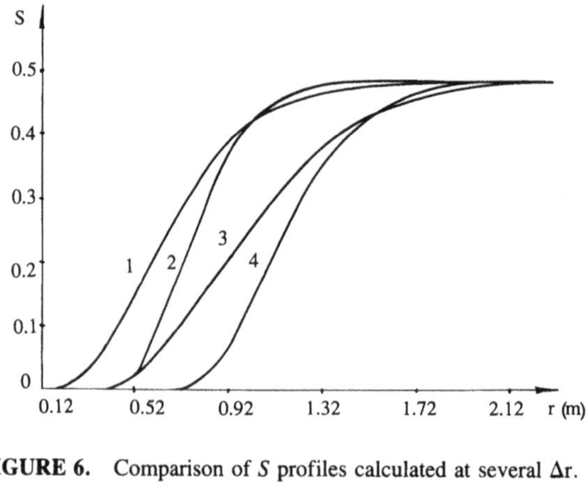

FIGURE 6. Comparison of S profiles calculated at several Δr.

pressure and saturation distributions. This means that one can use the exact solution for comparison with the numerical one until the displacement front approaches the boundary of the solution region.

TABLE 2

0.081	0.081	0.081	0.081	0.081	0.081	0.081	0.081	0.081	0.081	0.081	0.081	0.081	0.081	0.081	0.081	0.081	0.081	0.081	0.081	0.081
0.081	0.081	0.081	0.081	0.081	0.081	0.081	0.081	0.081	0.081	0.081	0.081	0.081	0.081	0.081	0.081	0.081	0.081	0.081	0.081	0.081
0.081	0.081	0.081	0.081	0.081	0.081	0.081	0.081	0.081	0.081	0.081	0.081	0.081	0.081	0.081	0.081	0.081	0.081	0.081	0.081	0.081
0.081	0.081	0.081	0.081	0.081	0.081	0.081	0.081	0.081	0.082	0.083	0.082	0.081	0.081	0.081	0.081	0.081	0.081	0.081	0.081	0.081
0.081	0.081	0.081	0.081	0.081	0.081	0.081	0.081	0.082	0.086	0.091	0.086	0.082	0.081	0.081	0.081	0.081	0.081	0.081	0.081	0.081
0.081	0.081	0.081	0.081	0.081	0.081	0.081	0.083	0.089	0.102	0.118	0.102	0.089	0.083	0.081	0.081	0.081	0.081	0.081	0.081	0.081
0.081	0.081	0.081	0.081	0.081	0.081	0.084	0.092	0.111	0.140	0.170	0.140	0.111	0.092	0.084	0.081	0.081	0.081	0.081	0.081	0.081
0.081	0.081	0.081	0.081	0.081	0.083	0.092	0.115	0.155	0.195	0.220	0.195	0.155	0.115	0.092	0.083	0.081	0.081	0.081	0.081	0.081
0.081	0.081	0.081	0.081	0.082	0.089	0.111	0.155	0.204	0.231	0.239	0.231	0.204	0.155	0.111	0.089	0.082	0.081	0.081	0.081	0.081
0.081	0.081	0.081	0.082	0.086	0.101	0.140	0.193	0.231	0.240	0.240	0.240	0.231	0.193	0.140	0.101	0.086	0.082	0.081	0.081	0.081
0.081	0.081	0.081	0.083	0.091	0.118	0.169	0.220	0.239	0.240	0.240	0.240	0.239	0.220	0.169	0.118	0.091	0.083	0.081	0.081	0.081
0.081	0.081	0.081	0.082	0.086	0.101	0.140	0.195	0.231	0.240	0.240	0.240	0.231	0.195	0.140	0.101	0.086	0.082	0.081	0.081	0.081
0.081	0.081	0.081	0.081	0.082	0.089	0.111	0.153	0.204	0.231	0.239	0.231	0.204	0.153	0.111	0.089	0.082	0.081	0.081	0.081	0.081
0.081	0.081	0.081	0.081	0.081	0.083	0.092	0.113	0.155	0.195	0.220	0.195	0.155	0.113	0.092	0.083	0.081	0.081	0.081	0.081	0.081
0.081	0.081	0.081	0.081	0.081	0.081	0.084	0.092	0.111	0.140	0.170	0.140	0.111	0.092	0.084	0.081	0.081	0.081	0.081	0.081	0.081
0.081	0.081	0.081	0.081	0.081	0.081	0.081	0.083	0.089	0.102	0.118	0.102	0.089	0.083	0.081	0.081	0.081	0.081	0.081	0.081	0.081
0.081	0.081	0.081	0.081	0.081	0.081	0.081	0.081	0.082	0.086	0.091	0.086	0.082	0.081	0.081	0.081	0.081	0.081	0.081	0.081	0.081
0.081	0.081	0.081	0.081	0.081	0.081	0.081	0.081	0.081	0.082	0.083	0.082	0.081	0.081	0.081	0.081	0.081	0.081	0.081	0.081	0.081
0.081	0.081	0.081	0.081	0.081	0.081	0.081	0.081	0.081	0.081	0.081	0.081	0.081	0.081	0.081	0.081	0.081	0.081	0.081	0.081	0.081
0.081	0.081	0.081	0.081	0.081	0.081	0.081	0.081	0.081	0.081	0.081	0.081	0.081	0.081	0.081	0.081	0.081	0.081	0.081	0.081	0.081
0.081	0.081	0.081	0.081	0.081	0.081	0.081	0.081	0.081	0.081	0.081	0.081	0.081	0.081	0.081	0.081	0.081	0.081	0.081	0.081	0.081

TABLE 3

0.081	0.081	0.081	0.081	0.081	0.081	0.081	0.081	0.082	0.083	0.084	0.083	0.082	0.081	0.081	0.081	0.081	0.081	0.081	0.081	0.081
0.081	0.081	0.081	0.081	0.081	0.081	0.081	0.081	0.082	0.083	0.084	0.083	0.082	0.081	0.081	0.081	0.081	0.081	0.081	0.081	0.081
0.081	0.081	0.081	0.081	0.081	0.081	0.081	0.082	0.084	0.088	0.092	0.088	0.084	0.082	0.081	0.081	0.081	0.081	0.081	0.081	0.081
0.081	0.081	0.081	0.081	0.081	0.081	0.082	0.086	0.093	0.162	0.113	0.102	0.093	0.086	0.082	0.081	0.081	0.081	0.081	0.081	0.081
0.081	0.081	0.081	0.081	0.081	0.083	0.087	0.098	0.115	0.134	0.152	0.134	0.115	0.098	0.087	0.083	0.081	0.081	0.081	0.081	0.081
0.081	0.081	0.081	0.081	0.083	0.088	0.101	0.124	0.154	0.180	0.198	0.180	0.154	0.124	0.101	0.088	0.083	0.081	0.081	0.081	0.081
0.081	0.081	0.081	0.082	0.087	0.101	0.128	0.165	0.199	0.220	0.230	0.220	0.199	0.165	0.128	0.101	0.087	0.082	0.081	0.081	0.081
0.081	0.081	0.082	0.086	0.098	0.124	0.165	0.205	0.229	0.237	0.239	0.237	0.229	0.205	0.165	0.124	0.098	0.086	0.082	0.081	0.081
0.082	0.082	0.084	0.093	0.115	0.154	0.199	0.229	0.238	0.240	0.240	0.240	0.238	0.229	0.199	0.154	0.115	0.093	0.084	0.082	0.082
0.083	0.083	0.088	0.102	0.134	0.180	0.219	0.237	0.240	0.240	0.240	0.240	0.240	0.237	0.219	0.180	0.134	0.102	0.088	0.083	0.083
0.084	0.084	0.092	0.113	0.152	0.198	0.230	0.239	0.240	0.240	0.240	0.240	0.240	0.239	0.230	0.198	0.152	0.113	0.092	0.084	0.084
0.083	0.083	0.088	0.102	0.134	0.180	0.219	0.237	0.240	0.240	0.240	0.240	0.240	0.237	0.219	0.180	0.134	0.102	0.088	0.083	0.083
0.082	0.082	0.084	0.093	0.115	0.154	0.199	0.229	0.238	0.240	0.240	0.240	0.238	0.229	0.199	0.154	0.115	0.093	0.084	0.082	0.082
0.081	0.081	0.082	0.086	0.098	0.124	0.165	0.205	0.229	0.237	0.239	0.237	0.229	0.205	0.165	0.124	0.098	0.086	0.082	0.081	0.081
0.081	0.081	0.081	0.082	0.087	0.101	0.128	0.165	0.199	0.220	0.230	0.220	0.199	0.165	0.128	0.101	0.087	0.082	0.081	0.081	0.081
0.081	0.081	0.081	0.081	0.083	0.088	0.101	0.124	0.154	0.180	0.198	0.180	0.154	0.124	0.101	0.088	0.083	0.081	0.081	0.081	0.081
0.081	0.081	0.081	0.081	0.081	0.083	0.087	0.098	0.115	0.134	0.152	0.134	0.115	0.098	0.087	0.083	0.081	0.081	0.081	0.081	0.081
0.081	0.081	0.081	0.081	0.081	0.081	0.082	0.086	0.093	0.102	0.113	0.102	0.093	0.086	0.082	0.081	0.081	0.081	0.081	0.081	0.081
0.081	0.081	0.081	0.081	0.081	0.081	0.081	0.082	0.084	0.088	0.092	0.088	0.084	0.082	0.081	0.081	0.081	0.081	0.081	0.081	0.081
0.081	0.081	0.081	0.081	0.081	0.081	0.081	0.081	0.082	0.083	0.084	0.083	0.082	0.081	0.081	0.081	0.081	0.081	0.081	0.081	0.081
0.081	0.081	0.081	0.081	0.081	0.081	0.081	0.081	0.082	0.083	0.084	0.083	0.082	0.081	0.081	0.081	0.081	0.081	0.081	0.081	0.081

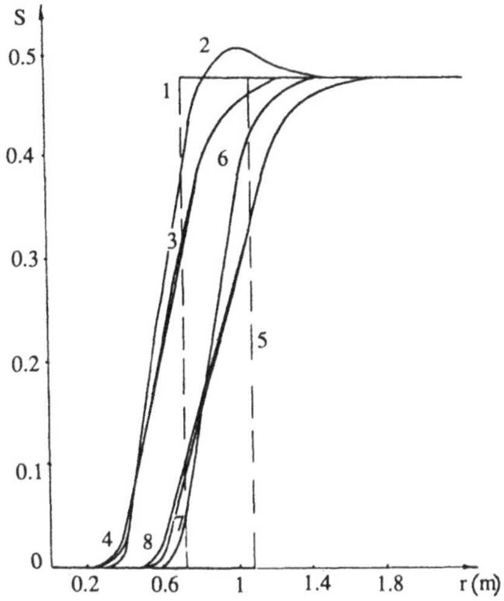

FIGURE 7. Comparison of exact and numerical solutions—two-dimensional model.

Figure 7 shows the saturation profiles for the numerical two-dimensional solution and for the corresponding exact one at 0.555 day (curves 1 to 4) and 1.11 day (curves 5 to 8). Owing to the central symmetry, the saturation distribution along the ray issuing from the center and parallel to the abscissa axis is presented. The designation of the exact solution coincides with Figure 4; curves 2 and 6, 3 and 7, and 4 and 8 correspond to the time steps 0.0308 day, 0.0154 day, and 0.0077 day, respectively. In this case the composition and saturation discontinuities also are calculated most precisely for the maximum Δt. On the other hand, calculation near the stability limit can lead to the oscillating saturation profiles. To decrease the processor time for the calculations one should use the variable Δt, depending on the concrete values of P and z_i at a given moment of time. Notice that the observed maximal value of Δt in the two-dimensional problem is three to four times less than in the one-dimensional one.

Table 4 demonstrates the characteristics of the method obtained by solving the two-dimensional problem. The disposition of the results in the table and the numeration of the calculations are equivalent to Table 1. The calculation required 160 K of computer memory. One time step was calculated in 50 sec of processor time.

It follows from Tables 1 and 4 that, for the schemes presented, approaching the stability limit is characterized more by deviation of the maximum $\sum_{\kappa=1}^{I} z_i$ value from 1, than by the integral balance error.

Lastly, let us estimate the value of the artificial diffusion coefficient D_{num} for the one- and two-dimensional cases. The order of D_{num} can be determined from the relationship

$$D_{num} \cong \frac{(\Delta L_D)^2}{\Delta t_*} \tag{4.4.1}$$

where $\Delta L_D = L_D(t_2) - L_D(t_1)$, $\Delta t_* = t_2 - t_1$; $L_D(t)$ is the width of the saltus

TABLE 4

Number of Calculation	Relative Error in Mixture Integral Balance (%)	Relative Error in CH_4 Integral Balance (%)	Relative Error in C_{10+} Component Integral Balance (%)	Maximum Deviation of $\sum_{\kappa=1}^{l} z_\kappa$ from 1 Over All Sites
1	0.157	0.6	3.7	30.0
2	0.156	0.9	1.5	5.2
3	0.1	1.1	1.1	0.1

smoothing zone at a moment t. Calculating $L_D(t_1)$ and $L_D(t_2)$ from curves 4 and 8 of Figure 4, we obtain for the one-dimensional case:

$$D_{num} = \frac{(0.18)^2}{0.555 \cdot 86400} = 0.68 \cdot 10^{-6} \frac{m^2}{s}$$

In the two-dimensional case, calculating $L_D(t_{1,2})$ from curves 4 and 8 of Figure 7, we have

$$D_{num} = \frac{(0.3)^2}{0.555 \cdot 86400} = 1.88 \cdot 10^{-6} \frac{m^2}{s}$$

For compositions close to the gaseous phase of the mixture used in the above calculations, one has obtained experimentally the value 10^{-6} m^2/s of the effective diffusion coefficient (the case of small Pecle numbers was considered).[9] The value is of the same order as D_{num} calculated by using Equation (4.4.1). It follows that in some cases of flow modeling the unavoidable artificial diffusion can be considered as a kind of analogue of the real diffusion processes that were neglected in the model under discussion.

Thus, as a result of numerical analysis it is shown that the proposed finite-difference schemes can be used in the calculations of real multicomponent reservoir flow processes. The efficiency of the schemes is connected mainly with the absence of iterations by P and z_i calculation.

REFERENCES

1. **Murkes, M. N., Rozhdestvenskii, V. A., and Shovkrinskii, G. Yu.,** *Zh. Vychisl. Mat. Mat. Fiz.*, 17(3), 696, 1977; *USSR Comp. Math. Math. Phys.*, 17(3), 133, 1977.
2. **Mitlin, V. S.,** Method of numerical solving one- and two-dimensional problems of multicomponent reservoir flow, in *Numerical Methods of Continuum Mechanics*, ITPM SO AN SSSR, Novosibirsk, 17(4), 110, 1986 (in Russian).
3. **Yanenko, N. N.,** *The Method of Fractional Steps*, Springer-Verlag, New York, 1971.
4. **Aziz, K. and Settari, A.,** *Petroleum Reservoir Simulation*, Elsevier, London, 1980.
5. **Mitlin, V. S.,** New Methods of Calculation of Enriched Gas Action Upon a Gas-Condensate Seam, Ph. D. thesis, VNIIGAS, Moscow, 1987 (in Russian).
6. **Van Quy, N., Simandoux, P., and Corteville, J.,** A numerical study of diphasic multicomponent flow, *Soc. Pet. Eng. J.*, 12(2), 171, 1972.
7. **Levi, B. I. and Shakirov, H. G.,** *Izv. Akad. Nauk SSSR, Mekh. Zhid. Gaza*, 4, 101, 1985; *Fluid Dynamics*, 20(4), 581, 1985.
8. **Barenblatt, G. I., Entov, V. M., and Ryzhik, V. M.,** *Motion of Fluids in Natural Rocks*, Klumer Academic Publ., Dordrecht, 1990.
9. **Ter-Sarkisov, R. M., Makeev, B. V., and Guzhov, N.A.,** Experimentally determining diffusion coefficients of hydrocarbon gas-condensate mixtures, in *Scientific and Technical Problems of Development of Gas, Gas-Condensate, and Gas-Oil Deposits*, VNIIGAS, Moscow, 1983, 33 (in Russian).

5

Numerical Study of the Miscible Displacement of Multicomponent Systems

5.1. PHYSICAL ASPECTS OF MISCIBLE DISPLACEMENT

The mathematical modeling of multicomponent flow through porous media is necessary in connection with modern methods of displacing various kinds of hydrocarbon mixtures with solvents. The efficiency of the displacement process is determined by the degree of miscibility of the injected and formation fluids.[1,2] Complete miscibility is achieved either by increasing the pressure of the formation fluid above the critical value P_{cr}, or by increasing the content in the formation or injected fluid of the component which during mixing reduces the value of P_{cr}. For hydrocarbon deposits in a late stage of exploitation it is technically impracticable to achieve complete miscibility by increasing the pressure. Achieving complete miscibility by changing the chemical composition requires the use of a large amount of expensive solvent. Accordingly, it is worth investigating the partial miscibility variant.

For fluids that can be represented in the form of three-component systems, displacement under conditions of partial miscibility was investigated.[3-6] It was shown that for the injection of "dry" and "enriched" gases the structures of the transition flow zone are quantitatively different. In the first case the component concentrations and phase saturations vary monotonically through the formation. The main difference between the second case and the first resides in the nonmonotonicity of the space variation of the concentrations and saturations (formation of a "bank" in the liquid phase). In the transition zone there may be a considerable decrease in interphase tension thought to be connected with the "softening" of the phase interface. The system obtained as a result of mixing the formation mixture with "enriched" gas approached the critical point at which the phases are indistinguishable and is characterized by an increased content of the components with equilibrium constants in the order of unity.

In order to properly describe the physical properties of multicomponent multiphase systems it is necessary to know the law of corresponding states in the form of the dependence of the system properties on the parameters of the critical point. At the same time, the convergence of the densities and viscosities of the gaseous and liquid phases with equalization of the phase compositions near the critical point is not decisive for the reservoir flow calculations.

More important are the actual flow characteristics of the process — the phase permeability functions $f_g(S_g)$ and $f_w(S_w)$. Owing to the decline in the role of surface effects as the critical state is asymptotically approached, the relationship

$$f(S_g) = S_g,\ f_w(S_w) = S_w$$

apparently first mentioned by Nikolaevskii et al.[1] must be satisfied. This implies the need to introduce an explicit dependence of $f_{g,w}$ on the phase compositions.[7,8]

It is necessary to take note that concentration dependence of the phase permeabilities has been considered in several publications. For instance, the experimental measurements of the oil and water phase permeabilities vs. the surfactant concentration are presented in book by Babalian et al.[9] The results of the experiments were used to construct a mathematical model of the displacement process. The influence of surface tension on phase permeability changes in an oil-gas system was studied by Amaefule and Handy.[10] Interpolation functions for phase permeabilities dependent on surface tension and phase saturations were reported on the basis of two-phase flow experiments in a paper by Fulcher et al.[11] However, in many cases the functions $f_{g,w}$ are assumed to depend only on the phase saturations. In part, this is associated with the difficulty of carrying out the experiments to determine the actual dependence of $f_{g,w}$ on the composition. A second reason appears to be the preconception that the dependence of the phase permeabilities on the composition is important only in a small neighborhood of the critical point. Below it is shown that, for displacement processes using gaseous solvents enriched in intermediate components, a change in interphase tension, leading to convergence of the phase mobilities, is very significant. It is important that the change in the structure of the interphase surface can be taken into account within the framework of a large-scale approximation, i.e., without including into the mathematical model dissipative (capillary and diffusive) effects.

Let us consider the system of equations of a one-dimensional multicomponent equilibrium reservoir flow (2.1.9). In order to check the adequacy of model (2.1.9) and of the real flow process, numerical solutions of the corresponding problem were compared with the results of experiments on the linear plane displacement of a gas-condensate mixture by a gas enriched in intermediate components.[12] In the experiments, the model of the formation was filled with a real reservoir liquid hydrocarbon phase, which was then displaced by an equilibrium gas phase until the liquid disappeared from the output flow. Then, in a second stage of the experiment enriched gas was pumped through the model. The parameters of the process were measured only at the model outlet.

In the calculations in this section we numerically reproduced only the second part of the experiments, when the enriched gas acted on the mixture at the initial condensate saturation is close to the residual saturation S_w^*.[13]

For a mathematical description of the experiments it is necessary to solve the system of ten equations (2.1.9) for $\alpha = 0$ on the interval $(0, L)$, where L is the length of the formation model. The initial conditions

$$P(r,0) = P^o, \; z_i(r,0) = z_i^o, \qquad i = 1, \ldots, l, \; r \epsilon (0,L)$$

correspond to the initial values of the pressure and composition in the beginning of the second part of the experiment.

At the inlet to the formation the composition of the pumped gas and the injection rate are supposed to be constant

$$z_i(0,t) = \delta_i, \qquad i = 1, \ldots, l$$

$$\left.\frac{k\beta}{N}\frac{\partial P}{\partial r}\right|_{r=0} = v, \qquad N = N(\epsilon_1, \ldots, \epsilon_{l-1}, P), \; \epsilon_i = \frac{\beta_i}{\beta}$$

Here ϵ_i is the portion of the i-th component in the flow.

At the outlet constant pressure is maintained

$$P(L,t) = P^o$$

The phase densities and viscosities were determined from Equation (2.1.10) with $\lambda_1 = \lambda_2 = \ldots = \lambda_8 = 1$. The equilibrium constants were approximated by quadratic dependences on the pressure and a single composition parameter R defined by Equation (2.1.11), and the initial information for constructing the interpolation polynomials was the result of experiments on a PVT bomb. The quantity R in Equation (2.1.11) must be determined in such a way that the compositions of the gas and liquid phases of the initial formation system correspond to the same value of

$$\frac{y_{int}}{y_{int} + y_{he} + d_{tr}} = \frac{x_{int}}{x_{int} + x_{he} + d_{tr}} \tag{5.1.1}$$

Here x_{int}, x_{he} and y_{int}, y_{he} are portions of intermediate and heavy components in liquid (gaseous) phase. It is not hard to see that in this case the quantity R is invariant along the conode (for mixing the gas and liquid phases in any proportion).

The phase permeability functions were taken in the form

$$f_w = \begin{cases} \left(\dfrac{S_w - S_w^*}{1 - S_w^*}\right)^{\tilde{\gamma}}, & S_w \geq S_w^* \\ 0, & S_w < S_w^* \end{cases} \tag{5.1.2}$$

$$f_g = \begin{cases} \left(\dfrac{S_g - S_g^*}{1 - S_g^*}\right)^{\tilde{\gamma}}, & S_g \geq S_g^* \\ 0, & S_g < S_g^* \end{cases}$$

The following initial data were used in the calculations: $L = 5$ m, $k = 10^{-14}$ m^2, $m = 0.222$, $P^o = 10.5$ MPa, $T = 335$ K, $v = 1$ m/day. The initial composition was 0.0001 (N_2), 0.56 (CH_4), 0.08905 (C_2H_5), 0.05365 (C_3H_8), 0.03565 (C_4H_{10}), 0.02535 (C_5H_{12}), 0.0333 (C_6H_{14}), 0.0813 (C_7H_{16}), 0.5975 (C_9H_{20}), and 0.06005 ($C_{10}H_{22}$). The composition of the enriched gas was 0.50 (CH_4), 0.24 (C_2H_6), 0.17 (C_3H_8), and 0.09 (C_4H_{10}).

Two groups of calculations were carried out. In the first calculations the phase permeability parameters were assumed to be constant ("frozen"):

$$S_w^* = 0.36, \; S_g^* = 0.12, \; \tilde{\gamma} = 3$$

In the second group the phase permeabilities were considered to be "moving", i.e., to depend on the phase property proximity parameter, for which we chose the surface tension σ, which vanishes at the critical point of the multicomponent system. The quantity σ was derived by Equation (2.1.12), the residual saturations and the exponent in Equation (5.1.2) were defined in such a way that at $\sigma \rightarrow 0$ the phase permeabilities become straight lines

$$S_w^* = \frac{0.6\sigma}{0.0032 + \sigma}, \; S_g^* = \frac{1}{3} S_w^*, \; \tilde{\gamma} = 1 + 5S_w^* \qquad (5.1.3)$$

and the coefficients in the expression for S_w^* being determined from laboratory experiments by the injection of gas mixtures of various compositions through the formation model up to the establishment of phase equilibrium.[12]

It should be noted that near the critical point the definition of the residual saturations as constants is physically contradictory. On the one hand, molecules of phases similar in composition should possess practically the same mobility, and such phases should differ very little by their motion through porous media (almost complete absence of capillary forces). On the other hand, as a result of the "rigid" definition of $S_{g,w}^*$, a part of the formation system (liquid) may be immobile, although in physical properties it differs little from the mobile part of the system (gas). The necessity of the phase permeabilities transformation to straight lines also is dictated by the requirement for continuity of the transition through the critical point from a two-phase state to a one-phase one in TMRF, which is equivalent to continuity of the conductivity coefficients β_i in Equation (2.1.9).

The system of Equations (2.1.9) was solved by a finite-difference method considered in the previous chapter. In Figures 8 and 9 we have plotted the distributions of pressure (curves 1), liquid saturation (curves 2), mixture fractions of CH_4 (curves 3), C_3H_8 (curves 4), and C_7H_{16} (curves 5) for "moving" and "frozen" permeabilities, respectively, at the dimensionless time $t_u = 0.5$. Here and further on in this chapter changing the parameters is considered versus t_u, equal to the ratio of the injected gas volume to the pore volume of the model;

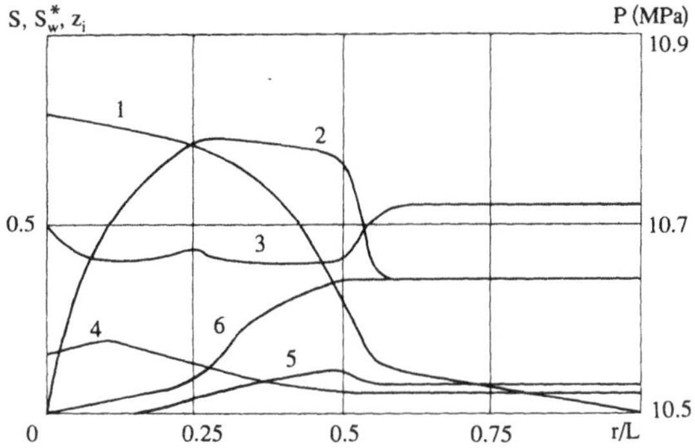

FIGURE 8. Spatial distribution of characteristics of the reservoir mixture. Case of "mobile" permeabilities.

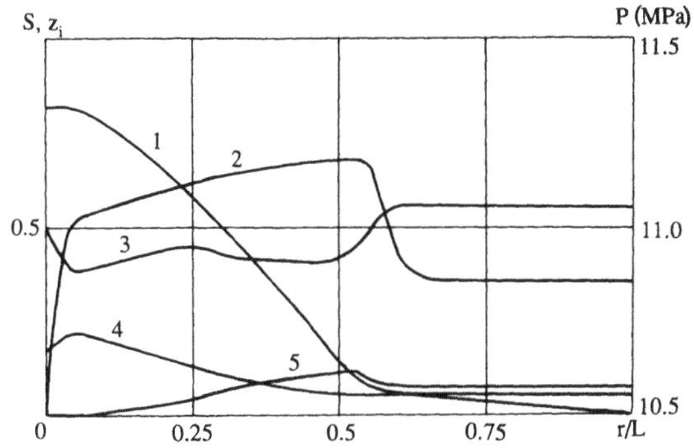

FIGURE 9. Spatial distribution of characteristics of the reservoir mixture. Case of "frozen" permeabilities.

$$t_u = \frac{vt}{mL}$$

for the plane linear flow. In the case of "moving" permeabilities the pressure difference is mostly expended in the front of the bank, while in the case of "frozen" permeabilities it is expended uniformly over the whole bank. The front of the bank is enriched in heavy components, the rear in intermediate components. The maxima of the intermediate and heavy components contents lie closer together in the case of "moving" permeabilities. In the bank zone the methane content has two minima and one maximum. In the first case the rear of the bank contains more heavy components, which leads to substantial differences in the saturation distributions in the calculations. Curve 6 in Figure 8 corresponds to the distribution of the residual saturation of the liquid phase. Clearly, at the rear of the bank S_w^* decreases practically to zero.

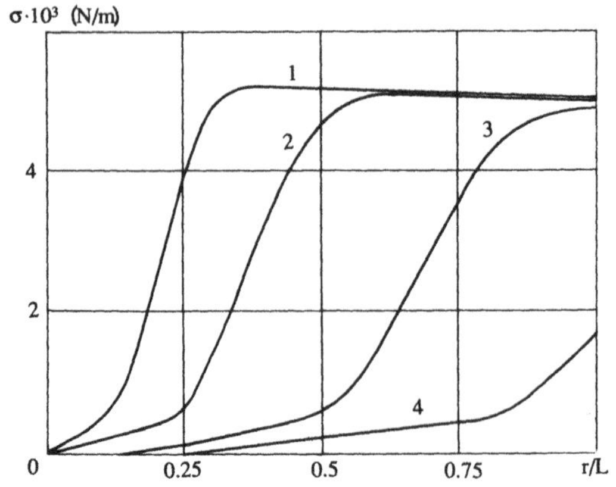

FIGURE 10. Evolution of the surface tension profile.

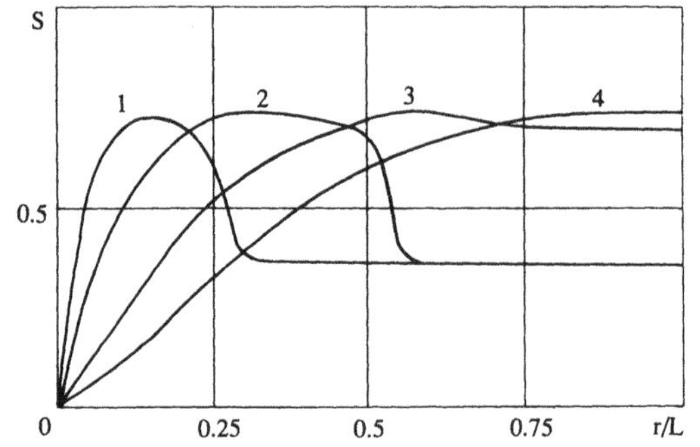

FIGURE 11. Evolution of the liquid saturation profile.

The moment of arrival of the bank to the outlet of the formation model corresponds to the injection of one pore volume. The fraction of light components in the output flow thereupon falls sharply, the intermediate fraction increases monotonically, and the heavy fraction first increases strongly and then decreases. The pressure difference between the ends of the formation model behaves similarly with the time: at first it increases reaching the maximum, and then decreases. In the case of "frozen" permeabilities the difference is sufficiently greater than in the case of "moving" permeabilities (discussed thoroughly in the next section).

The surface tension distributions are shown in Figure 10. Curves 1 to 4 correspond to the injection of 0.25, 0.5, 1, and 1.5 pore volumes. Corresponding profiles of S_w are shown in Figure 11, and profiles of S_w^* are shown in Figure 12. The intervals of rapid growth of the interphase tension correspond to the front of the bank; at the lightened rear zone it varies little and is close to zero.

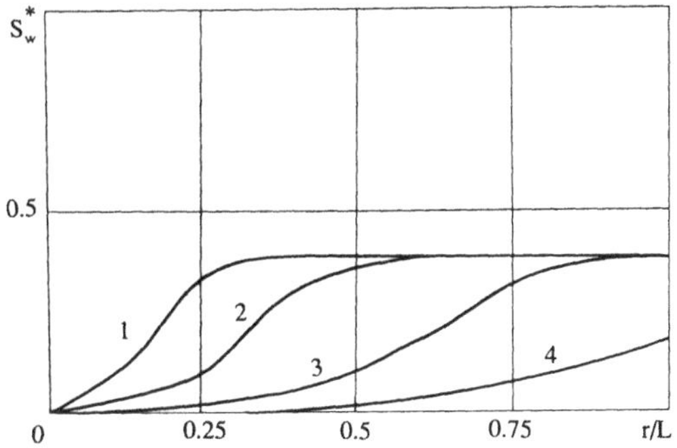

FIGURE 12. Evolution of the liquid residual saturation profile.

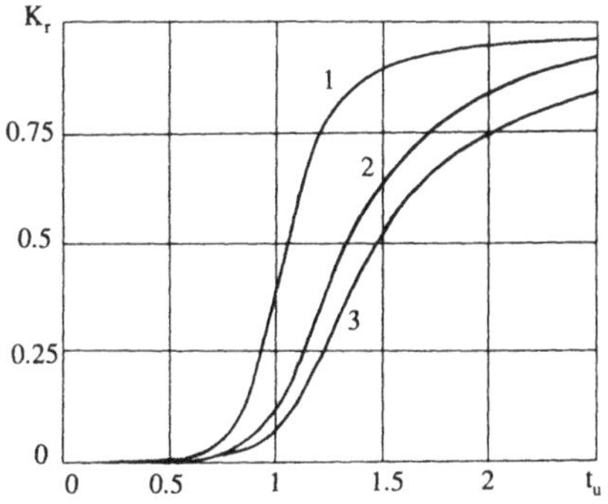

FIGURE 13. Comparison of the measured curve of the C_{5+}-components recovery coefficient and the calculated ones (simulation of the second stage of the experiment).

After passage of the bank the tension rapidly decreases. The formation-average liquid saturation $\langle S_w \rangle$ decreases much more slowly. Thus, at $t_u = 1.5$, $\sigma_{max} = 0.00178$ N/m, $\langle S_w \rangle = 0.505$, and at $t_u = 2.5$, $\sigma_{max} = 0.0041$ N/m, $\langle S_w \rangle = 0.325$. The decrease of the surface tension shows that the composition of the system remaining in the formation is near-critical.

In Figure 13 we have plotted the calculated (curves 2 and 3) and experimental (curve 1) dependences of the heavy components (C_{5+}) recovery coefficient K_r. A greater closeness to the experimental dependence and a greater recovery effect is achieved with "moving" permeabilities (curve 2). In this case the difference of the liquid mobility in the front and in the rear of the bank is sufficiently more than the difference in the case of "frozen" permeabilities. The conditions for the formation of a stable bank are therefore more favorable in the case of "moving" permeabilities (the difference in velocities before and after the discontinuity can be regarded as a measure of its stability[15]).

In order to estimate the effect of the geometry of the problem on the displacement parameters, system (2.1.9) was solved not only for $\alpha = 0$, but also

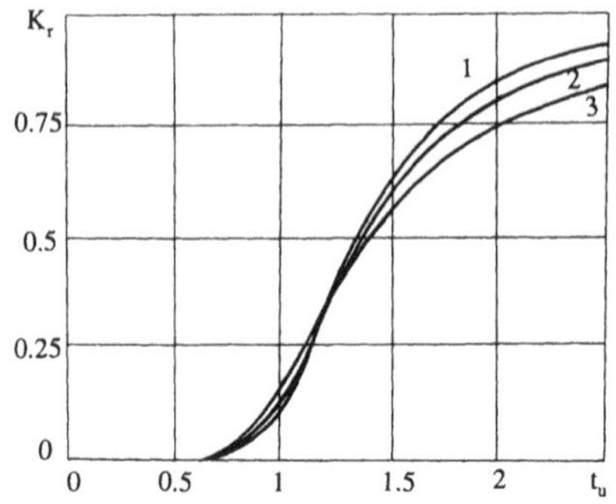

FIGURE 14. Behavior of the C_{5+}-components recovery coefficient by different geometries of the flow region. Curve 1: linear flow, 2: plane-radial flow, and 3: sphere-radial flow.

for the plane-radial and spherically radial cases, i.e., for $\alpha = 1$ and $\alpha = 2$, in the region (r_b, R_c), $= 0.127$ m, $R_c = r_b + L$. The initial and boundary conditions with respect to the concentrations at the well and with respect to pressure on the contour are the same as in the linear case. The condition of constancy of the rate of injection of "enriched" gas into the well has the form

$$-\frac{k\beta}{N}\frac{\partial P}{\partial r}\bigg|_{r=r_b} = v$$

The solution also depends on the dimensionless time

$$t_u = \frac{2r_b tv}{m(R_c^2 - r_b^2)}, \quad \alpha = 1$$

$$t_u = \frac{3r_b^2 tv}{m(R_c^3 - r_b^3)}, \quad \alpha = 2$$

The initial data are equivalent to the linear ($\alpha = 0$) case.

In Figure 14 we have plotted the dependence of the heavy components recovery coefficient for "moving" phase permeabilities in the cases $\alpha = 1$ (curve 2) and $\alpha = 2$ (curve 3). Curve 1 corresponds to the case $\alpha = 0$ and is equivalent to curve 2 in Figure 13. Should we have the piston displacement, the recovery coefficient K_r would not depend on the geometry of the region, and the injected amount of fluid would coincide with the obtained one. In the case of partially miscible displacement a more complicated dependence $K_r(t_u)$ on α was observed. Namely, in the first stage of passage of the bank out of the formation model ($t_u < 1.25$) the increase in the dimensionality of the problem leads to an increase in the recovery coefficient. However, in the later stages of the process the recovery coefficient decreases with an increase in α. It is interesting to notice that all the curves in fact have a common point of intersection, approximately for the time $t_u = 1.25$. Similar behavior was also observed in the case of "frozen" permeabilities.

According to the calculations considered above, taking into account the phase permeabilities changes near the critical point can have sufficient bearing on the flow structure description by the miscible displacement. Notice that despite the fact that the introduction of "moving" permeabilities brings the calculated and experimental values of the recovery coefficient closer together, the divergence of the curves still remains considerable. As will be shown in the next section, the complete modeling of the process, including the first stage, enables one to decrease the divergence of the calculated and experimental curves. We also will discuss below the results of comparing some other calculated and experimental parameters.

5.2. COMPARING THE RESULTS OF PHYSICAL AND NUMERICAL MODELING

The calculations discussed in the previous section were based on the assumption that after the displacement of the formation liquid phase by the gas, which is in equilibrium with respect to the formation liquid, the space distribution of concentrations and saturations in the formation is close to homogeneous. In accordance with this assumption the invariable initial conditions determined from experimental data of the amounts of injected gas and the obtained product were used in the calculations. As a matter of fact, after ending the equilibrium gas injection the space distributions of composition and saturations is essentially inhomogeneous. Namely, near the inlet to the formation a mixture poor in heavy components is collected, but the formation mixture near the outlet is more enriched by C_{5+} components. The liquid saturation near the outlet can exceed the formation-average value, which usually is considered as the residual saturation in the "frozen" models of phase permeabilities; the liquid saturation is close to zero near the inlet. More accurate description of the process requires mathematical modeling for both the first and the second stages of the experiment.[16]

Mathematical description of the process requires solving system (2.1.9) for $\alpha = 0$ on the interval $(0, L)$ with the initial conditions

$$P(r,0) = P^o, \; z_i(r,0) = x_i^o \,, \quad i = 1, \ldots, l \,, \quad r \in (0,L)$$

where x_i^o is the composition of the reservoir liquid phase. At the outlet the constant pressure P^o is maintained, at the inlet the constant injection rate v and the following conditions with respect to the composition of the pumped solvent are imposed

$$z_i(0,t) = y_i^o, \; t \leq t_1$$
$$z_i(0,t) = \delta_i, \; t > t_1 \,, \qquad i = 1, \ldots, l$$

Here y_i^o is the composition of the equilibrium (with respect to x_i^o) gaseous phase, t_1 is the time of ending the equilibrium gas injection, which was equal in the experiment to three pore volumes of the formation model.

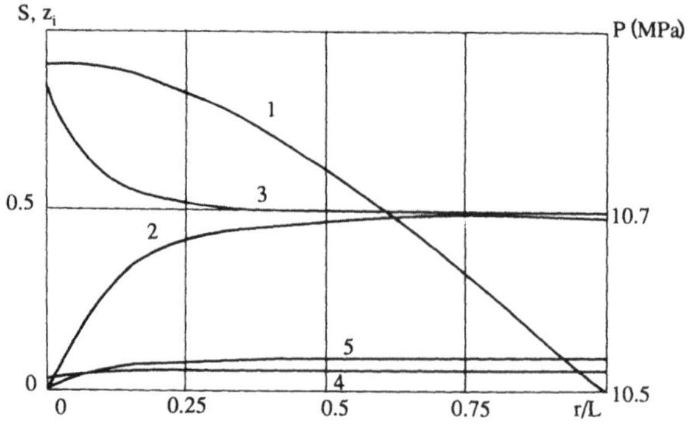

FIGURE 15. Spatial distribution of characteristics of the reservoir mixture after completing injection of the equilibrium gas.

Figure 15 shows the space distributions of the pressure (curve 1), liquid saturation (curve 2), the overall fractions of CH_4 (curve 3), C_3H_8 (curve 4), and C_7H_{16} (curve 5). One can see that the main changing of S and z_i is concentrated in the first quarter of the formation model. The pressure change per unit of length is very small at $r/L < 0.1$ (small liquid saturation), increases at $0.1 < r/L < 0.3$ (sufficient increase of amount of liquid), and achieves the constant and maximal value at $r/L > 0.3$ (practically invariable liquid saturation). One can see that the C_7H_{16} fraction changes essentially (from 0.005 at $r = 0$ to $r = L$). The fraction of heavy components is equal to 0.02 at $r = 0$, and to 0.33 at $r = L$. The results presented in Figure 15 correspond to the calculation with "frozen" phase permeabilities. Similar results are obtained in the case of "moving" phase permeabilities; we will not discuss them because at $t < t_1$ the results of both calculations differ little.

In Figure 16 the experimental (curve 1) and calculated (curve 2) recovery coefficients of components C_{5+}, and the experimental (curve 3) and calculated (curve 4) values of the formation-averaged liquid saturation vs. the number of pore volumes of the injected gas are given. At $t_u > 1$ the recovery of the heavy components is practically zero, and the liquid saturation changes little. From Figure 16 one can see the good coincidence of the experimental and calculated results.

In Figure 17 we show the calculated dependences of the liquid phase density at the outlet section of the formation model vs. time for "moving" (curve 3) and "frozen" (curve 4) phase permeabilities. The corresponding experimental data (shown by circles) were obtained by collection of the liquid in a special "pocket" connected with the outlet section of the formation model. At $t_u = 0.5$ (the moment of arrival of the equilibrium gas to the outlet in the first stage of the experiment) ρ_w decreases somewhat, which can be explained by a small deviation of injected gas composition used in the calculations from the equilibrium composition with respect to the formation liquid. In the stage of the enriched

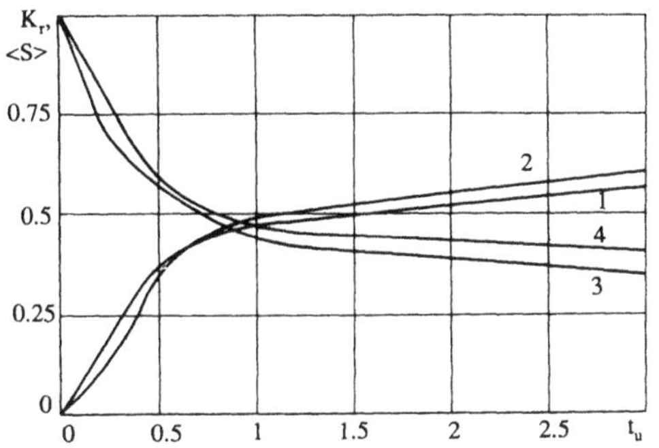

FIGURE 16. Comparison of the experimental and computational data at the stage of the equilibrium gas injection.

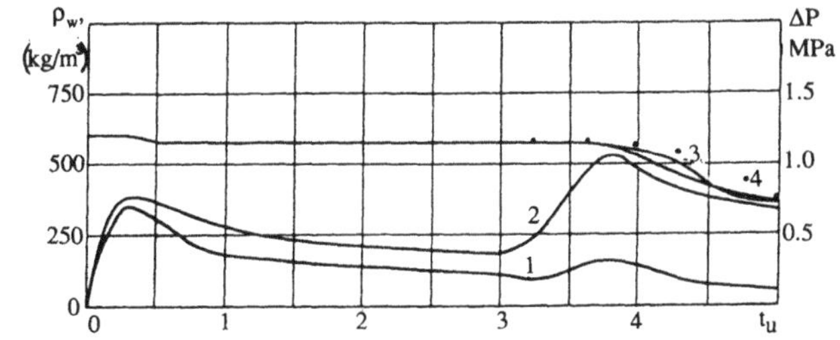

FIGURE 17. Temporal dependence of the pressure difference between the ends of the model and the density of liquid phase at the exit of the model, for different forms of $f_{w,g}$. The experimental data for the liquid density are shown by circles.

gas injection ($t_u > 3$) the calculated values of ρ_w first coincide exactly with the experimental ones, then (after passage of the bank through the formation model) a divergence appears, which decreases gradually with time. For instance, at $t_u = 2.5$ the experiment gives $\rho_w = 340$ kg/m^3, and the corresponding calculated value of ρ_w is 330 kg/m^3. Notice, that the curve 3 more precisely represents the shape of changing ρ_w in experiments, compared with curve 4.

Figure 17 shows the time-dependences of the pressure difference ΔP between two ends of the formation model, respectively, for the case of "moving" (curve 1) and "frozen" (curve 2) permeabilities. The first maximum of ΔP is reached at $t_u = 0.25$ immediately before the arrival of the first portions of the injected equilibrium gas to the outlet of the formation model. Then ΔP decreases until beginning the injection of the enriched gas. The later increase of ΔP is connected with flow resistance growth in the process of the forming and movement of the liquid phase bank. After passage of the bank, ΔP decreases again. The calculated value of ΔP in the second stage of the process is approximately three times greater in the case of "frozen" permeabilities than in the case of "moving" permeabilities. The fact that in the experiments the pressure difference did not exceed 1 MPa also speaks in favor of the model with "moving" permeabilities.

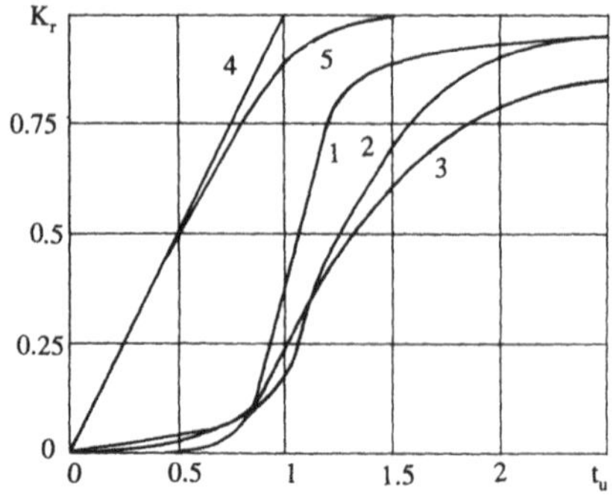

FIGURE 18. Comparison of the measured curve of the C_{5+}-components recovery coefficient and the calculated ones (complete simulation of experiment).

In Figure 18 the dependences of the recovery coefficient of components C_{5+} are represented; the reading of time in the figure starts from the moment the equilibrium gas injection ends. The experimental curve is designated by number 1, calculation in the case of "moving" and "frozen" permeabilities are represented by curves 2 and 3, respectively. As in Section 5.1, recovery of heavy components is greater in the case of "moving" permeabilities. At $t_u > 1$ the slopes of the calculated curves are larger than the corresponding slopes of the curves 1 and 2 in Figure 13. The agreement of the curves 1 and 2 in Figure 18 is better than the agreement of the corresponding curves in Figure 13. Since the calculations before beginning injection of enriched gas show the value of the liquid saturation at the outlet exceeds the residual saturation value, at $t_u < 0.5$ a small recovery of the heavy components is observed, which was practically absent in the experiment (see curve 1). This divergence may be connected with the existence of the end effect in the experiment, which is not taken into account in the framework of the mathematical model used (accumulation of the liquid phase near the outlet of the formation model because of the capillary "jump" there, which alternates by liquid movement out of the formation by reaching sufficient mobility).

The additional reason leading to the decrease of the calculated recovery coefficients slope is the influence of the numerical (artificial) diffusion. This effect is not connected with the concrete mathematical model used in the calculations, and is inherent in any finite-difference scheme.

In order to investigate the effect of artificial diffusion on the results of the computations let us consider the shape of the recovery coefficient of components C_{5+} in the problem of piston displacement discussed in the Section 4.3. As was shown, this problem has a self-similar solution, from which it follows that the heavy-component recovery coefficient coincides with t_u (curve 4 in Figure 18). Curve 5 is obtained from the numerical solution of the same problem. From the comparison of curves 4 and 5 it follows that the artificial diffusion has the most influence on the integral characteristics of the process in the stage of the front approach to the outlet. The decrease of the slope of the calculated recovery coefficient is connected with "smoothing" the numerical solution in a vicinity of discontinuity. However, as follows from Figure 8, by the calculations of enriched gas injection with the nonlinear functions of phase permeabilities, some (at least, two) discontinuities of concentrations and saturations profiles appear. The known analytical solutions for three-component models confirm this point.[3-6] Therefore, it follows that, due to the numerical diffusion, the slope of the recovery coefficient in the case of calculations of enriched gas injection can decrease sufficiently more than in the case of piston displacement. This is because of the "smoothing" of the numerical solution near each discontinuity. That can explain, in our opinion, the divergence of curves 1 and 2, which is essentially more than the divergence of curves 3 and 4. At the same time, such a fast change of the heavy-component recovery coefficient, which was observed in the experiment, may speak about the slowing down of the capillary and diffusive relaxation processes in the near-critical state. That should mean there is a necessity to take into account sufficient properties of multicomponent systems, such as the strong spatial correlation of fluctuations and slowing down of all the relaxation processes in a vicinity of the critical point, in the mathematical model of the process.

To conclude this chapter, we will consider some properties of phase permeabilities connected with the closeness of the reservoir system to the thermodynamic critical point and based on the scaling concept of the physical properties in the near-critical state. The case of the weak molecular interaction of fluid and pore boundaries will be discussed. The case of strong dependence of the phase state on the interactions "fluid-porous surface" will be considered in Section 9.4.

5.3. ASYMPTOTICAL BEHAVIOR OF PHASE PERMEABILITIES NEAR THE CRITICAL POINT

For complete mathematical models it is necessary to know the functional dependences of $S^*_{w,g}$ and $\widetilde{\gamma}$ on the parameters of the two-phase systems and the porous media for the situation of almost total mutual solubility of the fluid phases. This requires further theoretical and experimental study. The latter presupposes measurements of the residual saturations of the phases when solvents of different compositions are pumped through porous media of different structures, which, when combined with an analysis of the composition of the formation mixtures after pumping a sufficient amount of solvent, will enable one to determine the coefficients in expressions of the type of Equation (5.1.3) in each specific case. Near the critical point, obtaining reliable experimental data requires extremely high measuring accuracy, and in particular thermostating. At the same time, from the standpoint of a theoretical study, the situation may be simplified because of the universality of the behavior of different systems near the critical point. As an example, we will show how it is possible to determine the form of the dependences $S^*_{w,g}(\sigma)$ by using a scaling approach.[17]

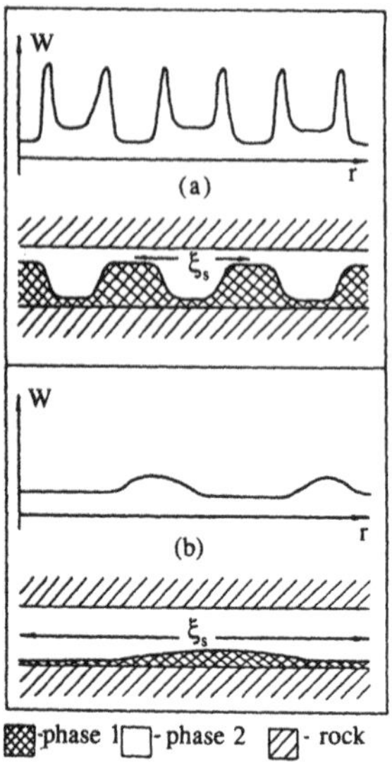

FIGURE 19. Changes in the distribution of a two-phase mixture and the free energy density W profile when approaching the critical point.

As noted above, the need to take into account the change in the properties of the interphase surface in determining $f_{g,w}$ follows from the physics of the mixing process in a porous media. In fact, the establishment of complete phase equilibrium is limited by the small sizes of the pore channels

$$\xi_p \sim \left(\frac{k}{m}\right)^{1/2}$$

Therefore, the distribution of a two-phase mixture in an isolated channel is a collection of microphases separated by surface layers.

For the purposes of discussion it is convenient to use scaling considerations. If the multicomponent system is far from a state of total mutual solubility of both phases, then the wetting phase will take the form of isolated drops. The spatial scale of the drop will be small owing to the smallness of the correlation length: actually, the quantity ξ_s is related to the vanishing determinant of the matrix of the second derivatives of the free energy with respect to the independent thermodynamic variables, and becomes infinite at the critical point.[18] The free energy density in the surface layer far exceeds that in the phases (Figure 19[a]) and is localized on distances in the order of the diameter of a single molecule.[1]

As is known, if the mobility of a phase of a multiphase system is to be ensured, its parts must be connected in the pore space. This can be achieved not only by increasing the phase saturation but also by adding components that bring the system closer to the state of total phase solubility. In the last case the free energy density in the surface layers decreases and approaches the values in the phases. The size of the drops and the thickness of the surface layer, which are proportional to ξ_s, increase substantially (Figure 19[b]). The residual phase saturations $S^*_{g,w}$ decrease, since the motion of a phase can take place through the connected part, which forms a thin layer and, owing to the low curvature of the interphase surface, occupies a much smaller part of the pore space than in the case of mutually insoluble fluids. Thus, in the case in question an extended cluster of drops is formed, not owing to an increase in their number but owing to an increase in their spatial sizes and a corresponding decrease in the interphase contact angle between the interfacial surface and the pore surface. A substantial increase in phase mobilities is to be expected when $\xi_s \geq \xi_p$. From this there follows the dependence of the phase permeabilities on the parameters of the porous medium. A similar spatially periodic structure of the microphases of a binary mixture, described by means of a free energy functional of the Ginzburg-Landau type, has been studied[20] (also see Chapter 8).

Since at the microlevel in the porous medium there is a single space scale ξ_p, one can represent the residual saturations in the form

$$S^*_{w,g} = \widetilde{F}(\xi_p/\xi_s)$$

where $\widetilde{F}$ is a monotonically increasing function at the small values of its argument; the specific form of $\widetilde{F}$ depends on the structure of the porous media. For sufficiently small $S^*_{w,g}$ this relation can be represented in the form

$$S^*_{w,g} \sim (\xi_p/\xi_s)^{\widetilde{\nu}_{w,g}}, \quad \widetilde{\nu}_{w,g} > 0 \tag{5.3.1}$$

Expanding the free energy density in a series about the value in a phase where it reaches a minimum (see Figure 19), we find that the excess energy density in the surface layer, which is proportional to σ, is determined by the matrix of the second derivatives of the free energy and, hence, is proportional to the square of the reciprocal correlation length.[14] We have

$$\sigma \cong k_B T \xi_s^{-2} \tag{5.3.2}$$

where T is the temperature, and k_b is the Boltzmann's constant. A relation analogous to Equation (5.3.2) was obtained for a semidiluted polymer solution.[17] Using Equations (5.3.1) and (5.3.2), we find

$$S^*_{w,g} \sim \left(\frac{\sigma \xi_p^2}{k_B T}\right)^{\frac{\widetilde{\nu}_{w,g}}{2}} \tag{5.3.3}$$

Notice that the $S^*(\sigma)$ dependence was investigated experimentally in an oil-water system.[10] For the oil $S^* \sim \sigma^{0.5213}$, and for the water $S^* \sim \sigma^{0.1534}$. However, the values of the water residual saturation were in all cases fairly large, and therefore representing it by a power dependence may be incorrect, whereas in the experiments the oil residual saturation reached fairly small values. Comparison of the experimental value of the exponent in the oil residual saturation dependence with Equation (5.3.3) gives $\tilde{\nu} = 1.04$. Experimental investigations of systems of the gas-condensate type with similar mobilities of both phases at small σ are required. Matching relationships of type of the Equation (5.3.3) suitable in the immediate vicinity of the critical point, with reliable experimental data obtained far from that point, should give a fairly complete description of the corresponding processes for systems of the gas-condensate type.

The above consideration is related to the static critical phenomena in porous media. Taking into account critical dynamics requires using nonequilibrium models of flow in porous media. Within the bonds of the model used in the above calculations, "adding" of nonequilibrium effects assumes to take into account the relaxation of surface tension or another function of composition,[1,19] with the relaxation time increasing unboundedly near the critical point.[18] According to S^* dependence on σ, the rearrangement of phase permeabilities in the near-critical state also will slow down.

REFERENCES

1. **Nikolaevskii, V. N., Bondarev, M. I., Mirkin, M. I., et al.,** *Motion of Hydrocarbon Mixtures in Porous Media,* Nedra: Moscow, 1968 (in Russian).
2. **Rozenberg, M. D., Kundin, S. A., Kurbanov, A. K., at al.,** *Flow of a Gasified Liquid and Other Multicomponent Mixtures in Oil Reservoirs,* Nedra, Moscow, 1969 (in Russian).
3. **Afanas'ev, E. F., Nikolaevskii, V. N., and Somov, B. E.,** Problem of the displacement of a multicomponent hydrocarbon mixture by gas injection, in *Theory and Practice of Petroleum Production,* Nedra, Moscow, 1971, 107, (in Russian).
4. **Entov, V. M.,** *Usp. Mekh.*, 4(3), 41, 1981 (in Russian).
5. **Bedrikovetskii, P. G.,** *Dokl. Akad. Nauk SSSR,* 262(1), 49, 1982; *Sov. Phys. Dokl.*, 27(1), 14, 1982.
6. **Zazovskii, A. F.,** *Izv. Akad. Nauk SSSR, Mekh. Zhid. Gaza,* 5, 116, 1985; *Fluid Dyn. USSR,* 20(5), 765, 1985.
7. **Nghiem, L. X., Fong, D. K., and Aziz, K.,** *Soc. Pet. Eng. J.*, 21(6), 687, 1981.
8. **Vafina, N. G. and Batalin, O. Yu.,** in *Int. Conf. Development of gas-condensate fields, Rep. Sect. 6. Fundamental and Research Scientific Investigation,* Krasnodar, 1990, 188.
9. **Babalian, G. A., Levi, B. I., and Tumasian, A. B.,** *Development of Oil Reservoirs by Using Surfactants,* Nedra, Moscow, 1983 (in Russian).
10. **Amaefule, J. G. and Handy, L. L.,** *Soc. Pet. Eng. J.*, 22(3), 371, 1982.
11. **Fulcher, R. A., Ertekin, J., and Stahl, C. D.,** *J. Pet. Technol.*, 37(2), 249, 1985.
12. **Ter-Sarkisov, R. M.,** *Gazov. Prom.*, 10, 26, 1982 (in Russian).
13. **Guzhov, N. A. and Mitlin, V. S.,** *Izv. Akad. Nauk SSSR. Mekh. Zhid. Gaza,* 4, 83, 1986; *Fluid Dyn. USSR,* 21(4), 576, 1986.
14. **Landau, L. D. and Lifshitz, E. M.,** *Statistical Physics,* Vol. 2, 3rd ed., Pergamon Press, Oxford, 1980.

15. **Rozhdestvenskii, B. L., and Yanenko, N. N.,** *Systems of Quasilinear Equations and Their Applications to Gas Dynamics,* Nauka, Moscow, 1987 (in Russian).
16. **Mitlin, V. S.,** *New Methods of Calculation of Enriched Gas Injection into a Gas-Condensate Seam,* Ph. D. thesis, VNIIGAS, Moscow, 1987 (in Russian).
17. **de Gennes, P.G.,** *Scaling Concepts in Polymer Physics,* Cornell University Press, Ithaca, NY, 1979.
18. **Ma, S. K.,** *Modern Theory of Critical Phenomena,* W. A. Benjamin, New York, 1976.
19. **Barenblatt, G. I., Entov, V. M., and Ryzhik, V. M.,** *Motion of Fluids in Natural Rocks,* Klumer Academic Publ., Dordrecht, 1984 (in Russian).
20. **Mitlin, V. S., Manevich, L. I., and Erukhimovich, I. Ya.,** *Zh. Eksp. Teor. Fiz.,* 88(2), 495, 1985; *Sov. Phys. JETP,* 61(2), 290, 1985.

6

Numerical Modeling Enriched Gas Action Upon a Reservoir

In connection with the essential necessity of increasing the efficiency of our use of the natural resources, solving the hydrocarbon recovery problem takes on great meaning. Especially, we are concerned with recovery of the condensate, which remains in a formation by the pressure decrease during the gas-condensate reservoirs exploitation. Some time ago this condensate was supposed to be irretrievably lost and its recovery seemed to be impossible in the U.S.S.R. However, experimental investigations carried out in the last few years by researchers at VNIIGAS (National Scientific Research Institute of Natural Gases) have showed, in principle, the possibility of recovering the remaining condensate from a formation.[1,2] One of the prospective methods for recovery consists in injecting a natural gas enriched by intermediate components into a gas-condensate formation. Some necessary data for preliminary estimation of the reservoir process should be based on laboratory studies (in PVT chamber) of the mixtures of reservoir fluid and enriched gas, and on the laboratory physical reservoir flow modeling. However, using these methods it is impossible to estimate the multidimensional behavior of the pressure and composition fields, and this leads to the necessity of computer compositional modeling. In this chapter we demonstrate some results of such modeling that were obtained using the numerical method presented above. The corresponding reservoir experiments were carried out at the gas-condensate formation Vuktyl (former U.S.S.R.).[3]

6.1. INJECTION OF ENRICHED GAS INTO A GAS-CONDENSATE SEAM WITH POSTERIOR PRODUCT RECOVERY

In this section some results of computer modeling the injection of enriched gas into a well and the posterior product recovery from the same well are presented. The one-dimensional model of multicomponent reservoir flow is used.[4,5]

Mathematical description of the process consists in the solving the plane-radial ($\alpha = 1$) problem (Equation 2.1.9) in a homogeneous seam with the initial conditions

$$P(r,0) = P^o, z_i(r,0) = z_i^o \ , \ r \in (r_b, R_c)$$

Here r_b is the well radius, R_c is the drain zone radius.

At the well changing the volume flow rate measured by normal conditions is given up:

$$2\pi rh \frac{k\beta}{N}\frac{\partial P}{\partial r}\bigg|_{r=r_b} = \begin{cases} Q_1, \ 0 < t \le t_1 \\ Q_2, \ t_1 < t \le t_2 \\ Q_3, \ t_2 < t \end{cases} \tag{6.1.1}$$

where $N = N(\epsilon_i, P_a, T_a)$;h is the seam thickness; ϵ_i is the fraction of the i-th component in the flow. At the drain contour the constant pressure is maintained

$$P(R_c,t) = P^o \tag{6.1.2}$$

Here t_1 is the time duration of the initial extraction stage, and t_2 is the time of completing the enriched gas injection stage. The compositional boundary conditions are imposed according to the direction of the mixture flow, i.e.,

$$\begin{aligned} &z_i(r_b,t) = \delta_i \text{ at } Q < 0 \text{ (injection)} \\ &z_i(r_b,t) \text{ are not imposed at } Q > 0 \text{ (extraction)} \end{aligned} \tag{6.1.3}$$

The following initial data were used in calculation: $k = 10^{-13}$ m^2, $m = 0.079$,$T = 335$ K , $r_b = 0.127$ m, $R_c = 40.127$ m, $h = 20$ m, $t_1 = 7$ days, $t_2 = 14$ days, $Q_1 = 100{,}000$ m^3/day, $Q_2 = -36{,}000$ m^3/day, $Q_3 = 100{,}000$ m^3/day, and $P^o = 10.62$ MPa. The initial reservoir composition was 0.001 (N_2), 0.56 (CH_4), 0.08905 (C_2H_6), 0.05365 (C_3H_8), 0.03565 (C_4H_{10}), 0.02535 (C_5H_{12}), 0.0333 (C_6H_{14}), 0.0813 (C_7H_{16}), 0.05975 (C_9H_{20}), and 0.06005 ($C_{10}H_{22}$). The composition of the enriched gas was 0.648 (CH_4), 0.092 (C_2H_6), 0.098 (C_3H_8), 0.16 (C_4H_{10}), and 0.002 (C_6H_{14}).

The radius of the drain contour was chosen for use in the calculations so that the composition there did not change after completing the injection. Calculation of the initial extraction stage was carried out to create a typical pressure and composition distribution before the injection began.

As was shown in the previous chapter, for the correct description of multicomponent mixtures in porous media, imposing the requirement of the equalization of phase mobilities by approaching the critical point is very important. We demonstrated that this was equivalent to holding the conditions

$$f_{w,g}(S_{w,g}) \longrightarrow S_{w,g}$$

For example, for the enriched gas action, changing the phase densities and

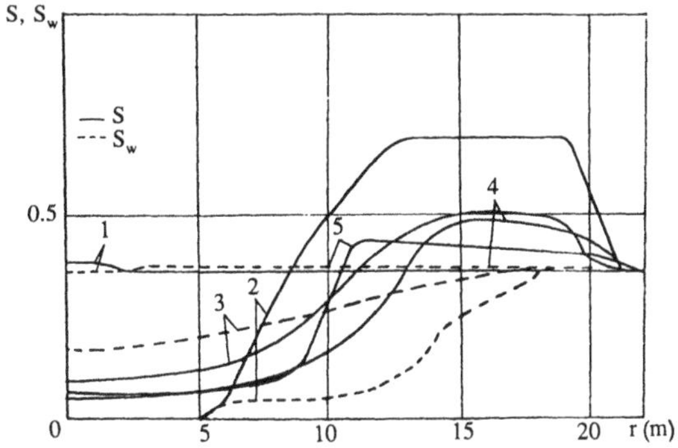

FIGURE 20. Plane-radial problem. Profiles of S and S_w^* at different moments of time: 1: t = 7 days, 2: t = 14 days, 3: t = 16.6 days, 4: t = 21 days, and 5: t = 140 days.

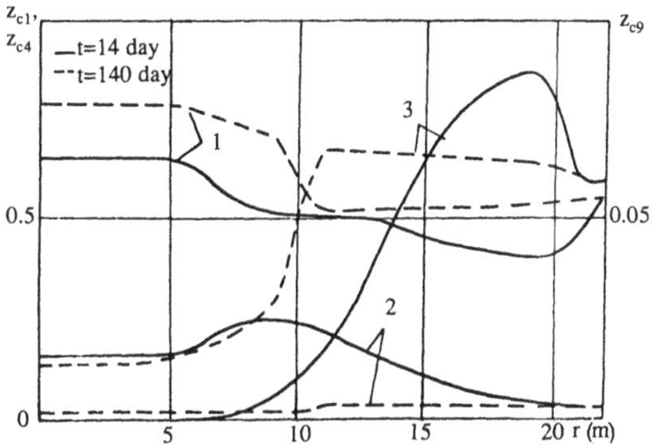

FIGURE 21. Profiles of reservoir concentrations (1: methane, 2: buthane, 3: nonane).

viscosities in the "bank" zone is negligible compared to changing the surface energy (Figure 10) and, correspondingly, to changing the phase mobilities regarding each other. In the modeling process described in this section we used this approach to determining the phase permeabilities. The quantities f_g and f_w were given by Equation (5.1.2), the residual saturations $S_{w,g}^*$ and the exponent $\tilde{\gamma}$ were obtained from Equation (5.1.3). The form of functions ρ_g, ρ_w, η_g, η_w, $\mathcal{K}_i$ is the same as that in the previous chapter.

Figures 20 to 23 show the results of computation of a ten-component reservoir flow model with boundary conditions (6.1.1) to (6.1.3).

During the first 7 days the fluid extraction from the well is carried out, the pressure difference between the well and the contour being 0.15 MPa (Figure 22). The reservoir liquid saturation does not differ basically from the residual saturation, only at $r < 2$ m $S > S_w^*$ (curves 1 in Figure 20). This can explain the increase in the fraction of heavy components and the decrease in the fraction of light components in the near-zone of the well (Figure 23). Then the enriched

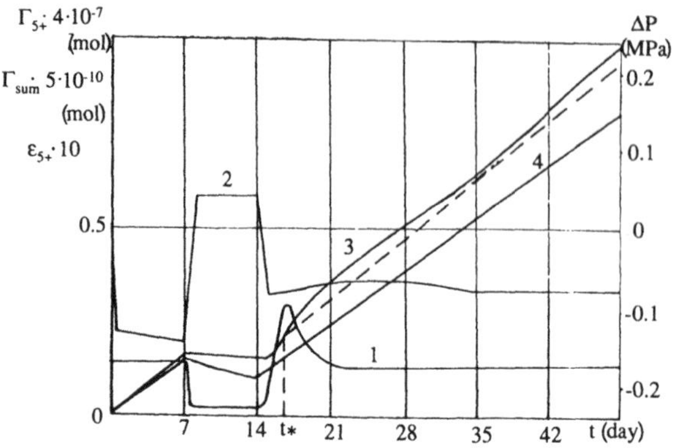

FIGURE 22. Temporal dependences of production parameters — 1: fraction of components C_{5+} in the flow at the well, 2: pressure difference between the well and the contour ΔP, 3: production of components C_{5+} (Γ_{5+}), 4: total production (Γ_{sum}). The moment of time t_* corresponding to condition $\Gamma_{sum}(t_*) = \Gamma_{sum}$ (7) is designated by an asterisk.

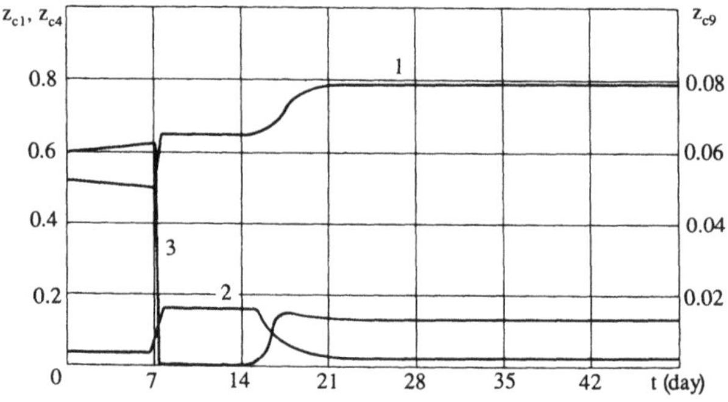

FIGURE 23. Temporal changes in the reservoir mixture composition in the near-zone of the well (1: methane, 2: buthane, 3: nonane).

gas injection begins, the condensate "bank" being formed as a result of the interaction of the injected gas and formation mixture. The bank moves along the seam and "shovels up" the heavy components. The distributions of S and S_w^* in the seam at several moments of time are presented in Figure 20. The component structure of the condensate bank is shown in Figure 21; the heavy components (continuous curve 3) are collected in the front zone of the bank, the intermediate components (continuous curve 2) are collected in its rear zone. The amount of light components (continuous curve 1) reaches a minimum in the front zone of the bank. The residual saturation S_w^* vanishes behind the bank, is small in its rear zone, and reaches the initial value in its front zone (dashed curve in Figure 20).

After one week of injection the extraction begins again.

The condensate bank redistributes in the seam so that the maximum liquid saturation in the bank zone decreases and the width of the bank increases. Figure 22 shows the time dependence of the fraction of the heavy (C_{5+}) components in flow at the well (curve 1). The moment the bank arrives at the well corresponds to the maximum fraction of components C_{5+}. At some time the heavy components move near the well both in the liquid and gaseous phases. However, after extracting the product from the rear bank zone, which is the richest in the intermediate components, the liquid phase becomes immobile near the well, and the later recovery of C_{5+} components is provided by the inflow of the gaseous phase only.

The immobile liquid phase collected near the well is poorer in heavy components compared to the liquid phase of the unperturbed reservoir system. This is because it was formed from the rear zone of the bank. The heavy components in the mobile gaseous phase are condensed when reaching the near-well zone; the C_{5+} fraction in the flow at the well decreases and becomes less than its initial value, and the C_{5+} fraction in the near-zone of the well increases (curve 3 in Figure 23). Gradually as the reservoir system approaches equilibrium, the value of the C_{5+} fraction in the flow becomes equal to the initial one. The S and z_i profiles have a sharp break at $r = 10$ m (continuous curve in Figure 20 and dashed lines in Figure 21). In the region $r < 10$ m the reservoir mixture contains less of the components C_{5+} compared to the initial one. In the region $10 \text{ m} < r < 25$ m the liquid phase, being a part of the remaining front bank zone, is heavier than the liquid of the initial system, and the saturation there is a little larger than S_w^*. The motion of the heavy components toward the well happens extremely slowly and could provide an additional long-time increase of the heavy components recovery. One can see (Figure 22) that the C_{5+} components recovery dependence goes higher than the dashed line, which would correspond to the recovery of heavy components starting from the moment t_1 in the extraction regime (the dashed line goes parallel with the initial ($t < 7$ day) section of the C_{5+} recovery curve). The difference between the continuous and dashed curves determines the full recovery of heavy components connected with the treatment of the near-well zone by enriched gas, and this additional recovery equals approximately 20 ton.

6.2. INJECTION OF ENRICHED GAS INTO A WELL WITH PARALLEL PRODUCT EXTRACTION FROM ANOTHER WELL

The obvious troubles of numerically solving the multicomponent reservoir flow problems increase additionally if the flow region has a complicated shape, and the reservoir parameters are space-inhomogeneous. In some cases a simplification of the problem is possible — namely, for a formation, if the sets of wells act as parallel galleries, then the problem is reduced to the corresponding one-dimensional one. If one has five-pointlike disposition of the wells in a typical rectangular seam element, sometimes it is possible to use a system of rigid (i.e., weakly changing) flow pipes and to carry out the calculations for the flow pipes with a variable section. However, many real processes of multicomponent reservoir flow cannot be satisfactorily described by one-dimensional models. Reliable determination of the flow pipes is often impossible, even for homogeneous regions with simple geometry, and one should use the multidimensional simulation.

In this section we present the results of numerical solving of the two-dimensional problem of injection of enriched gas in a well with the parallel extraction from another well.[5,6] The description of the process is based on Equations (2.1.8) of the two-dimensional multicomponent reservoir flow.

The initial conditions are

$$P(\mathbf{r},0) = P^o, \; z_i(\mathbf{r},0) = z_i^o , \qquad i = 1, \ldots, l , \quad \mathbf{r} \in \mathcal{G}$$

Here $\mathcal{G}$ is the flow region with the boundary $\partial\mathcal{G}$.

The boundary conditions for the pressure are

$$P(\mathbf{r},t) = P^o, \; \mathbf{r} \in \partial\mathcal{G}$$

and the composition boundary conditions are imposed according to the flow direction at the boundary: namely, at the sections where the flow aims outside the boundary, the concentrations equal initial ones; at the remaining sections of the boundary they are calculated. The rectangular element of the seam is considered.

The initial data correspond to conditions of the Vuktyl deposit in 1985. In the calculations $k = 4.13 \cdot 10^{-15}$ m^2, $m = 0.081$, $T = 335$ K, and $h = 100$ m, the sizes of the rectangular seam element were 580 m and 360 m, the initial pressure was 10 MPa, the initial liquid saturation was 0.142, the mass flow rate at the injecting and extracting wells were -3 kg/s and 3 kg/s, respectively, the distance between the wells was 160 m, the space step of finite-difference grid was 20 m, and the time step of the scheme was 0.6 day. The initial reservoir composition was the following: 0.6803 (CH_4), 0.0367 (N_2), 0.087 (C_2H_6), 0.0421 (C_3H_8), 0.0219 (C_4H_{10}), 0.0145 (C_5H_{12}), 0.0187 (C_6H_{14}), 0.0364 (C_7H_{16}), 0.0254 (C_9H_{20}), and 0.037 ($C_{10}H_{22}$). The composition of the enriched gas was 0.50 (CH_4), 0.24 (C_2H_6), 0.17 (C_3H_8), and 0.09 (C_4H_{10}).

The phase permeabilities were determined in the form of Equation (5.1.2) for liquid phase and in the form

$$f_g = \begin{cases} \left(\dfrac{S_g - S_g^*}{1 - S_g^*}\right)^{\tilde{\gamma}} (4 - 3S_g) , & S_g \geq S_g^* \\ 0 & , \; S_g < S_g^* \end{cases}$$

for the gaseous phase; $S_g^* = 0.1$, $S_w^* = 0.3$, and $\tilde{\gamma} = 3$. The phase densities and viscosities were determined by Equation (2.1.10) at $\lambda_1 = \lambda_2 = \lambda_3 = \lambda_4$

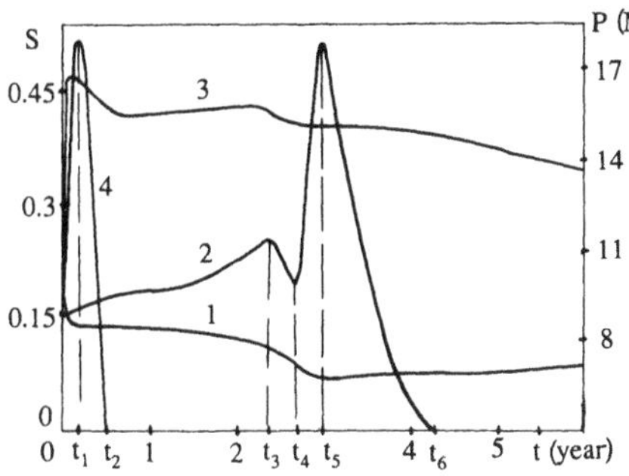

FIGURE 24. Two-dimensional problem. Changes in the pressure and liquid phase saturation in the near-zones of extracting and injecting wells.

$= \lambda_5 = 1$, $\lambda_6 = \lambda_8 = 0$, $\lambda_7 = 1$. The approximation of the equilibrium constants is the same as that in the previous chapters.

The results of the calculations are presented in Figures 24 through 33. The extracting and injecting wells are designated by the digits 1 and 2 correspondingly.

Figure 24 shows the time dependences of the pressure and liquid saturation in the near-zones of well 1 (curves 1 and 2) and well 2 (curves 3 and 4). First the bank of liquid phase forms around well 2. Soon after the liquid phase becomes mobile and its saturation in the near-zone of well 2 falls to zero (curve 4). While the bank forms, the pressure at well 2 reaches its maximum. The deviation of the pressure at a well from the pressure at the boundary of flow region should be considered as a criterion of the changing of the well near-zone resistance.

The sharp fall in the pressure at well 2 ($t_2 > t > t_1$) changes to a slow increase ($t_3 > t > t_2$). The liquid saturation near well 1 increases, first because of the pressure fall, then because of the inflow of gas enriched by the intermediate components and moving more quickly than the condensate bank itself; the pressure in the near-zone of well 2 decreases.

At $t_4 > t > t_3$ the liquid saturation at well 1 falls sharply: the liquid phase moving from the well 2 collects near the well 1, and the gas, which is in equilibrium with the "heavy" front zone of the bank and which is bereft of heavy components compared to the unperturbed formation gaseous phase, comes to well 1. As a result, the condensate in the near-zone of well 1 evaporates. This behavior is essentially connected with the two-dimensional character of the problem under consideration.

At $t_6 > t > t_4$ the condensate bank reaches well 1. The maximum of the liquid saturation at the well corresponds to the moment $t = t_5$. At $t > t_6$ the processes in the formation establish, the liquid saturation at well 1 vanishes, and the deviations of the pressure at the wells become close to each another.

Figure 25 shows the time dependences of the fractions of CH_4, N_2, C_2H_6, C_3H_8, C_4H_{10}, and C_{5+} components in the flow at well 1 (curves 1 to 6, correspondingly). The fraction of the intermediate components in the flow increases

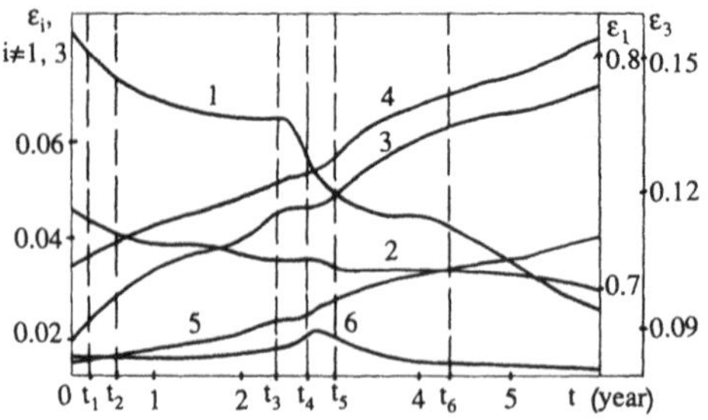

FIGURE 25. Dynamics of changes in the production composition of an extracting well.

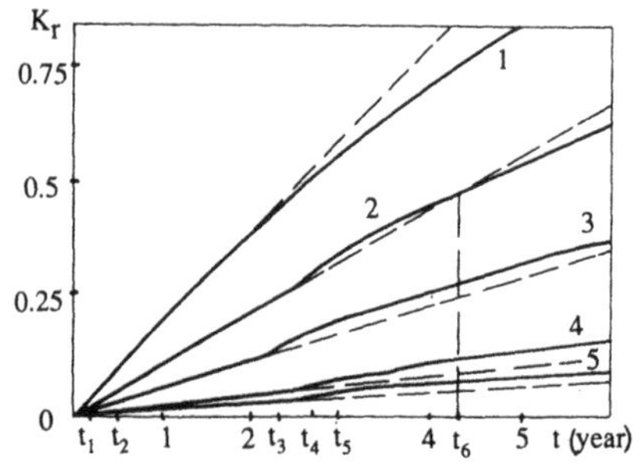

FIGURE 26. Temporal dependence of the recovery coefficients of heavy components.

with time. Its sharp growth, as well as the sharp fall of the fraction of the light components during the first year are connected with the pressure decrease at well 1. During the second year the process in the near-zone of well 1 establishes and the product composition changes little. In the beginning of the third year the composition changes more sharply because of the enriched gas inflow. At $t_4 > t > t_3$ the product composition is reestablished again because of the "locking" of the near-zone by the condensate bank, which approaches well 1 from three sides. The moment of the bank arrival to the well corresponds to the maximum of the fraction of C_{5+} components in the flow. Approximately at $t = t_6$ the value of the C_{5+} fraction becomes equal to the original, and then it decreases. Therefore, the heavy components recovery by means of enriched gas action is effective until the disappearance of the liquid phase in the near-zone of well 1.

Figure 26 illustrates the difference in the character of the heavy components recovery. Here we show the time dependences of the C_5H_{12}, C_6H_{14}, C_7H_{16}, C_9H_{20}, and $C_{10}H_{22}$ recovery coefficients normalized on their initial resources in the flow region (curves 1 to 5 correspondingly). The dashed lines show the recovery coefficients, which would correspond to the reservoir exploitation without injecting the enriched gas. One can see that as a result of the enriched gas action, the additional C_5H_{12} extraction falls, the additional C_6H_{14} extraction

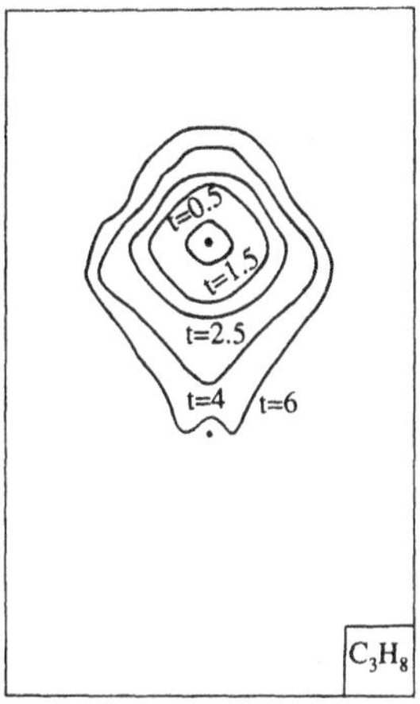

FIGURE 27. Position of the line of maximum values of the propane fraction at different moments of time.

increases until t = t_5 and then falls, and the additional extraction of the components C_{7+} increases. Thus, the positive effect of the enriched gas action on the heavy components recovery is connected generally with the recovery of C_7H_{16} and the heavier components. This shows that the representation of a multicomponent reservoir system as a binary or treble mixture of conditional components, when all the components C_{5+} are replaced by one, can be incorrect.

The dynamics of the composition and saturation fields is shown in Figures 27 to 30. The curves on Figures 26, 27, and 28 are the lines of the maximum values of the overall fraction of C_3H_8, liquid saturation, and the overall fraction of C_9H_{20}, correspondingly, which are shown at several moments of time. The curves in Figure 30 are the boundaries of the unperturbed zone for nitrogen, which concentrates generally in the gaseous phase and is displaced by enriched gas most quickly. The lines of the maximum fractions of the heavy components correspond to the motion of the ''heavy'' front zone of the bank, they move ahead of the lines of the maximum fractions of intermediate components corresponding to the ''light'' rear zone of the bank. The maximum liquid saturation is disposed in the rear zone of the bank and is a little behind the position of maximum fractions of the heavy components.

Figures 31 to 33 show the isobar maps calculated at several moments of time. One can see that changing the structure of the pressure field is quite essential and is characterized by changing the shape of isobar P = 10 MPa: the isobar is convex downward at the times corresponding to forming the bank at well 2 ; its convexity in fact disappears at the moment of the bank approach to well 1; and after passage of the bank the isobar becomes convex upward.

The same figures show the form of a flow pipe. The angle between its boundaries near well 2 is the same for all the figures. We have fixed the angle especially to demonstrate that the deformation of the flow pipe is quite essential

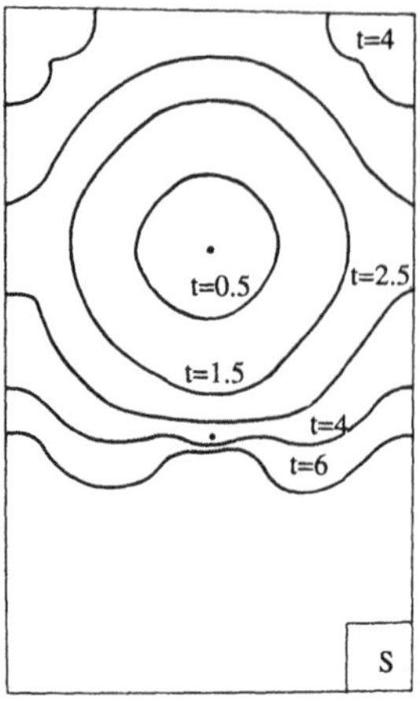

FIGURE 28. Position of the line of maximum values of liquid saturation at different moments of time.

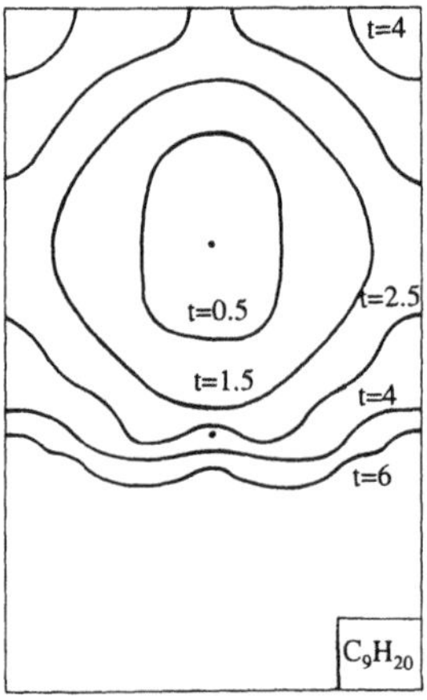

FIGURE 29. Position of the line of maximum values of the nonane fraction at different moments of time.

(though this also could be expected from the temporal behavior of isobar P = 10 MPa, discussed above). The important thing is that the changes are not connected with the formation geometry complexity or with the mutual disposition of wells, but with changing the component structure of the reservoir flow during the miscible displacement.

The analysis of the results obtained enables us to make the following conclusions.

The two-dimensional reservoir flow of multicomponent systems is characterized by some peculiarities connected with the mutual influence of the composition, saturations, and pressure fields.

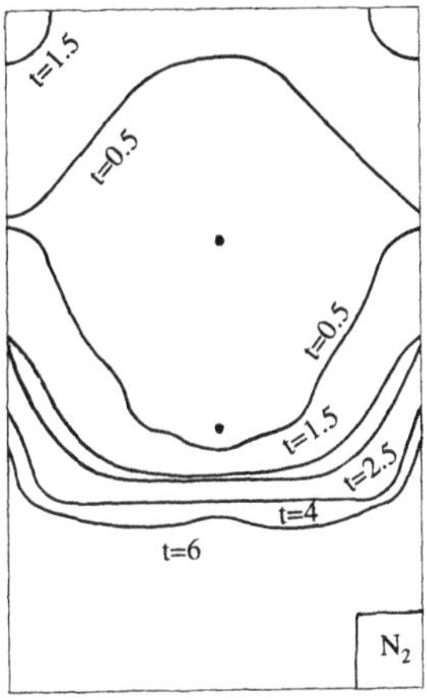

FIGURE 30. Position of the front of nitrogen displacement at different moments of time.

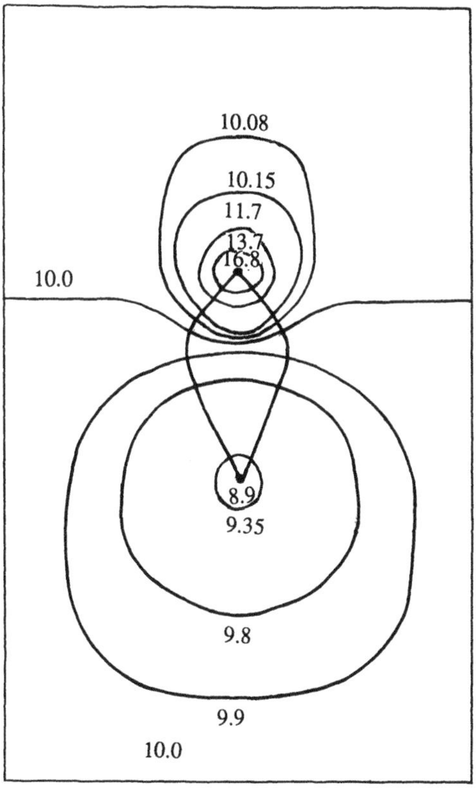

FIGURE 31. Two-dimensional pressure distribution at t = 0.2 years.

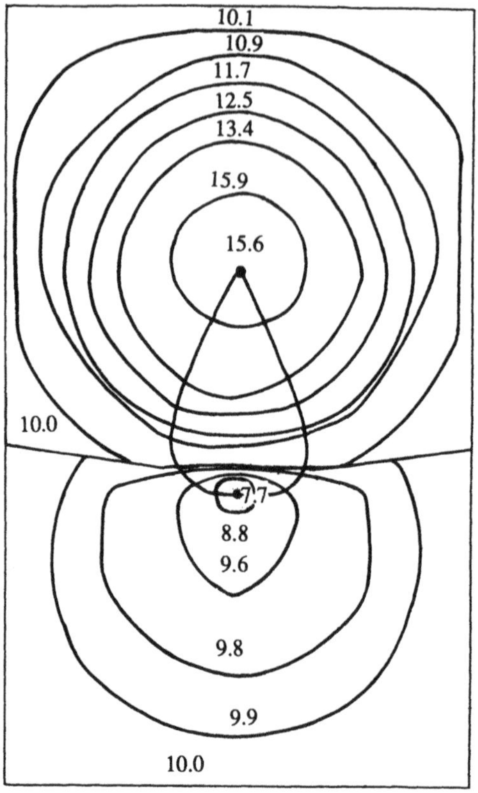

FIGURE 32. The same as that in Figure 31, but t = 2.4 years.

In the case of the absence of a reliable criterion for replacing the multicomponent system by a conditional one (with a small number of components), using the binary or treble models in the components recovery calculations can lead to incorrect results.

In conditions of strong changes of the pressure field, using the flow pipe method for the practical calculations may not be reliable. An interesting theoretical problem, which as far as we know has not been solved, would be to obtain equations for the sufficiently slow changing of the flow pipes in the mathematical model under consideration using small parameter methods[7] (i.e., obtaining the equations of the first approximation, if we are to consider the method of the rigid flow pipes as the zero approximation[8,9]).

We have considered above only a few examples of practical applications of the numerical methodics that have been worked out.[5] Currently, the program complex described is being used in several research institutes in the former U.S.S.R. without interference of author, for carrying out projects concerned with the estimation of possible variants in the development of oil–gas–condensate formations.[10,11] Actual applications of the methodics are connected with the deposits in the Ukraine, Komi Republic, Astrakhan province, and so on.

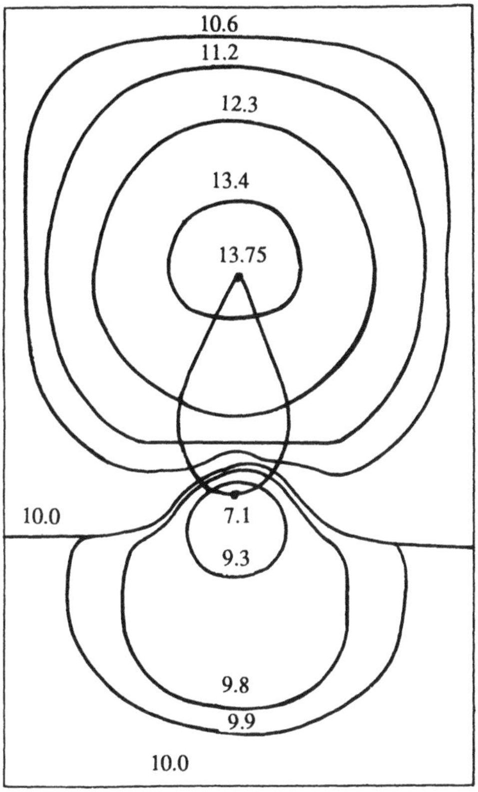

FIGURE 33. The same as that in Figure 31, but t = 6 years.

REFERENCES

1. **Ter-Sarkisov, R. M.,** *Gazov. Prom.*, 10, 26, 1982 (in Russian).
2. **Stepanova, G. S., Ter-Sarkisov, R. M., and Zenkina, L. D.,** *Gazov. Prom.*, 2, 31, 1984 (in Russian).
3. **Ter-Sarkisov, R. M., Solov'ev, O. N., Shmyglya, O. P., et al.,** Results of experimentally-industrial using enriched gas for enhanced condensate recovery on the gas-condensate deposit, Vuctyl, in *Increase of Efficiency of Development Systems of Natural Gas Deposits*, VNIIGAS, Moscow, 1985, (in Russian).
4. **Guzhov, N. A., Mitlin, V. S., Ter-Sarkisov, R. M., et al.,** Calculation of near-zone parameters of a well of a gas-condensate deposit in process of enriched gas injection, in *Development of Gas-Condensate Deposits with Anomalous High Reservoir Pressure*, VNII-GAS, Moscow, 1985, 80 (in Russian).
5. **Mitlin, V. S.,** *New Methods of Calculation of Enriched Gas Action Upon a Gas-Condensate Seam*, Ph.D. thesis, VNIIGAS, Moscow, 1987 (in Russian).

6. **Mitlin, V.S. and Tsybul'skii, G. P.,** Comparative analysis of displacement calculations using one-dimensional and two-dimensional multicomponent reservoir flow models, in *Mechanics of Multiphase Multicomponent Systems,* MING, Moscow, 200, 58, 1986 (in Russian).
7. **Nayfeh, A.-H.,** *Perturbation Methods,* Wiley-Interscience, New York, 1983.
8. **Aziz, K. and Settari, A.,** *Petroleum Reservoir Simulation,* Elsevier, London, 1980.
9. **Basniev, K. S. and Bedrikovetskii, P. G.,** Multiphase displacement of miscible fluids in porous media, in *Advances in Science and Technology of VINITI* (Itogi Nauki i Tekhniki VINITI). Ser. Complex and Special Sections of Mechanics, Vol. 3, VINITI, Moscow, 1988, 81 (in Russian).
10. **Ter-Sarkisov, R. M., Nikolayev, V. A., Makeev, B.V., et al.,** in Int. Conf. Development of Gas-Condensate Fields Rep. Sect. 3, Development of Oil-Gas-Condensate Fields, Krasnodar, 1990, 72.
11. **Gritsenko, A. I., Nikolaevskii, A. V., Peshkin, M. A., et al.,** *Izv. Akad. Nauk SSSR, Mekh. Zhid. Gaza,* 2, 185, 1990; *Fluid Dyn. (USSR),* 25(2), 323, 1990.

7

Autowave Regimes of Two-Phase Multicomponent Reservoir Flow: Theory, Computation, Experiments

In this chapter we consider one of the possible scenarios according to which the stationary regimes of two-phase multicomponent reservoir flow may lose their stability and, as a consequence, according to which autowaves may appear. Pulsatory reservoir flow regimes are constantly encountered by engineers when they study the exploitation of oil-gas-condensate deposits. The observed pulsations may be caused by the action of nonmonotonic or periodic external forces (the moon, earthquakes, microseismic action, etc.). Such effects will not be considered here. We shall be interested in a situation where there is a time-monotonic (e.g., constant) inflow of energy into an open system, which, owing to its own nonlinearity, transforms the incoming energy into oscillations. We shall also be interested in the conditions (bifurcational values of the parameters) under which branching of the solution occurs and the stationary regime is replaced by the autowave. Such effects are studied by synergetics.[1] This approach is rather general: the specific aim of our present study is to find the relationship between the observed instabilities and typical features of phase diagrams for multicomponent mixtures. In our scenario the autowaves are associated with the cyclic changing of the following processes: the increase of the volume occupied by the liquid phase (condensation), the flow of the liquid as it becomes sufficiently mobile in the porous medium, and the diminishing of the volume occupied by the liquid phase (evaporation). Correspondingly, the conditions of instability are equivalent to the conditions of the system being in the region of retrograde condensation. We shall discuss in detail the relationship between retrograde phenomena and the so-called effect of negative volume of heavy components. It will be shown that there is a direct connection between these phenomena and the negative compressibility of an individual volume of the two-phase continuum moving in a porous medium, and that this is what leads to instability. It is shown that, mathematically, the loss of stability occurs via Hopf bifurcation and that the autowaves are the relaxation oscillations in a distributed system. Finally, we discuss the results of recent experiments to check the predictions of the considered theory. These experiments have confirmed the correctness of our theoretical concepts.[2-5]

7.1. STABILITY CONDITIONS

Let us again consider a system of the local equilibrium reservoir flow equations for a two-phase l-component mixture

$$m\frac{\partial N z_i}{\partial t} = \nabla(k\beta_i \nabla P)\,, \qquad i = 1, \ldots, l$$
$$\beta_i = \frac{f_g \rho_g}{\eta_g M_g} y_i + \frac{f_w \rho_w}{\eta_w M_w} x_i,\ N = S_w \frac{\rho_w}{M_w} + S_g \frac{\rho_g}{M_g} \qquad (7.1.1)$$

All designations in Equation (7.1.1) are equivalent to designations introduced in the previous sections.

Let k and m be constant. Let us consider the linear stage of evolution of the perturbation of a spatially homogeneous solution in the region that has the form of a rectangular parallelepiped. As it is usually done when one studies hydrodynamical fluctuations, let us impose periodic boundary conditions at the boundary of this region. Owing to the autonomy of system (7.1.1), this corresponds to spatially periodic initial conditions for the problem considered here. In the case of an arbitrary region, after one isolates the time derivatives of P and z_i in the initial equations, one should expand the solution in eigenfunctions of the differential operator on the right-hand sides of the linearized equations, which, in this case are complex exponents.

Let us present Equations (7.1.1) as

$$\nabla(k\beta_i \nabla P) = m\left(N\frac{\partial z_i}{\partial t} + z_i\frac{\partial N}{\partial t}\right), \qquad i = 1, \ldots, l \qquad (7.1.2)$$

By summing Equations (7.1.1) we get

$$\nabla(k\beta\nabla P) = m\frac{\partial N}{\partial t}\,, \qquad \beta = \sum_{i=1}^{l} \beta_i \qquad (7.1.3)$$

Substituting Equation (7.1.3) into Equation (7.1.2) and performing simple transformations, in the linear approximation we obtain

$$\frac{\partial P'}{\partial t} = l\nabla^2 P', \ \frac{\partial z_i'}{\partial t} = (\beta_i - z_i\beta)\frac{k}{mN}\nabla^2 P'\,, \quad i = 1, \ldots, l - 1 \qquad (7.1.4)$$

where $P' = P(r,t) - P^o$, $z_i' = z_i(r,t) - z_i^o$, and index '0' denotes the spatially homogeneous solution. In accordance with Equation (7.1.4), P' and z_i' are related as

$$z_i' = \frac{k(\beta_i - z_i\beta)}{mNI} P'$$

i.e., P is the leading variable. In what follows, the quantity

$$I = \frac{k}{m}\left(\frac{\partial N}{\partial P}\right)^{-1} \cdot \left[\beta - \sum_{j=1}^{l-1} \frac{1}{N}\frac{\partial N}{\partial z_j}(\beta_j - z_j\beta)\right] \qquad (7.1.5)$$

will be called the increment of stability of a spatially homogeneous solution.

The growth (or dying-out) of Fourier components of P′ and z_i' is determined by the eigenvalues of the linear response matrix of system (7.1.1) $\mathbf{A} = \{a_{ij}\}$:

$$a_{il} = -\frac{k}{mN} q^2 (\beta_i - z_i\beta), \quad 1 \le i \le l - 1;$$

$$a_{ll} = -q^2 I;$$

$$a_{ij} = 0, \quad j < l \qquad (7.1.6)$$

From Equation (7.1.6) it follows that matrix **A** has $l - 1$ identically zero eigenvalues and one non-zero eigenvalue $\lambda_l = -q^2 I$. Here

$$q^2 = 4\pi^2 \sum_{j=1}^{d} \left(\frac{n_j}{L_j}\right)^2, \quad n_j = 1,2, \ldots$$

is the wave vector squared, L_j is the size of the region in the j-th direction, and d is the dimension of the region. By virtue of Lyapunov's theorem, the sufficient condition of instability of a spatially homogeneous solution is[6]

$$I < 0 \qquad (7.1.7)$$

The possibility of losing stability essentially has to do with the multicomponent system having more than one phase. Indeed, in the one-phase case (for example, for pure gas) we have

$$\beta_i = z_i\beta, \qquad i = 1, \ldots, l$$

and Equation (7.1.7) is equivalent to the condition

$$\frac{\partial \rho_g}{\partial P} < 0$$

which corresponds to thermodynamic instability of the one-phase system and contradicts the initial assumption of the theory. Let us now show that when there are several liquids flowing in a porous medium without mixing with one another, condition (7.1.7) cannot be satisfied either. For simplicity, let us consider the system of equations for two-phase reservoir flow. For a larger number of liquids the calculations become more cumbersome but are performed in exactly the same way. We have

$$m\frac{\partial}{\partial t}(\rho_w S) = \nabla\left(\frac{k\rho_w f_w}{\eta_w}\nabla P\right),\ m\frac{\partial}{\partial t}(\rho_g(1 - S)) = \nabla\left(\frac{k\rho_g f_g}{\eta_g}\nabla P\right) \quad (7.1.8)$$

By linearizing the system (7.1.8) with respect to the spatially homogeneous solution and solving it with respect to P' and S', one finds that the stability of a nonperturbed solution is determined by the eigenvalues of the matrix $\mathbf{A} = \{a_{ij}\}$:

$$a_{11} = a_{21} = 0, \quad a_{22} = -\frac{k}{m}\frac{q^2\rho_w\rho_g(f_g/\eta_g + f_w/\eta_w)}{\rho_g S\partial\rho_w/\partial P + \rho_w(1 - S)\partial\rho_g/\partial P}$$

One of these eigenvalues is identically zero; the other is a_{22}. One can easily see, however, that a_{22} is always negative.

Let us finally note that when deriving condition (7.1.7) we made use of the dependence of N on P. When searching for analytical solutions of equations of multicomponent reservoir flow one often makes the assumption that the mixture is incompressible or that the sum of partial volumes of components is constant,[7] which results in $\partial N/\partial P \equiv 0$. In our case, if N is independent on P, the linearization gives $\nabla^2 P' \equiv 0$, the system of linearized equations becomes degenerate and the linear approximation becomes noninformative.

Condition (7.1.7) can also be used for determining the regions of instability of stationary solutions in the case where one introduces into Equation (7.1.1) terms that correspond to sources and sinks of sufficiently small power. This is a consequence of the fact that the positive eigenvalue λ_l changes little with a small variation of boundary conditions and of the equations themselves. Thus, the reservoir flow regimes for which the stationary solution of Equation (7.1.1) is unstable can be obtained when one 'turns on' sufficiently weak sinks against the background of initial conditions P^o and z_1^o such that $I(P^o, z_1^o, \ldots, Z_{l-1}^o) < 0$. So that condition (7.1.7) could hold, it is necessary that the multicomponent system has at least two phases, that its components mix, and that N depends on P.

In order to describe the instabilities that we have observed in the calculations, let us present the results of a numerical solution of the plane linear problem of multicomponent reservoir flow with the condition that the pressure at both end points is kept constant and equal to 46 MPa. The initial distribution of P and z_i was perturbed with respect to the spatially homogeneous solution. In the process of calculation, the concentrations at a boundary were not fixed if the flow came toward it; if the flow went away from it, the composition on the boundary was taken from the previous time layer. In the calculations we have taken $k = 10^{-15}$

m^2, m = 0.1, T = 353 K; the initial composition was 0.6926 (CH_4), 0.0616 (C_2H_6), 0.0334 (C_3H_8), 0.0055 (iC_4H_{10}), 0.0079 (nC_4H_{10}), 0.009 (N_2), 0.0625 (H_2S), 0.346 (CO_2), and C_{5+} (pentane and heavier) 0.0928. These data, and a set of fluxes in the next section of this chapter, correspond to conditions of the gas-condensate Karachganak (former U.S.S.R.) deposit in 1985.

The equilibrium constants were approximated by some functions of pressure; the density and viscosity of phases were approximated by functions of pressure and the molecular weights of the phases. The data necessary for constructing these dependences were obtained from preliminary calculations of phase equilibria for the initial mixture and different pressures according to the Peng-Robinson equation of state. The phase permeabilities were taken to be of the form

$$f_w = \begin{cases} 0, & S \le 0.2 \\ \left(\dfrac{S - 0.2}{0.8}\right)^3, & S > 0.2 \end{cases}, \quad f_g = \begin{cases} 0, & S \ge 0.8 \\ \left(\dfrac{0.8 - S}{0.8}\right)^3, & S < 0.8 \end{cases}$$

Figure 34 shows the profiles of P and S at different times. One can see that the solution is pulsating: the convexity of the profile changes. The pulsations of saturation are more interesting: at different times its profile has a different number of extrema. This means that the process involves waves of different wavelengths. By comparing Figures 34(b) and 34(f) we see that in the system, a successive change of spatial distributions of S occurs modulated and nonmodulated in amplitude. This effect is typical of nonlinear oscillations: for example, the Fermi-Pasta-Ulam return phenomenon in the theory of solitons.[8] The same phenomenon appears in equations with completely different structure, e.g., in the brusselator model with distributed parameters.[9] Figure 35 shows a fragment of the time variation of S in the middle of the solution region.

Calculations show that at each time the stability increment has a negative value at one or several sites on the different grids. At the same time, the perturbations with respect to the pressures 43 MPa and 48 MPa and the same composition died out: the system relaxed toward a spatially homogeneous state and at all points of the solution region the stability increment was positive. Thus, we have observed a 'window' in the range of P - variation such that inside that window autowaves are generated in the system. Direct calculations of I for the same composition when P varies show that $I < 0$ exactly in the region $43.5 < P < 47.5$ (see below, Figure 44). Thus, the test demonstrates that the appearance of autowaves is explained by hydrodynamic (but not numerical) instability.

7.2. PULSATORY FLOW REGIMES OF A MULTICOMPONENT MIXTURE

Below we present the results of the solution of the plane-radial problem of the operation of a gas-condensate well with constant pressure at the boundary

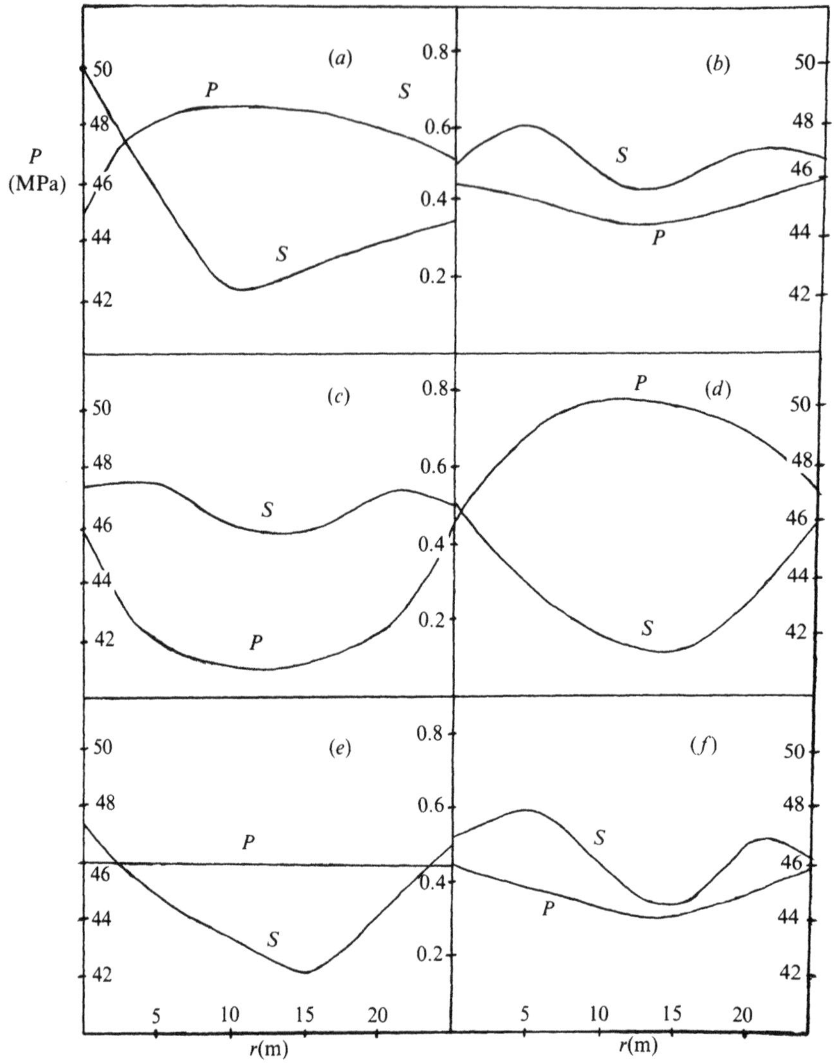

FIGURE 34. Evolution of perturbations in an unstable spatially homogeneous solution: profiles of P and S; a: t = 0, b: t = 2 days, c: t = 4 days, d: t = 10 days, e: t = 14 days, and f: t = 30 days.

of the sink zone. The solution is found in the region (r_b, R_c), $P(r,0) = P(R_c,t) = 55$ MPa and $z_i(r,o) = z_i(R_c,t)$; the initial composition is as presented in the previous section and is kept constant at the boundary of the drain zone. At the well, a constant total mass flux is fixed:

$$Q = 2\pi hk\beta M_f^{-1} r \left.\frac{\partial P}{\partial r}\right|_{r=r_b}$$

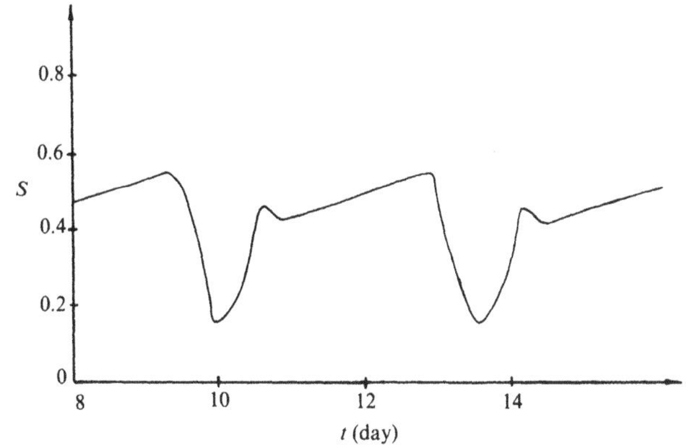

FIGURE 35. Fragment of evolution of S in the middle of the region of solution.

where $M_f = \sum_{i=1}^{l} M_i\epsilon_i$, and $\epsilon_i = \beta_i/\beta$ is the mole fraction of the i-th component in the flow. The values of k and m and the form of the functions $\rho_{g,w}$, $\eta_{g,w}$, $f_{g,w}$, and $\mathcal{K}_i$ are the same as in the previous section. The seam thickness is $h = 50$ m, $r_b = 0.12$ m, and $R_c = 25.12$ m.

The results of calculation are presented in Table 5 and in Figures 36 to 38. At $Q = 50$ ton/day and $Q = 70$ ton/day (Figure 36a) the problem has a stationary solution (the solution was considered stabilized if the pressure and saturation at the well were constant up to the fourth digit within 25 steps in time, i.e., 5 days). At $Q = 86.4$ ton/day the stationary solution lost its stability: after 160 days the smooth variation of parameters was replaced by pulsations. The pressure at the well at that time was 47.1 MPa. At $Q = 100$ ton/day the smooth lowering of pressure at the well was replaced by pulsations after 60 days. The amplitude of pulsations grew up to a certain magnitude, after which the solution transformed into a time-periodic one. The form of the stabilized pressure and saturation pulsations at the well, together with the fraction of gas in the product flow from the well, V_{gf}, and the fraction of C_{5+} components in the extracted gas, α_h, are shown in Figure 37a; the position of the time zero in Figure 37 is not important since the figure shows the behavior of the solution at large t. The profiles of S corresponding to its extreme values at the well at stabilized oscillations are shown in Figure 36b. In the cases where there are no stationary regimes, the stabilization time column, t_s, of Table 5 is left empty, while in the P, S, V_{gf} and α_h columns in these cases we present the values averaged over a period of stabilized oscillations for $r = r_b$. The quantities ΔP and ΔS are the doubled amplitudes of stabilized oscillations of pressure and saturation.

At $Q = 160$, 190, and 200 ton/day the stationary solution has existed (Figure 36c), but the approach to it was accompanied by pulsations which appeared when the pressure at the well reached 47.1 MPa. The pulsations of P at the well appeared between 44 and 38 MPa and gradually disappeared, after which P

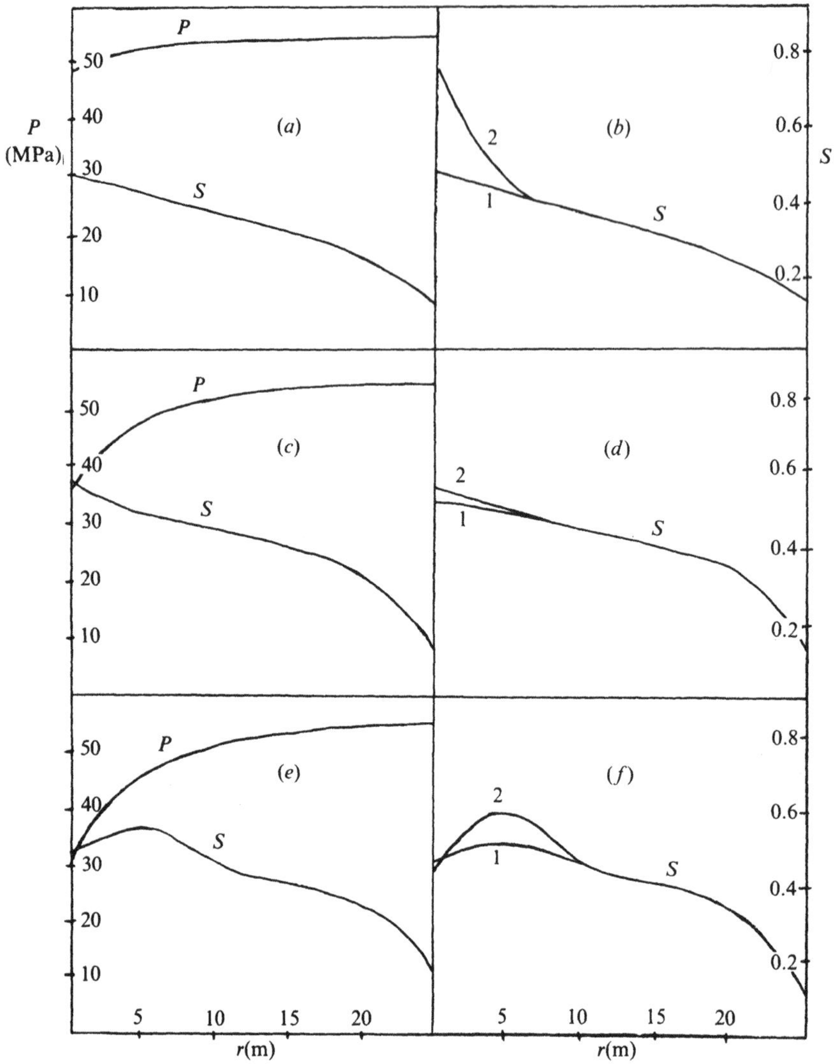

FIGURE 36. Profiles of P and S in the problem of the inflow of a gas-condensate mixture to a well (data of Karachganak deposit, U.S.S.R.); a, c, and e correspond to Q = 70, 200, and 420 ton/day, respectively, stationary regime exists, the stationary profiles are shown. In b, d, and f: Q = 100, 240, and 500 ton/day, respectively, stationary regime is absent, the profiles of S by its extremum values at the well are shown.

smoothly fell until the system reached a stationary regime. In these three calculations the pulsation time was maximum at minimum Q, the closest to the interval $86.4 < Q < 100$, within which the pulsations did not die out at all.

At Q = 210 ton/day, after the pressure at the well reached 34.9 MPa, a pulsatory regime appeared. At Q = 240 ton/day the stationary solution also

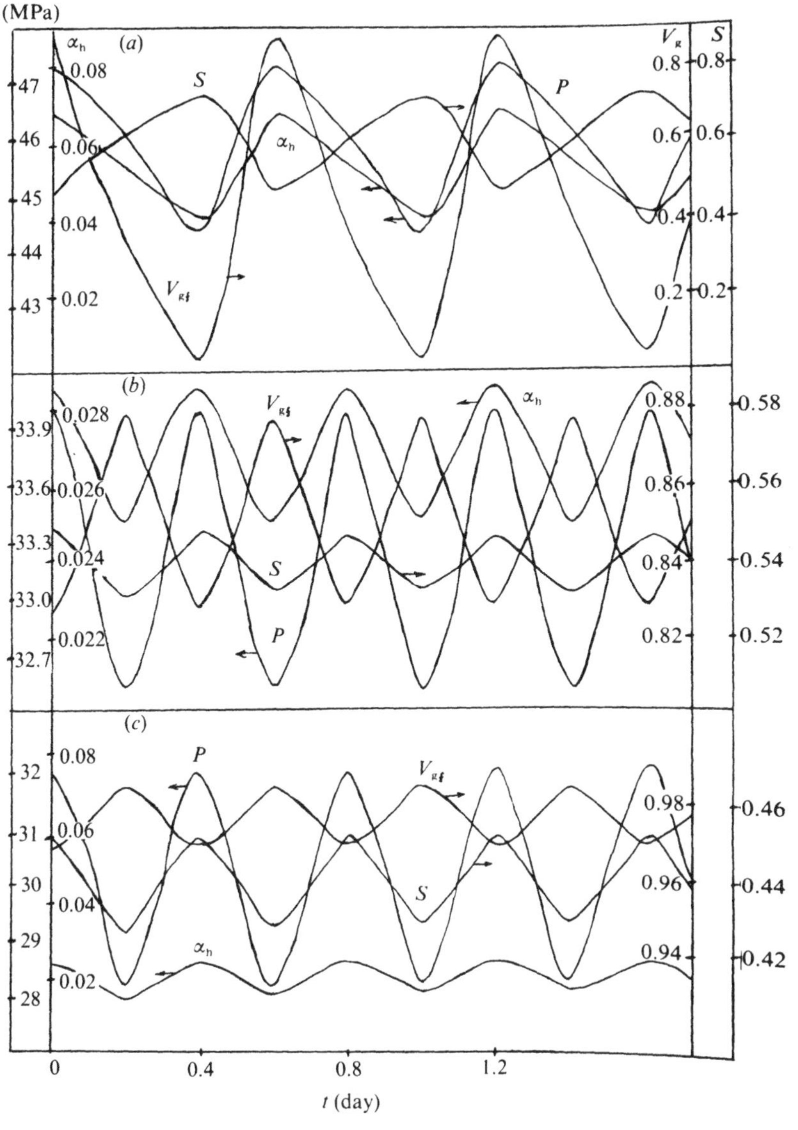

FIGURE 37. Oscillations of parameters at the large times; a: Q = 100 ton/day, b: Q = 200 ton/day, and c: Q = 500 ton/day.

existed. The form of stabilized pulsations at the well is shown in Figure 37b. The profiles of saturation at its extreme values at the well are featured in Figure 36d.

At Q = 324 and 420 ton/day the stationary solution appeared once again (Figure 36e). In this case, at the initial stage pulsations appeared, the amplitude

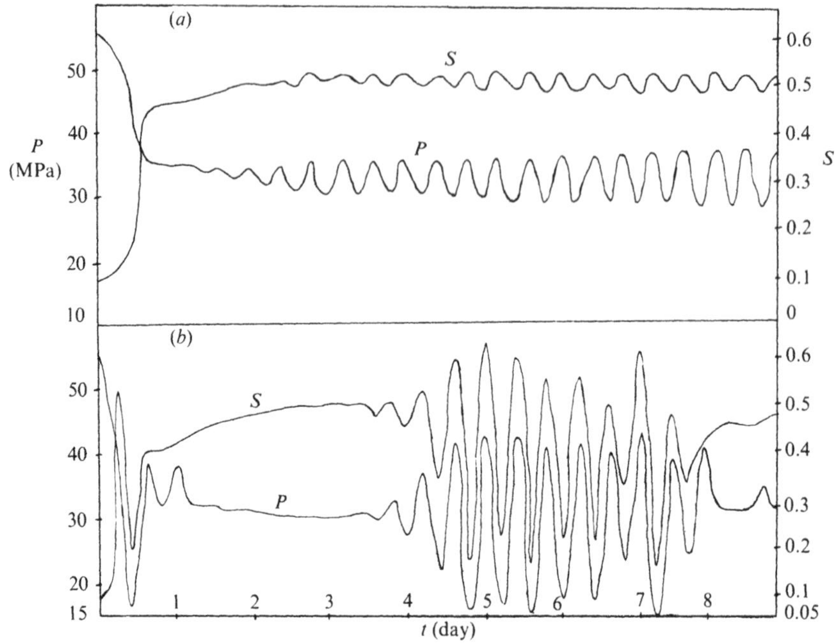

FIGURE 38. Dynamics of P and *S* at the well in the first 9 days: forming the oscillations-modulated (b: Q = 420 ton/day) and nonmodulated (a: Q = 324 ton/day) with regard to magnitude.

of which reached significant values: up to 10 MPa in pressure. The time dependence of P and *S* at the well during the first 9 days is featured in Figure 38. Compared to the pulsations in Figure 38a, those in Figure 38b have a noticeable modulation depth. In both of these cases, when the pressure at the well reached 47.1 MPa (the upper limit of the first zone of instability), the saturation sharply changed — the system was 'thrown out' into the stable zone. For some time after that the process went on smoothly and the pressure at the well fell, and then a new change occurred in the pulsatory regime (for Q = 324 ton/day this happened at 35 MPa; for Q = 420 ton/day at 32 MPa), but in the end the oscillations disappeared. It is interesting to note that at Q = 420 ton/day the system approached the stationary regime in a very complicated manner. Figure 39 shows the variation of pressure at the well at $t > 9$ days. One can see that in the calculations, the flashes of stochastic behavior (sharp peaks and gaps) alternated with dynamics that was almost periodic in time. As is well known, such effects have been observed in the scenario of turbulence generation via alternation.[10]

At Q = 500 and 700 ton/day, at large t one observed stabilized pulsations. As one can see in Figure 36f, they involved a larger portion of space than in the case of smaller fluxes. Moreover, periodically, in the vicinity of the well a billow of *S* forms, which subsequently flattens out.

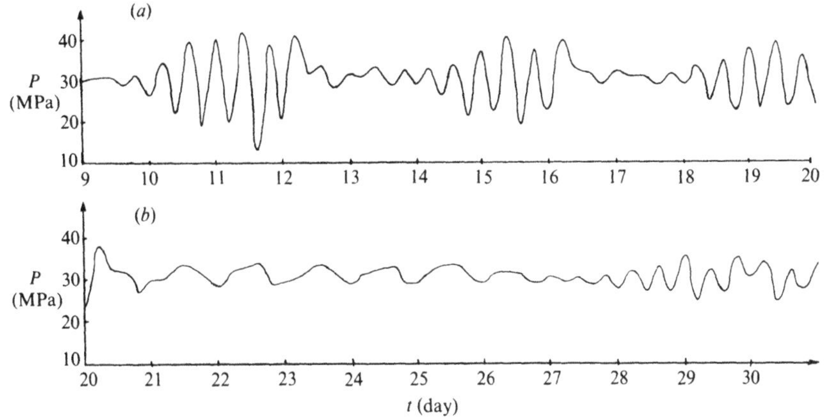

FIGURE 39. The same that in Figure 38b at $t > 9$.

It should be noted that pulsation generation does not depend on the geometry of the problem: this is shown by the calculations made in the previous section and, also, by the calculations for the planar linear problem with pressure fixed at one end and mass flux at the other. In this case, too, one has observed pulsatory regimes, and the values of parameters of the system at which the stationary solution became unstable were close to the corresponding values for the plane-radial problem.

Thus, at certain values of Q a self-sustaining regime is generated, with successive accumulation of the liquid phase in the zone surrounding the well and its leaving the seam. The pulsatory character of this process is due to the constant inflow of heavy components into the zone around the well and their condensation, owing to which the volume of the liquid phase becomes larger. When a sufficient amount of liquid is accumulated and it becomes sufficiently mobile, it comes to the well, the saturation profile flattens out, and so on. Let us note that in addition to the periodic regimes, at certain values of flux one observed nondamped oscillations which were not periodic in time, at least within the time considered in the calculation. The problem of stochastization of solutions of the multicomponent reservoir flow equations requires further study (for more about the stochastization of dynamic systems see, for example, Zaslavsky[11]).

From the fact that a spatially homogeneous solution may lose its stability and from what was noted in the previous section about the continuous dependence of the solution on the boundary conditions, it follows that in the inflow problems one may encounter profiles of P that are nonmonotonic in the spatial coordinate. In any case, if the initial composition and pressure are such that $I\,(P^o, z_1^o, \ldots, z_{l-1}) < 0$ and the pressure at the boundary of the sink zone equals the initial one, while the pressure at the well differs from $P(R_c)$ by a sufficiently small amount, quantity P will have nonmonotonic profiles. This is confirmed by the numerical calculations made in the previous section. The initial distributions of

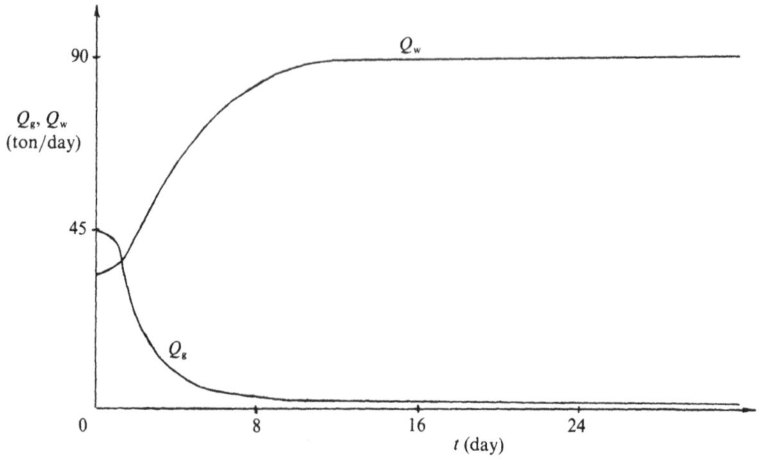

FIGURE 40. Dynamics of parameters in the problem with pressures at 45 and 55 MPa at the boundaries. The stationary regime exists.

pressure and composition were specially calculated to study the evolution of perturbations of a spatially homogeneous solution by preliminarily solving the problem with boundary condition for pressure of 45 and 47 MPa. As one can see in Figure 34a, the distribution of pressure is essentially nonmonotonic.

Figure 40 shows the time dependence of the mass flux for the gaseous and the liquid phases when the pressure at the well and at the boundary of the sink zone, respectively, is 45 and 55 MPa. There are no flux pulsations and the stationary regime sets in within 10 days. In the calculations where the flux at the well was fixed one observed a pulsatory regime, though the pressure at the well, averaged over a period of stabilized pulsations, was close to 45 MPa (see Table 5). This means that it is not at all necessary that for each stable stationary solution of the problem with fixed pressure at the well there should be a corresponding stable stationary solution of the problem with fixed flux. These two problems are different. In other words, the solution essentially depends on the fixed boundary conditions. Figure 41 shows the stabilized profiles of P and S for the problem with $P(r_b,t) = 45$ MPa and $P(R_c,t) = 55$ MPa. The saturation of the liquid phase at the well coincides with the mean saturation at the well in the problem with fixed flux of 100 ton/day. However, on the whole, the profiles of S in Figures 36b and 41 are different from each other. The stabilized value $V_{gf} = 0.35$ is different from the period-average value for the corresponding case in Table 5. The stabilized flux is $Q = 75$ ton/day and not 100 ton/day. Moreover, these values cannot be the same, since such a coincidence would mean that the stationary regime also exists for a fixed flux of 100 ton/day, and this has not been observed. In our calculations with pressures of 40 and 48 MPa at the well and 55 MPa at the boundary of the drain zone, there were also no flux pulsations, but at a pressure of 25 MPa at the well the pulsations did appear.

TABLE 5

Q (ton/day)	50	70	100	160	190	200	240	324	420	500	700
t_8 (day)	200	400	—	320	320	420	—	160	140	—	—
P (MPa)	52	48.5	46	38.5	36.1	35	33.2	31.3	30.3	30.1	27.8
α_h (%)	7.9	7.1	5.3	4.2	3.4	3.1	2.7	2.2	1.9	2.0	2.1
S (%)	37.5	46.2	62	55.1	55.5	56.3	53.9	50.5	47.5	44	34.1
V_{gf} (%)	98.1	90.5	45.8	75.8	77.8	77	85.3	92.8	96.1	97.8	98.6
ΔP (MPa)	—	—	2.9	—	—	—	1.6	—	—	3.8	17
ΔS (%)	—	—	26	—	—	—	1.7	—	—	2.6	18.9

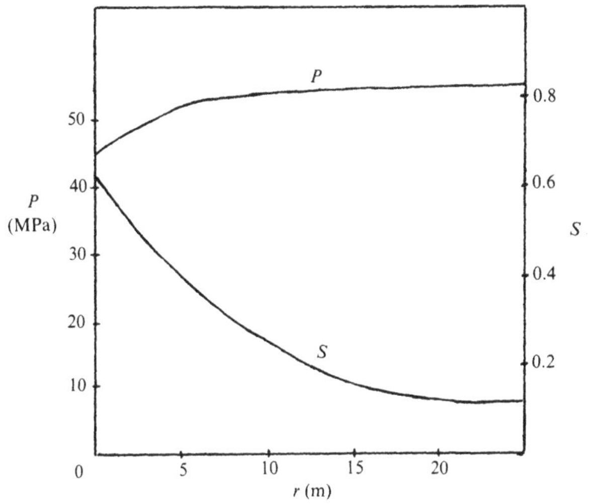

FIGURE 41. Profiles of P and S in the problem with pressures at 45 and 55 MPa at the boundaries.

In a series of calculations where the pressure at the well was kept equal to 46 MPa, while the pressure at the boundary of the sink zone varied, a reduction in $P(R_c)$ resulted in the appearance of pulsatory regimes. At $P(R_c,t) = 48$ MPa the stationary regime still existed, but at 47.5 MPa pulsations had already appeared. Figure 42 shows the time dependence of the flux for the gaseous and the liquid phases when the pressure at the boundary of the drain zone is 47.5 MPa. The fact that P has a nonmonotonic profile is indicated by the negative values of Q_g and Q_w at certain times (Figure 42b). As was already mentioned, the nonmonotonic character of the P-profile and the possibility of the flux changing its sign are both related to the fact that the problem solved here is very close to the problem of the evolution of a perturbation of a nonstable spatially homogeneous solution. Since model (7.1.1) makes no allowance for capillary effects at the boundary of the reservoir flow region, at $Q < 0$ it looks as if the seam 'sucks in' the mixture in the well. Taking into account the end effect in this model will apparently result in the sections in Figure 42, where Q is negative,

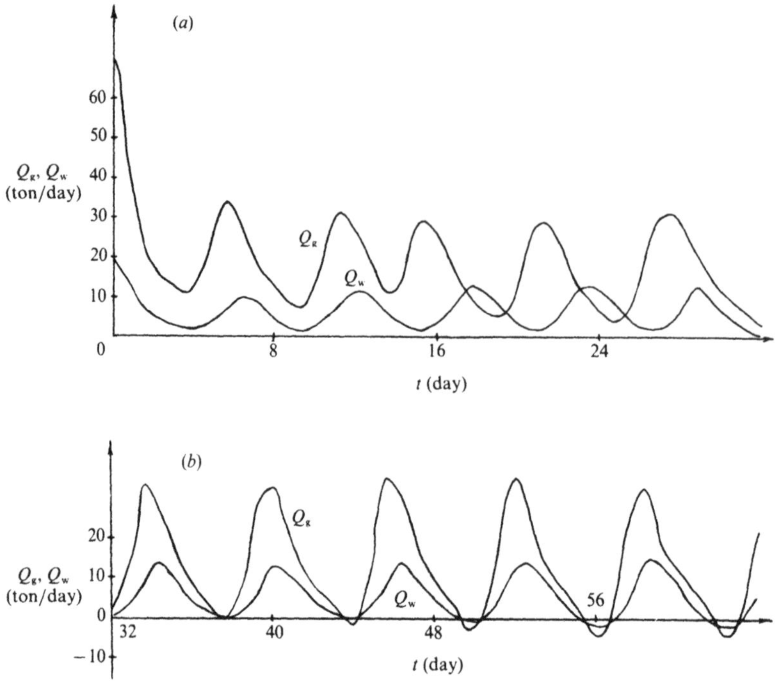

FIGURE 42. Oscillations of flow rates of gas and condensate at the well in the problem with pressures at 46 and 47.5 MPa at the boundaries; a: < 32 days, b: > 32 days.

being cut off. The well will become 'spitting': at one moment opening up and giving out products, at another, shutting down. Let us note that pulsatory regimes in the calculations with sufficiently large pressure difference (e.g., $P(R_c,t)$ = 55 MPa and $P(r_b,t)$ = 25 MPa) did not result in a negative flux and a non-monotonic profile of P.

Let us now examine the behavior of an important quantity, the stabilization time t_s (in those cases where the stationary solution exists). In the case of the simplest elastic reservoir flow regime of a homogeneous liquid, described by the piezoconductivity equation, the die-away time of pressure fluctuations about the spatially homogeneous solution depends only on the initial magnitude of the fluctuations and on the piezoconductivity coefficient. The die-away time is constant at equal initial magnitudes of fluctuations for all values of the unperturbed pressure. For more complicated flowing regimes, or in the case of sources or sinks, it seems natural enough to suppose that t_s increases as the flux becomes larger and, accordingly, the initial unperturbed state of the system moves farther away from its stationary state. The calculations made have shown that in the considered case t_s is a nonmonotonic function of Q. Namely, one observes a significant increase of t_s as one approaches an interval of Q-values where the stationary solution does not exist, $t_s \to \infty$ as $|Q - Q^*|^{-1}$. Here the asterisk labels the critical regime that separates the problems with a stable stationary solution from those whose stationary solution is unstable.

In the general case, the stability of a stationary regime is established by considering a certain eigenvalue problem for the operator on the right-hand side of the system of equations linearized with respect to the stationary solution. Suppose that at a certain value of the bifurcation parameter Q one of the eigenvalues λ crosses the imaginary axis, i.e.,

$$\operatorname{Re}(\lambda)\Big|_{Q=Q^*} = 0$$

To be definite, let $\operatorname{Re}(\lambda) < 0$ at $Q < Q^*$ and $\operatorname{Re}(\lambda) > 0$ at $Q > Q^*$. Then, at Q less than Q_* and close to Q_* we have

$$t_s \sim (\operatorname{Re}(\lambda))^{-1} = \operatorname{Re}^{-1}\left[\lambda(Q^*) + \frac{d\lambda}{dQ}\Big|_{Q=Q^*} \cdot (Q - Q^*)\right] + o\,(Q - Q^*)$$

$$t_s \sim \frac{1}{|Q - Q^*|} \tag{7.2.1}$$

The formula for the stationary inflow of a multicomponent mixture gives us[12]

$$Q \sim H_f(P(R_c)) - H_f(P(r_b))$$

$$|Q - Q^*| = \left|\frac{dH_f}{dP}(P(r_g) - P^*(r_b))\right| + o(P(r_b) - P^*(r_g))$$

where H_f is the flow potential (as is well known, for steady-state flow only one variable P is free: all the rest are expressed through P: $H_f = \int \beta dP$). Then t_s can also be estimated as

$$t_s \sim |\,P(r_b) - P^*(r_b)\,|^{-1} \tag{7.2.2}$$

Formulas (7.2.1) and (7.2.2) enable one to picture quantitatively the behavior of Q (Figure 43). As one approaches each interval of instability $(Q^*_{l,n}, Q^*_{r,n})$, hatched in the figure, t_s tends to infinity. Between $Q^*_{r,n}$ and $Q^*_{l,n+1}$ the quantity t_s is limited and passes through a minimum. Thus, in the problem considered, critical slowing down may occur before the stability is lost. Let us emphasize that the corresponding singular points in the parameter space are different from the purely thermodynamic critical point of a multicomponent liquid where all processes of fluctuation resolution slow down.[13] The new critical points are essentially related to the interaction of a multicomponent system with porous media.

One may notice that our results concerning the nonmonotonicity of the P-profile and the anomalously slow relaxation near an instability are independent of the specific form of system (7.1.1) and are due only to the very fact that

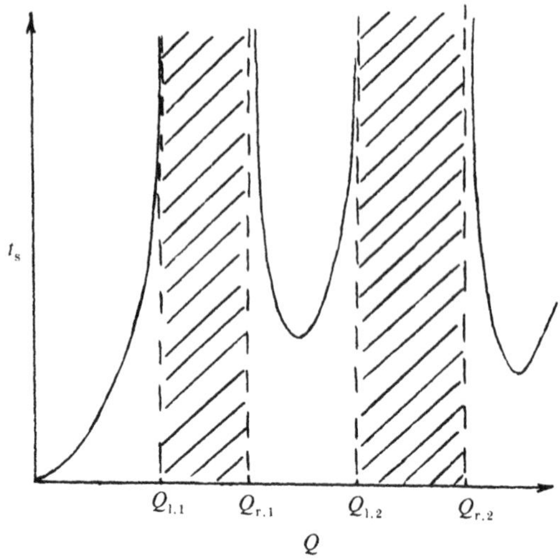

FIGURE 43. Qualitative behavior of the relaxation time of the reservoir system in dependence on an external parameter. The stationary regime is absent in the hatched regions.

unstable spatially homogeneous solutions do exist. This is why in the next section we discuss in detail the reasons why I can be negative.

7.3. RETROGRADE PHENOMENA AND NEGATIVE COMPRESSIBILITY OF AN INDIVIDUAL VOLUME

In order to understand what causes the generation of autowaves considered above, one should first examine the properties of I. Figure 44 shows the calculated P-dependences of the mole density N of the mixture (curve 1), of the saturation by liquid S (curve 2), and of I (curve 3) for the composition of the multicomponent system we have presented above. Despite the fact that conditions

$$\left.\frac{\partial \rho_w}{\partial P}\right|_{x, \ldots, x_{l-1}} > 0\,, \quad \left.\frac{\partial \rho_g}{\partial P}\right|_{y_1, \ldots, y_{l-1}} > 0 \tag{7.3.1}$$

were satisfied, function N(P) proved to be nonmonotonic, decreasing (except for a narrow interval 50 MPa $<$ P $<$ 51 MPa) in the pressure range where S grows with a decrease of P, i.e., in the region of retrograde condensation (the hatched region in Figure 44). Thus, the condition of negative compressibility of the multicomponent mixture does not contradict the conditions of positive compressibility of individual phases, and it is in these regions that $\partial N/\partial P$ was negative. This is not a coincidence. One may construct a hypothetical model of a multicomponent mixture with phase densities constant or only weakly depending on P (compared to S). The mass density of the mixture is

$$\tilde{N} = \rho_w S + \rho_g (1 - S) = (\rho_w - \rho_g) S + \rho_g \tag{7.3.2}$$

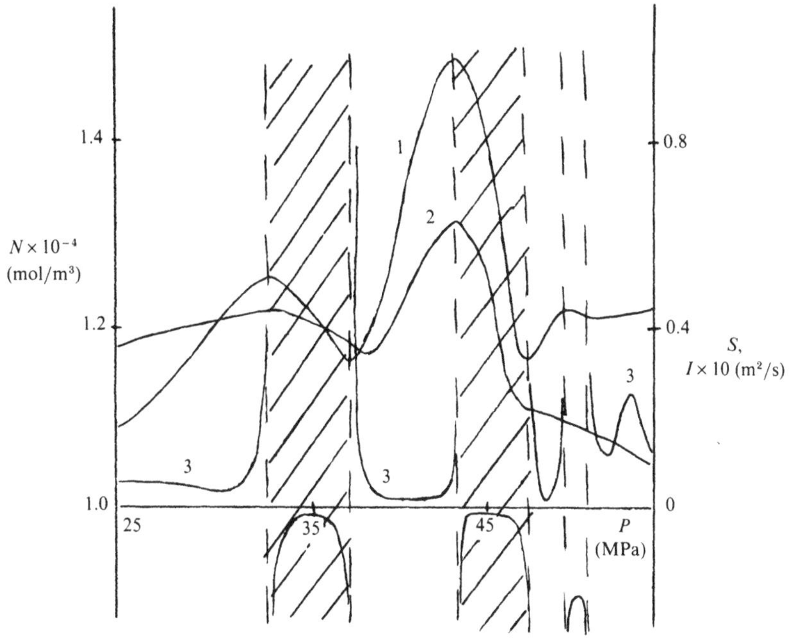

FIGURE 44. Dependences of N, S, and I on P calculated for the composition given in Section 7.1.

From this expression and the inequality $\rho_w > \rho_g$ it follows that in this model, at the fixed total composition, the condition $\partial\tilde{N}/\partial P < 0$ exactly corresponds to regions of retrograde condensation.

In order to understand the physical meaning of negative compressibility of a two-phase system moving in a porous medium one should turn to the known data on the thermodynamics of multicomponent systems. Figure 45a shows a typical pressure–temperature phase diagram for a mixture with constant composition. A characteristic geometric feature of such diagrams is that the section of the abscissa cut off by a vertical tangent to the boundary of the two-phase region is greater than the value of the abscissa for the critical point K and the size of the segment 0K′. The properties of the curve K′KK″ determine the configuration of lines of equal saturation (dashed lines) so that for temperatures above the critical point S(P) is nonmonotonic with a decrease of P (Figure 45b). The value of P at the point where S is maximum is called the maximum condensation pressure. The point at which the liquid appears with a decrease of P corresponds to the pressure at the beginning of condensation. The range of pressures below (above) the maximum condensation pressure corresponds to direct evaporation (to retrograde condensation).

In the context of the oil industry, the methods of finding the shape of S(P)-curves at fixed concentration of the mixture are called contact condensation

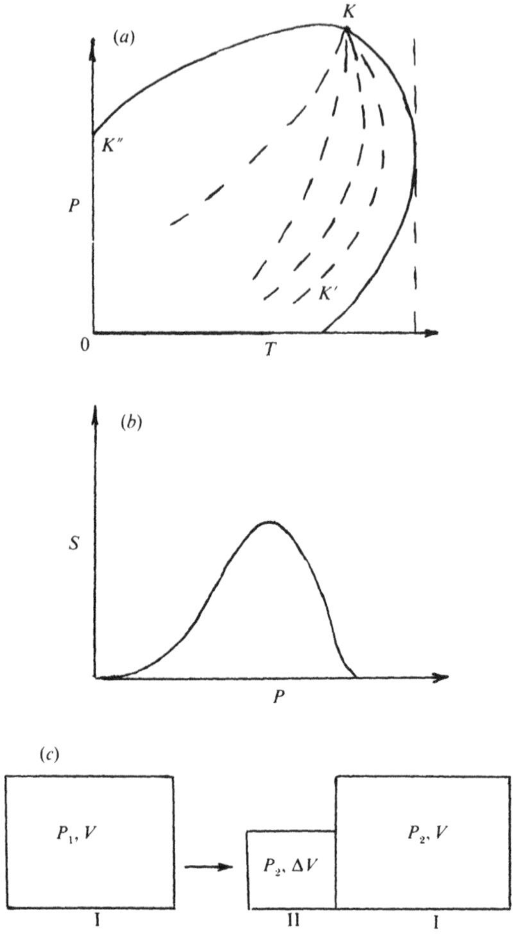

FIGURE 45. Diagram a: typical P-T phase diagram of a reservoir mixture, b: dependence $S(P)$ by retrograde condensation, and c: the evolution of an individual volume of mixture moving in a porous media. The effect of negative volume of heavy components leads to the negative compressibility of an individual volume ($P_2 > P_1$, $V + \Delta V > V$).

experiments. Another popular method of studying the properties of mixtures is the so-called differential condensation.[14] In this case, at each stage of condensation one lowers the pressure in the vessel containing the mixture and extracts the liquid that is condensed, and after that brings the rest of the mixture to the initial pressure or (in another variant of the experiment) to the initial volume. One can see that in this way one reproduces the effect of liquid condensation with the motion of an individual volume of the mixture through a porous medium. In any case, this is the generally accepted interpretation of experiments on differential condensation.

One of the most curious effects observed in experiments on differential condensation is the so-called negative volume of heavy components — the phenomenon whereby with condensation of liquid the volume of the remaining mixture increases instead of becoming smaller (if the pressure is the same) or (if the volume is kept constant) the pressure becomes higher. Apparently, this effect was first discovered by Katz and Sliepcevich.[15] In their experiments, which were made at fixed pressure, with successive extraction of C_7, C_6, C_5, C_4, C_3, and C_2 components, the volume of the mixture first increased and then (after all the C_4 and heavier components have been extracted) began to diminish.

Figuratively speaking, the heavy components had a sort of negative volume which, after they were extracted from the mixture, was added to the volume of the remaining system. Further experimental studies have shown that the effect of the negative volume of heavy components is observed (1) in the regions of retrograde condensation, and (2) near the critical point of the multicomponent system.[16] Let us note that it is the proximity to the critical point that may be the cause of a stronger dependence of S on P compared to $\rho_{w,g}$. This follows from the fact that in the near-critical state the lines of equal saturation in Figure 45a come infinitely close to one another, which means that a small variation of P can lead to a large variation of S. In this case the model of a mixture where the phase densities change very little compared to S predicts negative compressibility in the region of retrograde condensation. Thus, the effect of negative compressibility of a two-phase mixture (Figure 44) and the effect of a negative volume of heavy components should take place under similar thermodynamic conditions.

Let us now show that if the multicomponent mixture is selected in such a way that in experiments on differential condensation one observes the effect of negative volume of heavy components, then the properties of such a mixture, when it moves under the same conditions through a porous medium, are described by descending density. Indeed, in the case of differential condensation we can place the liquid deposited in vessel I into vessel II. If then the volume of vessel I is brought to its initial value V, the pressure P_2 in it, as was said above, will become higher than it was initially (P_1). Now let us change the volume ΔV of vessel II in such a way that the pressure in it will become equal to the new pressure P_2 in vessel I. We see that the pressure in system I + II is higher than the initial pressure in vessel I. At the same time the volume I + II is also greater than the volume of vessel I. Thus, by moving some of the molecules from one part of the system to another one can reproduce the effect of negative compressibility provided that the molecules were transferred inside the system according to their molecular weight (for instance, leaving the heavier molecules in their place).

The above two-chamber experiment was selected for a particular reason: it reflects the changes that occur in an individual volume of the mixture as it moves in the reservoir flow and comes into the region where the pressure is lower than it was at the beginning of condensation (Figure 45c). It is in this region that the phenomenon of the negative volume of heavy components occurs. Some of the heavy components are deposited into the initially immobile liquid phase while the lighter ones move on further. Thus, the part of the individual volume located higher upstream in the flow plays the role of chamber II, while the part downstream in the flow plays the role of chamber I. As was shown, for the individual volume as a whole, one will observe the effect of negative compressibility which is due to the fact that the volume moves in a porous medium that acts differently upon the gaseous and the liquid phases. In the two-chamber experiment the difference in action upon different phases is simply due to the gravitational separation of substances with different densities.

The above considerations show that when describing the hydrodynamics of a two-phase mixture in porous medium in the region of retrograde condensation one should take into account the possible lowering of the mixture density. This becomes especially evident when one describes the flow, not with Eulerian coordinates, which are secondary, but with Lagrange coordinates, which are primary in continuum mechanics and are designed for describing the evolution of an individual volume. It is also important that when we use a continuous description like Equation (7.1.1), we are dealing with quantities averaged over a large number of pore channels. In some of the channels belonging to the averaging volume differential condensation may occur, while the others will be filled with lighter components carried away along the flow. Thus, the appearance of a descending density branch in the thermodynamics of an individual volume of a two-phase mixture moving in a porous medium seems to be not only natural, but also inevitable. The scheme presented of the experiment with a two-chamber model gives one, in the first approximation, a method of determining the descending density branch under laboratory conditions.

7.4. FROM A NONEQUILIBRIUM MODEL OF MULTICOMPONENT RESERVOIR FLOW TO A MODEL WITH LOCAL EQUILIBRIUM

Let us now see how the results obtained above change as we turn to more complete models of reservoir flow. As is easily seen, the introduction of terms into system (7.1.1) that are responsible for capillary and diffusion effects, and are small compared to the terms describing convective transfer, will not affect the possibility of I changing its sign due to its denominator running through zero, if such a possibility existed initially in model (7.1.1). Taking into account the nonequilibrium character of interphase transfer is another matter. Let us examine this problem more closely and show that the account of nonequilibrium effects regularizes Equation (7.1.1) in a situation where the quantity $\partial N/\partial P$ may change its sign, and in the framework of model (7.1.1) we have relaxation oscillations discontinuous in time.

One may start with the system of equations of the form[17]

$$m\frac{\partial N z_i}{\partial t} = \nabla(k\beta_i \nabla P), \qquad i = 1, \ldots, l \tag{7.4.1}$$

$$m\frac{\partial N_g y_i}{\partial t} = \nabla(k\beta_i^{(g)}\nabla P) + \tau^{-1}\cdot\phi_i\,, \qquad i = 1, \ldots, l \qquad (7.4.2)$$

This system is a generalization of Equation (7.1.1) to the case of nonequilibrium multicomponent reservoir flow. Here $N_g = \rho_g(1 - S)/M_g$, $\beta_i^{(g)}$ is the part of β_i corresponding to the motion of the gas; and τ is the characteristic relaxation time for interphase mass exchange. In the general case we have a set of quantities τ_i for each component. Quantity ϕ_i measures the deviation from phase equilibrium in the i-th component and, according to the relations of nonequilibrium thermodynamics, is proportional to the difference of chemical potentials of the i-th components in the liquid and gaseous phases,

$$\sum_{i=1}^{l} y_i = \sum_{i=1}^{l} z_i = 1$$

In the zero-th approximation with respect to τ, Equations (7.4.1) and (7.4.2) give system (7.1.1) with the closing constraints

$$\phi_i = 0\,, \qquad i = 1, \ldots, l \qquad (7.4.3)$$

For system (7.4.1) and (7.4.2), the unknown quantities are $\mathbf{u} = (z_1, \ldots, z_{l-1}, P)$ and $\mathbf{v} = (y_1, \ldots, y_{l-1}, S)$. In the local-equilibrium model (7.4.1), (7.4.3), $\mathbf{v}$ can be expressed through $\mathbf{u}$ using relationships (7.4.3), and thus at $\tau = 0$ the unknown quantities are $\mathbf{u}$.

By considering the problem of the linear stage of the evolution of perturbations of a spatially homogeneous solution of system (7.4.1) and (7.4.3) in the region shaped like a rectangular parallelepiped with dimensions L_j, it is easily shown that conditions (7.3.1) are sufficient for the elements of the linear response matrix not to have poles. This holds for any $\tau > 0$, but at $\tau = 0$, as was shown above, despite the fact that conditions (7.3.1) are satisfied, the denominator of I can become zero. Since the transition from Equations (7.4.1) and (7.4.2) to Equations (7.4.1) and (7.4.3) occurs at $\tau \to 0$, it is natural to examine this transition using the method of perturbations.

By linearizing Equation (7.4.2) and considering the Fourier components of fluctuations, $\mathbf{u}_q'$ and $\mathbf{u}_q'$, instead of Equation (7.4.2) we get

$$\tau\left(\mathbf{A}\frac{d\mathbf{u}_q'}{dt} + \mathbf{B}\frac{d\mathbf{v}_q'}{dt}\right) = -\tau q^2\mathbf{C}\mathbf{u}_q' + \mathbf{E}\mathbf{u}_q' + \mathbf{F}\mathbf{v}_q' \qquad (7.4.4)$$

Here q^2 is the wave vector squared; **A, B, C, E,** and **F** are $l \times l$ matrixes depending on the spatially homogeneous solution. The structure of these matrices is determined by Equation (7.4.2). For instance, **E** and **F** are matrixes of the first derivatives of ϕ_i with respect to components of vectors **u** and **v**. By using Equation (7.4.4) one can show that, up to the second-order terms in τ, quantity $\mathbf{v}_q'$ is expressed through $\mathbf{u}_q'$ in the following way

$$\mathbf{v}_q' = -\mathbf{F}^{-1}\mathbf{E}\mathbf{u}_q' + q^2\tau\mathbf{F}^{-1}\mathbf{C}\mathbf{u}_q' + \tau(\mathbf{F}^{-1}\mathbf{A} - \mathbf{F}^{-1}\mathbf{B}\mathbf{F}^{-1}\mathbf{E})\frac{d\mathbf{u}_q'}{dt} \quad (7.4.5)$$

Let us note that if one makes the transformation to originals in Equation (7.4.5), one will get a relationship between **u** and **v** involving both time and space derivatives. The existence of matrix $\mathbf{F}^{-1}$ follows from the implicit function theorem of Equation (7.4.3) and from the fact that in models (7.4.1) and (7.4.3) **v** can always be expressed through **u** by using Equation (7.4.3).

After linearization and transition to Fourier components of fluctuations, system (7.4.1) acquires the form

$$\mathbf{D}\frac{d\mathbf{u}_q'}{dt} + \mathbf{M}\frac{d\mathbf{v}_q'}{dt} = -q^2\mathbf{G}\mathbf{u}_q' \quad (7.4.6)$$

where **D, M,** and **G** are $l \times l$ matrixes. By substituting Equation (7.4.5) into Equation (7.4.6) we obtain

$$\tau\mathbf{M}\mathbf{F}^{-1}\cdot(\mathbf{A} - \mathbf{B}\mathbf{F}^{-1}\mathbf{E})\frac{d^2\mathbf{u}_q'}{dt^2} + [(\mathbf{D} - \mathbf{M}\mathbf{F}^{-1}\mathbf{E}) + $$
$$+ \tau q^2\mathbf{M}\mathbf{F}^{-1}\mathbf{C}]\cdot\frac{d\mathbf{u}_q'}{dt} + q^2\mathbf{G}\mathbf{u}_q' = 0 \quad (7.4.7)$$

At $\tau = 0$, from Equation (7.4.7) one should obtain the linearized system (7.4.1) and (7.4.3) expressed in terms of Fourier components of fluctuations, or, equivalently, system (7.1.1). Transforming Equation (7.4.7) in such a way that the set of the matrix that stands in front of derivative $d\mathbf{u}_q'/dt$ and does not depend on q^2 becomes diagonal — this is achieved by the same transformations as those used in Section 7.1 for transforming the initial linearized system (7.1.1) into system (7.1.4) — we obtain

$$\tau\mathbf{R}^{I}\frac{d^2\mathbf{u}_q'}{dt^2} + (\mathbf{R}^{II} + \mathbf{R}^{III}q^2\tau)\frac{d\mathbf{u}_q'}{dt} + q^2\mathbf{R}^{IV}\mathbf{u}_q' = 0 \quad (7.4.8)$$

where $\mathbf{R}^{(n)} = \{r_{ij}^{(n)}\}$ are square matrixes of order l. Thus, instead of $2l$ Equations (7.4.7) and (7.4.8) we obtain l equations of the second order in t, with the initial conditions for $d\mathbf{u}_q/dt$ found from the initial conditions for systems (7.4.1), (7.4.2), and Equation (7.4.5). Comparing Equation (7.4.8) at $\tau = 0$ with Equation (7.1.4) we find

$$\mathbf{R}^{II} = \begin{pmatrix} 1 & & & 0 \\ & \cdot & & \\ & & \cdot & \\ & & & \cdot \\ & & & 1 \\ 0 & & & \dfrac{\partial N}{\partial P} \end{pmatrix}, \qquad \mathbf{R}^{IV} = \begin{pmatrix} 0 & \cdots & 0 & r^{IV}_{1l} \\ \cdot & & \cdot & \cdot \\ \cdot & & \cdot & \cdot \\ \cdot & & \cdot & \cdot \\ 0 & \cdots & 0 & r^{IV}_{ll} \end{pmatrix}$$

$$r^{IV}_{il} = \frac{1}{mN}(\beta_i - z_i\beta), \qquad i = 1, \ldots, l-1$$

$$r^{IV}_{ll} = \frac{k}{m}\left[\beta - \frac{1}{N}\sum_{j=1}^{l=1}\frac{\partial N}{\partial z_j}(\beta_j - z_j\beta)\right]$$

The form of matrixes $\mathbf{R}^{I}$ and $\mathbf{R}^{III}$ is unessential for what follows.

Now we notice that, up to the τ^2-terms, for obtaining the equation for $P'_q = u'_{q},l$ one should substitute the zero-th approximation of $\mathbf{u}'_{q,i}$, $i = 1, \ldots, l - 1$ into the last line in Equation (7.4.8). We obtain

$$\tau d_1 \frac{d^2P'_q}{dt^2} + \left(\frac{\partial N}{\partial P} + q^2\tau d_2\right)\frac{\partial P'_q}{dt} + q^2 d_3 P'_q = 0 \qquad (7.4.9)$$

where

$$d_1 = r^{I}_{ll},\ d_2 = r^{III}_{ll} - \sum_{i=1}^{l-1} r^{I}_{li}\, r^{IV}_{il},\ d_3 = r^{IV}_{ll}$$

The expression for d_2 is obtained by taking account of the fact that both $\mathbf{R}^{III}$ and $\mathbf{C}$ have non-zero elements only in their l-th column.

Thus, the proximity of $\partial N/\partial P$ to zero makes it necessary to take into account the second-order time derivatives in the equation for P, i.e., to take into account the nonequilibrium effects. By comparing Equation (7.4.9) and Equation (7.1.4), one can see that the condition $I^{-1} = 0$ in the local-equilibrium model at $d_1 \cdot d_3 > 0$ and

$$\frac{\partial N}{\partial P} + q^2\tau d_2 = 0$$

corresponds, in the nonequilibrium model, to the moment at which the time-periodic solutions of small amplitude appear. If, in addition, we have $d_1 \cdot d_2 > 0$, then all the conditions necessary for continuation of periodic solutions into the region of instability in the problem considered are satisfied. Indeed, from Equation (7.4.9) it follows that in the given problem the eigenvalues are found up to the τ^2 terms from the equation

$$\tau d_1\lambda^2 + \left(\frac{\partial N}{\partial P} + q^2\tau d_1\right)\lambda + q^2 d_3 = 0 \tag{7.4.10}$$

If a pair of eigenvalues corresponding to the minimum value of the wave vector squared,

$$q_*^2 = \frac{4\pi^2}{\sum_{j=1}^{d} L_j^2}$$

is purely imaginary — according to Equation (7.4.10), this will happen when $-\partial N/\partial P = q_*^2\,\tau d_2$ — then all the modes with smaller wavelength ($q^2 > q_*^2$) will have eigenvalues lying in the left half-plane. According to Hopf's theorem for distributed systems,[6] the periodical orbit corresponding to q_*^2 can be continued into the region $\partial N/\partial P < -q_*^2\,\tau d_2$.

The period of the auto-oscillations generated from Equation (7.4.9) equals

$$\tau_P = 2\pi\left(\frac{\tau d_1}{q_*^2 d_3}\right)^{1/2} \tag{7.4.11}$$

i.e., it is proportional to the spatial scale of fluctuations and to the square root of τ. By analyzing the structure of coefficients d_1 and d_3 in Equation (7.4.11) more closely one can show that the period of auto-oscillations is of the order of the geometric mean of the characteristic phase equilibrium time τ, and the characteristic time of transfer of a mixture particle in the reservoir flow along the solution region τ_f:

$$\tau_p \sim (\tau \cdot \tau_f)^{1/2}$$

Thus, if the auto-oscillations are generated in a region of size around 1 m (e.g., in the near-zone of a well) and have a period of about 1 h, the auto-oscillations at the scale of the formation (kilometers) will have a period of around a month.

In the one-mode approximation, the nonlinear equation for P_q' corresponding to the linearization of Equation (7.4.9) will have the form

$$\begin{gathered}\tau d_1 \frac{d^2P_q'}{dt^2} + \frac{dW(P_q')}{dt} + q^2 d_3 P_q' = 0\,, \\ W(P_q') = q^2\tau d_2 P_q' + \frac{\partial N}{\partial P} P_q' + \frac{1}{2}\frac{\partial^2 N}{\partial P^2}(P_q')^2 + \frac{1}{6}\frac{\partial^3 N}{\partial P^3}(P_q')^3\end{gathered} \tag{7.4.12}$$

This equation is obtained by keeping the terms nonlinear in P' in the function $N(P)$ in the vicinity of the spatially homogeneous solution with $\partial N/\partial P \approx 0$. It is close to the Van der Pol equation and at $\partial^2 N/\partial P^2 = 0$ coincides with it exactly.[18]

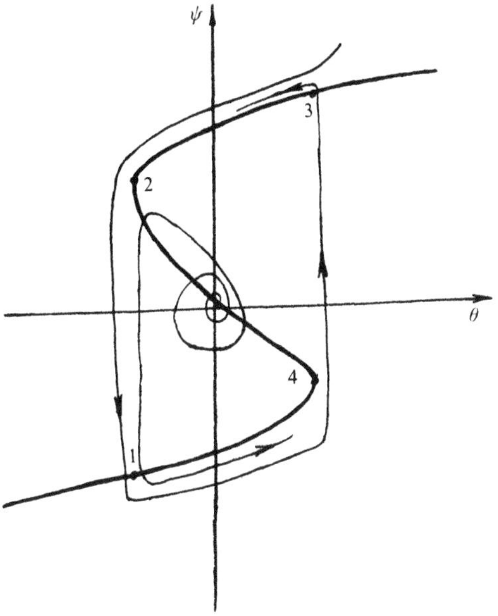

FIGURE 46. Relaxation oscillations in the dynamic system considered (one-mode approximation).

Like the Van der Pol equation, Equation (7.4.12) can be examined from the viewpoint of the theory of relaxation oscillations. Let us present Equation (7.4.12) in the form of equivalent systems

$$\begin{cases} d_1 \dfrac{d\psi}{dt_1} = \theta - W(\psi) \\ \dfrac{d\theta}{dt_1} = -\tau d_3 q^2 \psi \end{cases}, \quad \begin{cases} \tau d_1 \dfrac{d\psi}{dt} = \theta - W(\psi) \\ \dfrac{d\theta}{dt} = -\psi q^2 d_3 \end{cases} \tag{7.4.13}$$

where $\psi = P'_q$; $t_1 = t/\tau$ is the short timescale. Figure 46 shows the phase plane of Equation (7.4.13). The limit $\tau = 0$ in the right-hand side of the first system (7.4.13) describes the fast motion along sections 1–2 and 3–4 of the phase trajectory. The slow motion along 2–3 and 1–4 is described by equations of the second system in (7.4.13) at $\tau = 0$. Taking the limit $\tau = 0$ in the slow system in (7.4.13) leads to instantaneous relaxation of the solution on the vicinity of points 1, 2, 3, and 4, corresponding to zero compressibility, and the motion is described as being piecewise discontinuous with sections where parameters vary monotonically, alternating with discontinuities. At small non-zero τ the auto-oscillations will have the form of alternating fast and slow motions. It is noteworthy that here the fast motions correspond to being the system in the region of retrograde condensation in the space of thermodynamic parameters, while the slow motions are associated with the region of direct evaporation. Let us note that this picture corresponds to keeping only one mode in the nonlinear description. The calculations made in Sections 7.1 and 7.2 show that there may be more complicated autowave regimes corresponding to the interaction between several modes.

7.5. DISCUSSION OF EXPERIMENTAL RESULTS

Thus, we see that the possibility of autowave generation is related to the loss of stability of stationary solutions of equations of two-phase multicomponent reservoir flow. According to the above considerations, the loss of stability in a distributed system should occur in the region of retrograde condensation. The self-sustaining oscillations should be generated owing to the cyclic alternation of the following processes:

1. Accumulation of the liquid phase in the region of space corresponding to retrograde condensation,
2. Transport of the liquid in the reservoir flow into the region corresponding to evaporation, and
3. The diminishing of the volume occupied by the liquid phase in this latter region.

It is predicted that there may be nonmonotonic profiles of pressure and the anomalous increase of the time within which the stationary regime sets in as one of the parameters approaches the region of instability. This process may have the character of autowaves in an open system, i.e., under continuous inflow of energy from outside.

The results presented here have been mainly obtained in 1985, were reported at the 6th All-Union Congress on Mechanics in 1986, and were published as a short report in 1987. The next step was to obtain experimental confirmation of the theory. We do not refer here to the observation of fluid auto-oscillations with exploitation of the deposits — there have been plenty of such observations already. The principal idea was to do experiments under laboratory conditions in the range of parameters predicted by the theory. Such experiments were performed in 1988 and a complete account of the results obtained was presented in the literature.[19,20] Only the most important confirmations of the theory obtained in these experiments will be discussed here.

The experiments were done with a linear model of a seam with a porous medium poured on. The pressure at the inlet was kept constant and equal to 13.4 MPa (the system was open). The simulation was done at 296.5 K, with a multicomponent mixture for which the reservoir flow occurs both in the regime of retrograde condensation and in the direct evaporation regime, depending on the pressure at the given point of the linear model of a seam. The mixture is characterized by the pressure at the beginning of condensation, 12.7 MPa, and by the maximum condensation pressure, 12 MPa. The system was brought into the working regime by lowering the pressure in the model from 15 MPa.

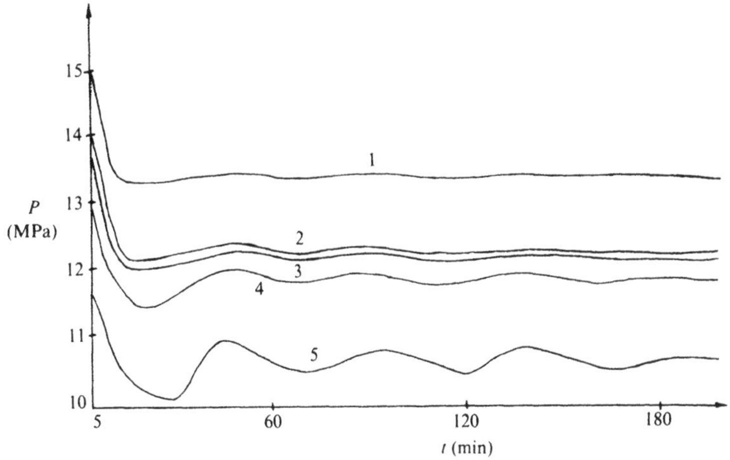

FIGURE 47. Pressure dynamics at different points of the seam model (experiment 1).

Two experiments were performed. In the first, the pressure at the inlet of the linear model was 9.4. MPa. Figure 47 shows the time variation of pressure at different points of the model. The i-th curve corresponds to measurements made at the point with coordinates $r_i = L \cdot i/6$. One can see that there are damped oscillations of pressure, their amplitude being the greater the closer the measurement point is to the exit of the model. At the point r_5 the amplitude of oscillations reached 0.45 MPa. The profiles of P change little along the seam when the pressure is around the pressure at the beginning of condensation or around the maximum condensation pressure.

A chromatographic analysis of the mixture coming out of the seam has shown that there are oscillations of composition in the flow.

Figure 48 shows the time variation of the flow velocity at the exit of the model (curve 1). The characteristics of the rate at the exit, Q, were measured here in units of m^3/min at T = 296.5 K and P = 0.1 MPa. The velocity oscillations have a period of 40 min. The maximum values of velocity corresponds to minima in curve 5 (Figure 47), and vice versa. The oscillations die away in about 3 to 3.5 h.

In the second experiment the pressure at the exit of the model was fixed at 10.6 MPa. All other parameters were the same. Figure 49 shows the time variation of pressure at different points of the model. The labeling of the curves corresponds to that in Figure 47. Despite the fact that the difference in pressure between the ends of the model is one and one-half times smaller than in the first experiment, the process now is much more complicated. One could very clearly observe that the pressure at the points located closer to the exit was uniformly higher than at the points located farther away from the exit (the overlapping of curves 2, 3 and 3, 4). The oscillations at the point r_5 had a clearly distinguished period of about 2 h. The amplitude of oscillations reached 0.4 MPa. At the times

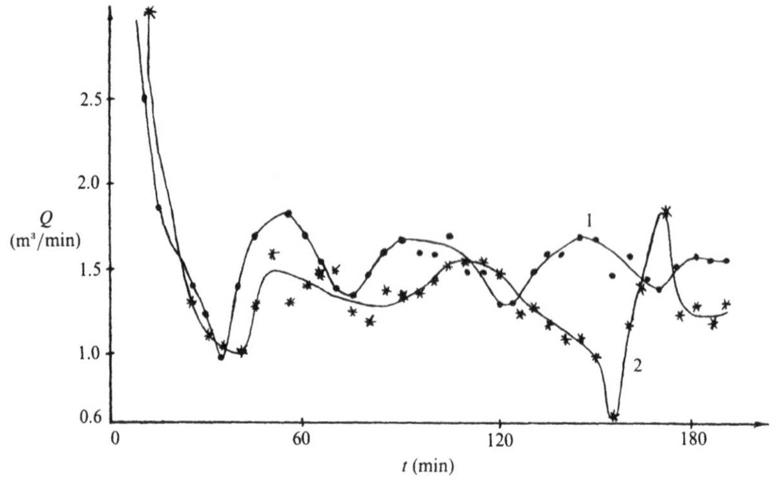

FIGURE 48. Flow rate at the exit of the seam model in the first (curve 1) and the second (curve 2) experiments.

when the oscillations in the first experiment died away, no damping of autowaves was now observed. According to our theory, this is an indication of the relaxation processes slowing down as the region of nonstable stationary solutions is approached.

Curve 2 in Figure 48 shows the time dependence of the velocity at the exit. The velocity reaches its minimum at $t = 40$ min. After that, against a background of small-scale oscillations, a slow increase in velocity occurs, which is then replaced first by a slow and then by a more rapid decrease. The minimum of velocity is reached at $t = 155$ min, and this minimum is much deeper than the first one. After that the velocity rises, which corresponds to cyclic resumption of auto-oscillations. Such behavior of the velocity is associated with a slow accumulation of the liquid phase at the exit replaced by an abrupt 'shutting down' of the reservoir flow, and then by the emergence of the liquid phase that has acquired sufficient mobility.

Thus, the autowaves in the case of a reservoir flow with phase transitions have been experimentally observed. The results of these experiments confirm the following predictions of the theory.

The autowaves can be observed in an open system if the region of the flow consists of areas corresponding to retrograde condensation and to evaporation processes.

In those regions of the flow that correspond to retrograde condensation, the pressure may have nonmonotonic profiles. According to the considerations of Section 7.3, this suggests an actual observation of the effect of negative compressibility of an individual volume of the moving mixture.

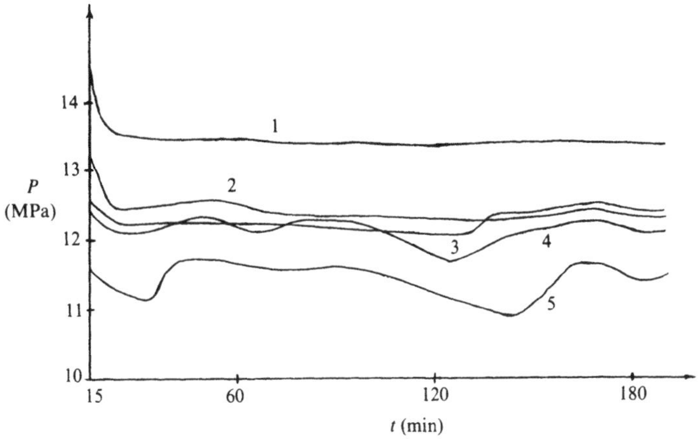

FIGURE 49. Pressure dynamics at different points of the seam model (experiment 2).

The nonmonotonic profiles were observed only when the pressure difference between the ends of the model was smaller (i.e., in the second experiment). This agrees remarkably well with the results of calculations presented in Section 7.2, where the pressure at one of the ends corresponded to an unstable spatially homogeneous solution, the pressure difference between the ends was varied, and nonmonotonic profiles of P were observed only for a sufficiently small pressure difference.

The relaxation time in the system grows as the instability is approached: in this case the parameter that determines the proximity of the system to the instability region is the deviation of pressure at the exit from the maximum condensation pressure. This result is very significant in confirming the theory, since in the second experiment it was natural to expect a more rapid damping of oscillations. This would have been the case if we were dealing with a one-phase liquid moving in the porous medium and the reservoir flow occurs without phase transitions. As one can see, the phase transitions in the flow process can significantly change the situation.

Thus, one more example of self-organization in dynamic systems has been discovered and described. Let us enumerate the succession of events that have led to these results. First, in calculations with the data from an actual deposit the autowave regimes of reservoir flow had been numerically observed; namely, in 1985 the author carried out calculations for a problem of stationary inflow into a well at a Karachaganak (former U.S.S.R.) deposit. However, it was found that for some initial conditions a stationary solution could not exist. Later, it was learned that the same effects impeding the use of standard methods of well testing had in fact been observed by Karachaganak. Then, the theory of these phenomena was constructed and, finally, confirmed experimentally. In our opinion, this course of events serves as an additional confirmation of the correctness of the theory.

It would be important to undertake a program of investigation of autowave reservoir flows. The program must include investigations of the thermodynamic properties of mixtures in retrograde conditions (in particular, experiments with a descending density branch were considered in Section 7.3), experimental investigations of the hydrodynamics of multicomponent fluids in porous media, and finally, improvement of numerical methods for solving multicomponent reservoir flow equations. The numerical method used above enables us to single out the regions of instability and to estimate the dynamics of autowaves. However, it was not especially intended to calculate the time-discontinuities (relaxation) processes; it was worked out for standard reservoir flow calculations. In a simple case that can be illustrated by a 'slow' Van der Pol equation

$$(1 - \psi^2)\frac{d\psi}{dt} = \psi \tag{7.5.1}$$

It is obvious that any direct numerical method leads to loss of accuracy near discontinuous points $\psi = \pm 1$. Two approaches to correction are possible. First, the term $\sim d^2\psi/dt^2$ should be added to Equation (7.5.1) and then one can solve the full Van der Pol equation. For the problem of a multicomponent reservoir flow it should mean solving general Equations (7.4.1) and (7.4.2). Secondly, Equation (7.1.1) can be solved with control of the value $\partial N/\partial P$ at any sites of the different grid ($|\partial N/\partial P| > \epsilon_{reg}$). In my opinion, these are possible actions to avoid the difficulties connected with mathematical modeling of the system with discontinuities considered in this chapter.

The theory presented above is directed to describe gas-condensate reservoir systems. A description of a reservoir flow of liquids with a newly developed gas phase, when the pressure in the system is higher than full saturation pressure, is available in the literature.[21] The investigation of stationary reservoir flow regimes stability shows the possibility of stability loss and appearance of time-periodical and stochastic oscillations. The bifurcation parameter here is the difference in pressure at the ends of the flow region, that is the increase of the flow rate, as by the description of the transition to the turbulence in usual hydrodynamics, leads to the nontrivial regimes.

Experimental investigation of a reservoir flow of oil saturated by gas above the phase transition pressure has been reported.[22] Oscillations of the flow parameters were observed for a reservoir flow of such an aerated liquid, which is connected, in the opinion of author of the experiment, with the presence of micro-embryos of gas in the liquid. This situation differs from those discussed in this chapter (no phase transition), but anyway one can arrive at a conclusion about the necessity to study reservoir flow oscillation regimes which can reflect not only the fact of a noise presence, which is inherent in any system, but the physical essence of the process.

REFERENCES

1. **Haken, G.,** *Synergetics. An Introduction,* Springer-Verlag, Berlin, 1983.
2. **Mitlin, V. S.,** in *Thesis of Talks, 6th All-Union Congr. Theor. Appl. Mech.,* Tashkent, 1986, 453 (in Russian).
3. **Mitlin, V. S.,** *Dokl. Akad. Nauk SSSR,* 296, 1323, 1987; *Sov. Phys. Dokl.,* 32(10), 796, 1987.
4. **Mitlin, V. S.,** in *Int. Conf. Development of gas-condensate fields, Rep. Sect. 6, Fundamental and Research Scientific Investigation,* Krasnodar, 1990, 26.
5. **Mitlin, V. S.,** *J. Fluid Mech.,* 222, 369, 1990.
6. **Marsden, J. E. and McCracken , M.,** *The Hopf Bifurcation and its Applications,* Springer-Verlag, New York, 1976.
7. **Barenblatt, G. I., Entov, V. M., and Ryzhik, V. M.,** *Motion of Fluids in Natural Rocks,* Klumer Academic Publ., Dordrecht, 1990.
8. **Yuen, H. C. and Lake, B. M.,** in *Solitons in Action,* Longren, K. and Scott, A., Eds., Academic Press, New York, 1978, 89.
9. **Nicolis, G. and Prigogine, I.,** *Self-Organization in Non-Equilibrium Systems,* John Wiley & Sons, New York, 1977.
10. **Landau, L. D. and Lifshitz, E. M.,** *Hydrodynamics,* Nauka, Moscow, 1986 (in Russian).
11. **Zaslavsky, G. M.,** *Stochasticity of Dymamical Systems,* Nauka, Moscow, 1984 (in Russian).
12. **Khristianovich, S. A.,** *Prikl. Mat. Mekh.,* 5, 277, 1941, (in Russian).
13. **Ma, S.,** *Modern Theory of Critical Phenomena,* W. A. Benjamin, New York, 1976.
14. **Amyx, J. W., Bass, D. M., and Whiting, J.R.,** *Petroleum Reservoir Engineering. Physical Properties,* McGraw-Hill, New York, 1960.
15. **Katz, D. L. and Sliepcevich, C. M.,** *Oil Weekly,* 116(13), 30, 1945.
16. **Velikovsky, A. S. and Stepanova, G. S.,** Negative volume of the less volatile component in methane mixtures with different hydrocarbons, in *Study of Gas-Condensate Deposits, Proc.,* VNIIGAS, No. 17/25, Moscow, 232, 1962 (in Russian).
17. **Nicolaevskii, V. N.,** *Mechanics of Porous and Fractured Media,* Singapore World Sci. Publ., 1990.
18. **Andronov, A. A., Vitt, A. A., and Khaikin, S. E.,** *Theory of Oscillations,* Nauka, Moscow, 1982 (in Russian).
19. **Makeev , B. V. and Mitlin, V. S.,** *Dokl. Akad. Nauk SSSR,* 310(6), 1315, 1990; *Sov. Phys. Dokl.,* 35(2), 147, 1990.
20. **Makeev, B. V. and Mitlin, V. S.,** in Int. Conf. Development Gas-Condensate Fields, Rep. Sect. 6, Fundamental and Research Investigation, Krasnodar, 1990, 137.
21. **Chasanov, M. M. and Yagubov, I. N.,** Int. Conf. Development of Gas-Condensate Fields, Rep. Sect. 6, Fundamental and Research Investigation, Krasnodar, 1990, 66.
22. **Stepanova, G. S.,** *Gazov. Prom.,* 4, 51, 1989 (in Russian).

8

Kinetically Stable Structures in the Flow Models with Phase Transitions

The dynamics of spatially distributed nonlinear systems is characterized by a set of features, of which the self-organization phenomenon is one. This means that, as a result of the nonlinear interaction in a system, some typical inhomogeneous structures can form, and they often derive the asymptotical behavior of solutions by arbitrary initial conditions. Such structures characterize specific properties of a given nonlinear medium and are currently the object of intensive study in the natural sciences. One of the possibilities of a nonlinear system asymptotical evolution, which is often observed, is the formation of nonstationary, for instance, autowave structures; a corresponding example was considered in the previous chapter. Another is the asymptotical formation of stationary dissipative structures corresponding to the local minima of energy functional in the system.[1]

However, other situations are also possible when a system slows down quickly near a structure which is unstable, but has long "lifetime". The time of the nonlinear retardation need not be only of the order of the characteristic time of the process, but can essentially exceed it. In this chapter, the results of investigation of kinetically stable structures (the term has been suggested by Erukhimovich[2] in the process of the preparation of this paper) in TMRF and in the phase transition theory are presented.

8.1. KINETICALLY STABLE STRUCTURES IN TWO-PHASE MULTICOMPONENT RESERVOIR FLOW THEORY

The consideration of this section is based on previous work.[3] Let us again consider the system of Equations (7.1.1). The problem of how the solution goes over into a steady-state solution, when the latter is fairly close to a spatially homogeneous solution and the fluctuations are fairly small, can be solved by analyzing the linearized system (7.1.4) with reference to the spatially homogeneous solution (P = const, z_i = const). This point was considered in Section 7.1 in connection with the problem of self-oscillation in multicomponent reservoir flow. The relaxation to the equilibrium is defined by increment I.

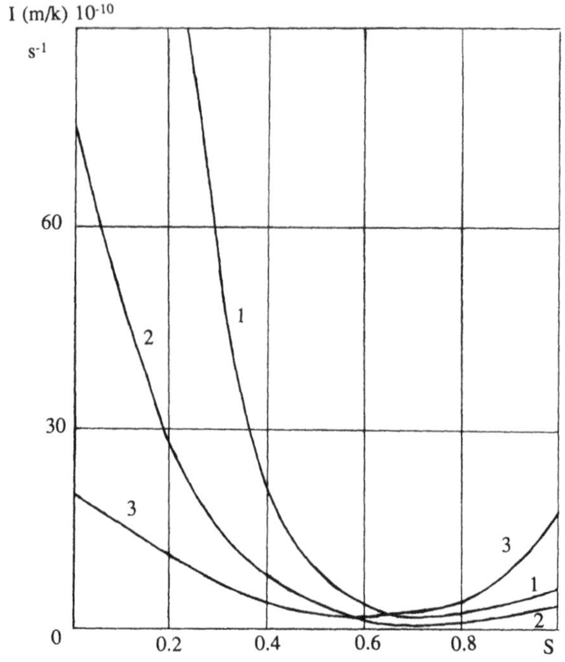

FIGURE 50. Dependence of the inverse relaxation time at the pore scales on S for different P and a given methane/n-hexane mixture.

For a binary mixture the phase compositions, densities, and viscosities depend only on the pressure. Taking P and S as the independent variables, instead of Equation (7.1.5) we obtain

$$I = \frac{k}{m}\left[\beta_1\left(\frac{\rho_w x_2}{M_w} - \frac{\rho_g y_2}{M_g}\right) - \beta_2\left(\frac{\rho_w x_1}{M_w} - \frac{\rho_g y_1}{M_g}\right)\right] \cdot$$

$$\left[\left(\frac{\rho_w x_2}{M_w} - \frac{\rho_g y_2}{M_g}\right)\left(S\frac{\partial}{\partial P}\left(\frac{\rho_w x_1}{M_w}\right) + (1 - S)\frac{\partial}{\partial P}\left(\frac{\rho_g y_1}{M_g}\right)\right) - \right.$$

$$\left. - \left(\frac{\rho_w x_1}{M_w} - \frac{\rho_g y_1}{M_g}\right)\left(S\frac{\partial}{\partial P}\left(\frac{\rho_w x_2}{M_w}\right) + (1 - S)\frac{\partial}{\partial P}\left(\frac{\rho_g y_2}{M_g}\right)\right)\right]^{-1} \quad (8.1.1)$$

Expression (8.1.1) can be written somewhat more simply in mass variables; in this case x_1, x_2, y_1, and y_2 are mass fractions, and the combinations ρ_w/M_w and ρ_g/M_g should be replaced by ρ_w and ρ_g.

In Figure 50 we have plotted the relations $I^* = mk^{-1}I(S)$ calculated from Equation (8.1.1) for various values of P. The mixture in question consisted of CH_4 (75 mol%) and n-C_6H_{14} (25 mol%) at a temperature of 335 K. The properties of the two-phase mixture were calculated from the Peng-Robinson equation of state. The phase permeabilities were taken in the form

$$f_w = \begin{cases} \left(\dfrac{S - 0.36}{0.64}\right)^3 , & S > 0.36 \\ 0 , & S \leq 0.36 \end{cases} \qquad f_g = \begin{cases} \left(\dfrac{0.88 - S}{0.88}\right)^3 , & S < 0.88 \\ 0 , & S \geq 0.88 \end{cases}$$

Curves 1 to 3 correspond to 8, 12, and 17 MPa. Clearly, the curves are convex downward, the minimum being reached at $S \sim 0.6$–0.8. With an increase in pressure, the relaxation rate for the pure gas, which exceeds by several orders of magnitude that for the pure liquid, decreases, while the relaxation rate for the liquid increases. The equalization of the relaxation times for the liquid and the gas is associated with the convergence of the phase properties at the upper pressure limit of the two-phase region and hence the critical point is approached. From Figure 50 it also follows that as the pressure increases the minimum of the relaxation rate is shifted towards the left. We note that I^* corresponds to the reciprocal of the relaxation time on scales of the order of the pore channels. The evolution on scales L is determined in the linear approximation by characteristic times of the order of L^2/I.

If the deviation from the steady-state solution is large, then in order to describe the relaxation it is necessary to solve the nonlinear system (7.1.1). Below we present the results of calculating the recovery of a gas-condensate mixture from one end of a reservoir with linear geometry after injecting enriched gas through the same end. A finite-difference algorithm for solving the problem was described in Chapter 4.

The initial and boundary conditions are as follows:

$$P(r,0) = P^o, \; z_i(r,0) = z_i^o , \qquad i = 1, \ldots , l , \quad 0 < r < L ;$$

$$t < t^*: z_i(0,t) = \delta_i , \qquad i = 1, \ldots , l, \; - \frac{k\beta}{N}\frac{\partial P}{\partial r}\bigg|_{r=0} = v ;$$

$$t \geq t^*: \frac{k\beta}{N}\frac{\partial P}{\partial r}\bigg|_{r=0} = v, \; N = N(\epsilon_1, \ldots , \epsilon_{l-1}, P), \; \epsilon_i = \frac{\beta_i}{\beta} ;$$

$$z_i(L,t) = z_i^o, \; P(L,t) = P^o$$

The quantity t^* corresponds to the moment of the dimensionless time $t_u = vt/mL = 0.75$, $mL/V = 0.96 \cdot 10^5$ s, and was so chosen that at the beginning of recovery there was practically no disturbance of the composition at $r = L$. The composition of the injected gas was 66% (CH_4), 10% (C_2H_6), 10% (C_3H_8), and 14% (C_4H_{10}). The other initial data completely correspond to previous investigation[4] (also see Section 5.1).

In Figure 51 we have plotted the profiles of the saturation S at various moments of time. Curve 1 corresponds to the end of injection, curves 2 to 6 to $t_u = 1$, 2, 30, 110, and 170, respectively. Clearly, the process of relaxation of the fluctuations proceeds in two stages. The first occupies the interval $t_u = 1$ to 3, when the tail of the liquid-phase bank formed flows towards the reservoir

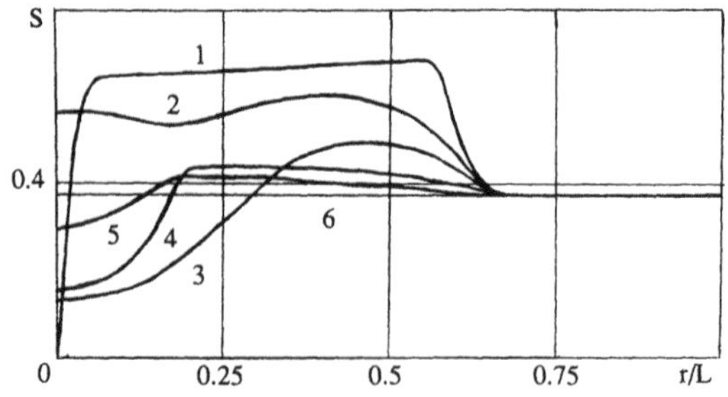

FIGURE 51. Profiles of S on the way to the stationary inflow regime.

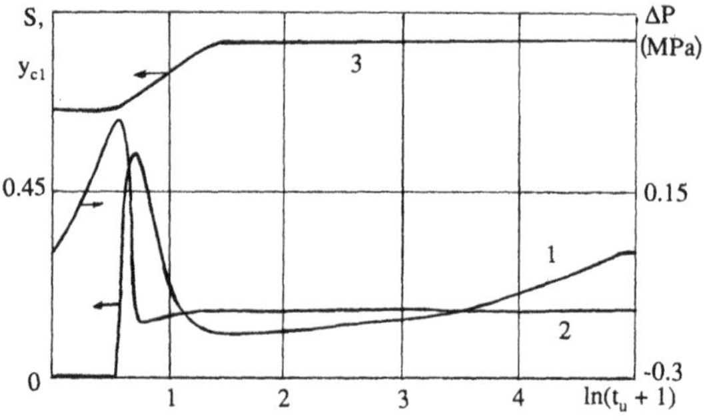

FIGURE 52. Dynamics of reservoir flow parameters: separation of fast and slow variables.

inlet. Then the saturation at the point r = 0 falls sharply, and when $t_u \approx 10$ a saturation profile with a sharp bend is formed (curve 4). The second stage is characterized by a considerable slowing down of the process. The steady-state regime is entered at time t_u = 160–170.

Figure 52 shows the process characteristics as a function of ln (t_u + 1). Curve 1 corresponds to $S(0,t)$, curve 2 to $\Delta P = P(0,t) - P(L,t)$. One can see that the evolution of the pressure, as distinct from saturation, takes place during the first stage. The phase compositions also reach their steady-state values in the course of several units of t_u. The calculations were made for a ten-component mixture; as an example we have reproduced the dependence of the CH_4 fraction in the gas phase (curve 3). Thus, the slow evolution of S takes place in the presence of almost steady-state distributions of pressure and phase compositions. The overall composition z_i then varies to the extent that S varies.

The time of formation of the fluctuations of S differs from their relaxation time by two orders. Such different behavior of the solution for the injection and subsequent recovery regimes is associated with the irreversibility of discontinuous flows.[5] The specifics of the approach to the steady-state regime are also determined by the finiteness of the solution domain, since in an infinite domain there is no steady-state solution of the problem under consideration.

The binary system considered in the beginning of this section (Figure 50) was chosen to serve as a conditional replacement of the multicomponent system described here and in Section 5.1. The characteristic relaxation time is determined from the linear theory on the basis of the following data: $I^* = 1.5 \cdot 10^{11}\ s^{-1}$ (which corresponds to the residual saturation value $S = 0.36$), $k = 10^{-14}\ m^2$, $L = 5$ m, and $m = 0.222$. One can see that in order of magnitude it amounts to only 10^3 s. Since the unit of t_u corresponds to the injection time of a fluid amount equal to the formation pore volume, the ratio of the relaxation time, calculated with the help of the linear theory, to the characteristic time of process is a small quantity in the order of 0.01 (the presence of the small parameter in the theory under consideration was used by asymptotical calculations). In the same time, the ratio of the "nonlinear" relaxation time and the characteristic process time in the order of 100. This speaks to the fact that the linear description of the process is not applicable in the case of a large deviation of the actual solution from the steady-state condition.

The calculations were carried out for various space and time steps, and in all cases after rapid changes in the pressure and phase compositions we observed a slow evolution of S during times in which ~160 pore volumes are injected. This suggests that when spatially highly inhomogeneous initial distributions of S and z_i have somehow been created in the reservoir, testing the composition of gas-condensate wells production over the usual periods of time accepted in well testing may not give correct results, since the composition may slowly change over much longer periods; for example, condensate may appear.

Though the experimental investigation of the discussed retardation by the instant changing of flow direction was not carried out, one can notice that at the stage of creating the bank (curve 1 in Figure 51) the results of calculations were comparable with the experiment, (see Chapter 5). Since the comparison showed the concordance of the numerical and the experimental results, one can hope that in the future when carrying out the experiments one will be able to observe the long-time relaxation.

It should be noted that quasi-steady solutions of the type described above have been previously postulated by Nikolaevskii et al.[7] Namely, when one of the phases (for example, the liquid phase) is stationary and $f_w = 0$, $f_g = 1$, system (7.1.1) goes over into a system of ordinary differential equations, if the phase compositions and pressure are assumed not to depend on time. Numerical calculations, carried out for $S(L,t)$ equal to the residual saturation, confirm the validity of the approach adopted by Nikolaevskii.[7] Moreover, similar slowing of the relaxation was also observed in calculations with $S(L,t) = 0.42$, i.e., for the case of mobility of both phases of the unperturbed reservoir mixture.

A problem similar to that considered above was solved for the plane-radial case.[8] However, the whole of the relaxation process was not traced owing to the very large computation time involved. Subject to computational constraints, the formation of a S profile with a sharp bend that differs from the known steady-state distribution and is extremely slowly transformed suggested that the system also might have a discontinuous steady-state solution. Let us consider this question in more detail.

In the context of the approach adopted by Khristianovich[9] in finding steady-state solutions, the reservoir phase compositions and saturations are determined from the pressure and overall composition in the flow (invariant along the streamlines), using the relationships

$$\epsilon_i^o = \epsilon_i(z_1, \ldots, z_{l-1}, P), \qquad i = 1, \ldots, l-1 \tag{8.1.2}$$

where ϵ_i^o is the composition in the flow at the boundary of the reservoir. If system (8.1.2) can be solved for z_i, all the reservoir characteristics are functions of the pressure, and the steady-state flow is determined from the Laplace equation for the flow potential. The question is as follows: is the system (8.1.2) always solvable for z_i and, consequently, is it possible for the steady-state solution not to be single-valued, with the result that it becomes necessary to introduce strong discontinuities?[10]

It should be noted that the question of the solvability of system (8.1.2) is a fundamental one, since the reserves of oil-gas-condensate deposits are estimated by analyzing the composition of the production, i.e., the quantities ϵ_i^o. The absence of a single-valued relation between the compositions in the flow and in the reservoir would mean that the problem of estimating the reserves is incorrectly defined and makes it necessary to introduce additional data. The question posed is part of the more general question of the number of possible solutions of system (8.1.2). For mixtures with $l > 2$ it remains unsolved, and in particular it is not known whether solutions of system (8.1.2) always exist. However, for binary mixtures it is possible to prove the single-valued solvability of Equations (8.1.2).

The solvability condition is written in the form

$$\det\left\{\frac{\partial \epsilon_i}{\partial z_j}\right\} \neq 0 \tag{8.1.3}$$

For a binary mixture, that can be described by two independent variables P and S, condition (8.1.3) takes the form

$$\frac{\partial \epsilon_1}{\partial S} \neq 0, \ \epsilon_1 = \left(\frac{f_g}{f_w} \cdot \frac{\rho_g y_1}{\eta_g M_g} + \frac{\rho_w x_1}{\eta_w M_w}\right)\left(\frac{f_g}{f_w}\frac{\rho_g}{\eta_g M_g} + \frac{\rho_w}{\eta_w M_w}\right)^{-1} \tag{8.1.4}$$

all the functions in Equation (8.1.4), apart from $f_{g,w}$, depend only on P. Differentiating, instead of Equation (8.1.4) we obtain

$$\operatorname{sign}\frac{\partial \epsilon_1}{\partial S} = \operatorname{sign}\left[\frac{d}{dS}\left(\frac{f_g}{f_w}\right) \cdot (y_1 - x_1)\frac{\rho_g \rho_w}{M_g M_w \eta_g \eta_w}\right] \neq 0 \tag{8.1.5}$$

From geometrical considerations it is clear that for a binary mixture in the plane "pressure-fraction of the first component" the branches corresponding to the gas and liquid phases close up only at the critical point, bounding the two-phase zone.[11] Thus, in the two-phase zone the quantity $(y_1 - x_1)$ preserves its sign. We did not consider the case of a one-phase state; it is obvious that in this case the compositions in the flow and in the reservoir simply coincide.

The derivative $d/dS\,(f_g/f_w)$ is expressed in the form

$$\frac{d}{dS}\left(\frac{f_g}{f_w}\right) = \frac{(df_g/dS)f_w - (df_w/dS)\cdot f_g}{f_w^2} \tag{8.1.6}$$

For saturations corresponding to the immobility of one of the phases there are no steady-state solutions.[7] If both phases are mobile, then the vanishing of Equation (8.1.6) is equivalent to the equality

$$\frac{\dfrac{df_g}{dS}}{\dfrac{df_w}{dS}} = \frac{f_g}{f_w} \tag{8.1.7}$$

However, the left side of Equation (8.1.7) is less, and the right side greater than zero, which also proves condition (8.1.5).

Thus, the observed anomalous retardation of the relaxation process by changing the flow rate sign seems not to be connected with the presence of any additional stationary structures and is defined by an unusually strong effect of irreversibility of a discontinuous flow structure which can not break down in the same manner as it has appeared: in this case it is impossible "to rewind the film back".

8.2. KINETICALLY STABLE STRUCTURES IN BINARY SYSTEMS DECOMPOSITION

The theory discussed below is applicable to many physical systems: binary liquid mixtures and solutions, binary metal melts, and solid solutions. Therefore it is necessary to state clearly that the verbal interpretation of the theoretical results in this section has a general geochemical orientation, i.e., their applications to the solid solutions of rock minerals are primarily considered.[12]

Many rock minerals (piroxenes, feldspars, etc.) are the solid solutions with complete or partial miscibility of components at temperatures in the order of the crystallization temperature. In the region of subsolidus temperatures (below the crystallization temperature) they can perform solid state phase transformations, for example, inversion transitions, ordering, and decomposition of a solid solution.[13,14] The minerals are sources of valuable information about the prehistory of rocks. The investigations of the crystal and phase structure of the mineral solid solutions, carried out by the electron microscopy method, precise X-ray structural analysis, and so on, enable us in many cases to use them as the natural geothermometers, geobarometers, and geospeedometers.[13,14] However, the reconstruction of the rock history strongly depends on the mathematical model of the structural transformations, which is used in interpretation of the experimental data.

Below we will consider the spinodal decomposition of a solid solution, that is the postcrystalization transformation controlled by diffusion and not connected with changing the crystal structure geometry. The spinodal decomposition equation was proposed and studied by Cahn in connection with the problem of the diffusion microstructure formation in binary metal alloys.[15-17] Imposing the crystal structure on the diffusion redistribution of the mixture components produces the most advantageous spatial directions of decomposition (the directions of the concentration waves motion) and, in particular, enables us to use one-dimensional mathematical models to describe the coherent (components having similar crystal lattices) microphase structures.[16] The nonlinear spinodal decomposition equation has been considered,[17] and it has been shown by numerical modeling that the influence of nonlinear terms after the initial (linear) stage of the process is the factor limiting the exponential growth of the composition fluctuations. It was pointed out that nonlinear interaction of the fastest growth mode and short-wave modes leads to forming sharper (nonsinusoidal) composition structures, which are typical for the two-phase state.[18] However, the importance of the role of nonlinear terms is not always realized in the practical applications of the theory. Particularly in geochemistry, a quite common point of view is that the spinodal description is applicable only at the early stages of the process,[14] and then one has to use the Lifshitz-Slyozov theory[19] where the dependence of the mean microphase scale on time is determined (modification of this theory for the case of dependence of the diffusion Onsager coefficient on the space fluctuation scale has been reported by De Gennes[20]).

Our attention is mainly concerned with a discussion of the nonlinear Cahn equation at the times exceeding the time of exponential fluctuations growth.[21,22] As will be shown, the nonlinear effects are reflected in the purely nonelastic interaction of the neighboring decomposition structures, which is of interest to compare with the purely elastic interaction of the soliton solutions of the Corteveg- de Vries shallow water equation. A nonlinear equation turns out to be quite effective to describe the process of spinodal decomposition. The process is seen as a sequence of fast rearrangements alternating with long retardations near the unstable steady-state periodic solutions of the Cahn equation. We will discuss the concept of dependence of the coherent equilibria of binary solid solutions on the overall composition and will attempt to give a possible explanation of the contradictions between existing theory and experimental data, considering the unstable and long-lived composition microstructures. The consideration of a more common equation of convective diffusion leads to spinodal decomposition against a background of oscillating motion. The stability properties of steady-state solutions are generalized for the case of convective diffusion; for the purely temporal dependence of the flow rate it is shown that the running wave solutions, which correspond to the steady-state solutions of the problem without convection, are unstable.

Main Equation

The evolution of the volume fraction of a component of a binary mixture is considered in the framework of the following equation[23,24]

$$\frac{\partial \phi}{\partial t} + \mathbf{V} \cdot \nabla \phi = \nabla \Lambda \nabla \frac{\delta F}{\delta \phi} \tag{8.2.1}$$

Here $\Lambda(\phi)$ is diffusion Onsager coefficient which can be expressed in terms of the self-diffusion coefficients of both components, F is the free energy functional in the form

$$F = \int (F_o(\phi) + K(\phi)(\nabla\phi)^2)\,d\mathbf{r} \tag{8.2.2}$$

the operator $\delta/\delta\phi$ designates the variational derivative, $\mu = \delta F/\delta\phi$ is the chemical potential[25]

$$\mu = \mu_o - 2K\nabla^2\phi - K_\phi(\nabla\phi)^2, \; \mu_o = \frac{dF_o}{d\phi}, \; K_\phi = \frac{dK}{d\phi} \tag{8.2.3}$$

$\mathbf{V}$ is the rate of the convective transfer.

Below we will basically consider the solutions of problem (8.2.1) with the periodic boundary conditions. This means that the evolution of spatially periodic initial conditions caused by the diffusion and convection will be studied. Other statements of the problem are also possible. Actually, at the boundary one can give up the condition

$$(\mathbf{n}, \Lambda\nabla\mu) = 0 \tag{8.2.4}$$

which is added by the condition

$$(\mathbf{n}, \nabla\phi) = 0 \tag{8.2.5}$$

(smooth changing ϕ, or vanishing the gradient part of energy (8.2.2) near the boundary). One can see that it is sufficient to derive $\phi(\mathbf{r},t)$ for a given $\phi(\mathbf{r},0)$: it can be shown, for example, by writing the difference analogue of Equation (8.2.1). The conditions (8.2.4) and (8.2.5) correspond to the case of small diffusion intermixing outside of the flow region compared to intermixing within it, i.e., both components move with approximately the same rates inside the region and with different rates outside of it.[26] For example, this can happen when the temperature in a region increases sharply, which leads to an increase in the diffusion coefficient.

Equation (8.2.1) at $\mathbf{V} = 0$ describes the spinodal decomposition in the solid solutions and high-viscous liquid mixtures. The case $\mathbf{V} \neq 0$ corresponds, for example, to the convection of a binary melt under conditions of the thermodynamic instability of its one-phase state.

Linear Analysis of Problem

The initial stage of the diffusion redistribution of the composition should be studied by the linearized Equation (8.2.1). Let $\mathbf{V} = 0$. Representing the deviation $\phi(\mathbf{r},t)$ from the mean concentration ϕ_o in form of Fourier series we obtain

$$\delta\phi = \phi(\mathbf{r},t) - \phi_o = \sum_{\mathbf{q}} c(\mathbf{q})\exp\left[-\frac{t}{\tau(\mathbf{q})} + j \cdot \mathbf{q} \cdot \mathbf{r}\right]$$
$$\tau^{-1}(\mathbf{q}) = q^2\Lambda(\phi_o)\left(\frac{\partial^2 F_o}{\partial\phi^2} + 2Kq^2\right),\ j^2 = -1 \tag{8.2.6}$$

Thus, the function $\delta\phi$ is represented in form of superposition of elementary concentration waves. It follows from Equation (8.2.6) that at $\partial^2F_o/\partial\phi^2 > 0$ all the modes die out; at $\partial^2F_o/\partial\phi^2 < 0$ there is the sphere with radius q_c in the wave vector space,

$$q_c = \left(-\frac{\partial^2 F_o}{\partial\phi^2}\Big/2K\right)^{1/2} \tag{8.2.7}$$

the wave vectors within which correspond to the exponentially growing modes. From Equation (8.2.6) one can see that the $\tau^{-1}(q^2)$ dependence has parabolic form. At $\partial^2F_o/\partial\phi^2 < 0$ it is convex upward, and the modes with the largest growth rate correspond to the sphere with radius $q_M = q_c/\sqrt{2}$.[15] The positive sign of the second derivative of F_o by ϕ is the condition of thermodynamic stability of a one-phase state of binary mixture. In the case of an opposite sign the space-homogeneous state is thermodynamically not advantageous, and one observes its breakdown.

The linear theory enables us to obtain the dependence of space scales of the fastest growth mode at the initial stage as a function of parameters of initial conditions.[2] For simplification let us consider the one-dimensional case. If one can solve the problem in a interval with the length L, then among all the modes with periods L/n, $n = 1,2,\ldots$, at $L > \xi_c = 2\pi/q_c$, there exists two competing ones in sense of maximum growth rate. Clearly, their periods are L/n and $L/(n + 1)$ for a certain n, and they are disposed from different sides of $2\pi/q_M$. By changing the length one has a discrete set of L_n values such that $\tau^{-1}(2\pi n/L_n)$ and $\tau^{-1}(2\pi(n + 1)/L_n)$ are equally close to the maximum amplification coefficient $\tau^{-1}(q_M)$. When L goes over the values, the period of the maximum

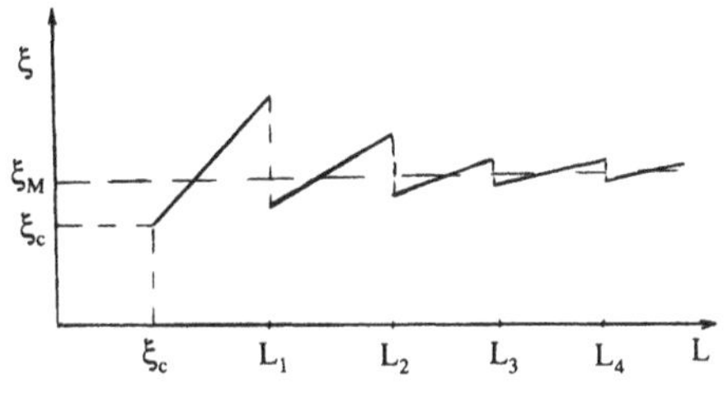

FIGURE 53. Dependence of the period of the fastest mode on the region scale at the initial stage of phase separation.

growth mode changes from L/n to $L/(\text{n} + 1)$. The bifurcation values are defined by the condition

$$\tau^{-1}\left(\frac{2\pi \text{n}}{L}\right) = \tau^{-1}\left(\frac{2\pi(\text{n} + 1)}{L}\right) \tag{8.2.8}$$

wherefrom after substituting Equations (8.2.6) and (8.2.7) into Equation (8.2.8) one obtains L_n, the values of period before and after a jump, and the value of the period jump

$$L_n = \xi_M(\text{n}^2 + \text{n} + 0.5)^{1/2},\ \xi(L_n - 0) = \frac{L_n}{\text{n}} = \xi_M\left(\frac{\text{n}^2 + \text{n} + 0.5}{\text{n}^2}\right)^{1/2},$$

$$\xi(L_n + 0) = \frac{L_n}{\text{n} + 1} = \xi_M\left(\frac{\text{n}^2 + \text{n} + 0.5}{(\text{n} + 1)^2}\right)^{1/2}, \tag{8.2.9}$$

$$\Delta\xi(L_n) = \xi(L_n + 0) - \xi(L_n - 0) = -\xi_M\left(\frac{\text{n}^2 + \text{n} + 0.5}{\text{n}^2(\text{n} + 1)^2}\right)^{1/2}$$

Formulas (8.2.9) permit drawing the dependence of the fastest mode period on the region scale (Figure 53). This dependence looks like a saw with the teeth decreasing with an increase of L. The period changes linearly between the bifurcation points. It follows from Equation (8.2.9) that at large L the period approaches to ξ_M, and the difference $L_n - \xi_M \cdot \text{n}$ approaches to $0.5 \cdot \xi_M$. Thus, for the large spatial scale of the initial fluctuations (essentially more than ξ_c) the period of the fastest mode equals ξ_M. The form of dependence presented in Figure 53 is connected with nonmonotonicity of $\tau^{-1}(q^2)$.

The results concerned with the bifurcation of the dominant mode period look more cumbersome. There have been extensive studies on this subject,[2] but here one can notice that the structures formed at the initial stage belong to three main types: sphere-like, layered, and cylindrical, depending on the space scales of initial fluctuation. The bifurcation conditions in a general form will be discussed in the next section. As was pointed out, the additional limitation of the geometric form of fluctuations is determined by the crystal structure of the solid solution.

There are a lot of natural minerals (feldspar cryptoperthites, moon clinopyroxenes, etc.[14]) with the layered coherent decomposition microstructure forming precisely because of that. Imposing the crystal structure also changes the growth exponent in the structures enlargement law (see section 9.2).

The linear analysis of Equation (8.2.1) at $\mathbf{V} \neq 0$, $\mathbf{V} = \mathbf{V}(t)$ leads to the appearance of an additional multiplayer of the type $\exp(j\mathbf{q} \int_0^t \mathbf{V}dt)$ in each term of the Fourier expansion (8.2.6). In the situation of spinodal decomposition this means that the growth of the long-wave modes is accompanied by oscillations. The case $\mathbf{V} = \mathbf{V}(\mathbf{r},t)$ is more complicated and is not considered here.

Steady-State Solutions

The stationary solutions of the one-dimensional problem (8.2.1) were considered for the case $\mathbf{V} = 0$.[2,21,27] The case $\mathbf{V} = \mathbf{V}(t)$ is reduced to the case $\mathbf{V} = 0$ by changing the variable x:

$$x \longrightarrow x - \int_0^t V dt \tag{8.2.10}$$

In the multidimensional case, Equation (8.2.10) is generalized as

$$\mathbf{r} \longrightarrow \mathbf{r} - \int_0^t \mathbf{V} dt$$

It is convenient to present the results in the phase plane.[28] The steady-state solution is determined from the equation

$$\Lambda \frac{d}{dx}\left(\frac{\delta F}{\delta \phi}\right) = 0 \tag{8.2.11}$$

but should the right-side of Equation (8.2.11) be non-zero, the periodic boundary condition cannot hold. By taking into account Equation (8.2.3) we obtain the integral of the steady-state problem in the form

$$F_o(\phi) - K(\phi)\gamma^2 = C\phi + h, \; \gamma = \frac{d\phi}{dx} \tag{8.2.12}$$

The dependence $F_o(\phi)$, for which the spinodal decomposition is possible, is given by a function with two inflection points (curve 1 in Figure 54a). Correspondingly the function $\mu_o(\phi)$ should have two extrema, which correspond to the spinodal (curve of the absolute thermodynamic instability) at a given temperature (curve 2 in Figure 54a). The potential of the conservative system (8.2.12) differs from F_o on the quantity $C\phi$.

Changing the constant C leads to the motion of the points of rest of the conservative system defined by the relation

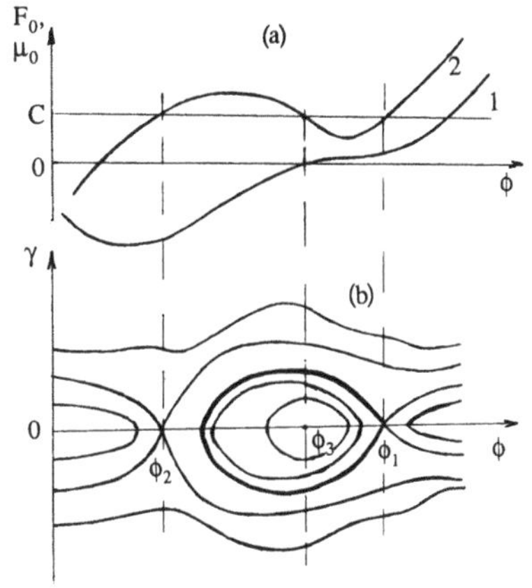

FIGURE 54. Construction of the phase plane of a stationary problem using the concentration dependence of free energy F_o and its derivative.

$$\frac{1}{K}\left(\mu_o(\phi) - C\right) = 0 \qquad (8.2.13)$$

As is clear from Equation (8.2.13), C is the value of chemical potential for the space-homogeneous solution. The appearance of the multiplayer K^{-1} in Equation (8.2.13) is connected with the fact that the function $K^{-1}(\phi)$, which is physically non-negative, can vanish for some models of spinodal decomposition (for example, in case of polymer mixtures it vanish at $\phi = 0$ and $\phi = 1$). We will not consider here the case $K^{-1}(\phi) = 0$.

The phase plane is presented in Figure 54b. The region of periodic solutions is bounded by a separatrix, which issues from the saddle point $(\phi_1,0)$, 0) and is shown by a bold line. The separatrix also can issue from the point $(\phi_2,0)$, which depends on the C value. The condition divides these two situations has the form

$$\frac{F_o(\phi_1^*) - F_o(\phi_2^*)}{\phi_1^* - \phi_2^*} = \mu_o(\phi_3^*),\ \mu_o(\phi_3^*) = \mu_o(\phi_2^*) = \mu_o(\phi_1^*)$$

The first of the relationships is obtained by equating the constants h for the separatrixes issuing from both saddle points. It is easily seen that the values ϕ_1^* and ϕ_2^* correspond to the binodal; in this case the boundary of the periodic solutions region corresponds to the solution of a type of wall with the values ϕ_1^* and ϕ_2^* at infinity.

The closer a periodic trajectory is to the separatrix, the larger is its period. The minimum period $\xi_c = 2\pi/q_c$ is determined from condition (8.2.7) and corresponds to the steady-state solutions with small magnitude disposed in the phase plane near the vortex point of rest.

Notice an important circumstance: problem (8.2.1) is solved in the region with the length L, owing to the conservativity of the equation considered the space mean value ϕ_o being invariant. Therefore the constants C and h are determined from the conditions

$$\int \frac{\phi\sqrt{K}d\phi}{\sqrt{F_o - C\phi - h}} = \phi_o \int \frac{\sqrt{K}d\phi}{\sqrt{F_o - C\phi - h}} \tag{8.2.14}$$

$$2\int \frac{d\phi\sqrt{K}}{\sqrt{F_o - C\phi - h}} = \xi = \frac{L}{n} \geq \xi_c\ ,\ n = 1,2, \ldots \tag{8.2.15}$$

The integration in Equations (8.2.14) and (8.2.15) is carried out between the denominator zeroes of expressions behind the integral signs. Conditions (8.2.14) and (8.2.15) have the following sense. For a phase plane with a arbitrary C there is a set of periodic trajectories whose periods are divisors of L. However, the solutions should not necessarily satisfy condition (8.2.14). That is why the phase curves with a certain ϕ_o and L are disposed in the phase planes, which correspond to discrete C values — for every n there is its own pair (C_n, h_n).

The analysis carried out above is correct also by imposing the impermeability conditions. Relationship (8.2.11) is equivalent to Equation (8.2.4). Condition (8.2.5) means that the appropriate phase curves have the starting and ending points at the ϕ-axis of the phase plane (Figure 54b). One can consider the case of C values corresponding to the existence of the only root of Equation (8.2.13); then the only steady-state solution satisfying the periodic conditions is the trivial homogeneous one.

"Turning on" the convective term in Equation (8.2.1) should lead to the following: the steady-state solutions of the problem with periodic boundary conditions should be translated with the instantaneous rate V, i.e., the solutions of a type of running waves. The role of the solutions for the problem considered cannot be understood without their stability analysis which will be discussed below.

Lyapunov Function

Let us show that the free energy functional (8.2.2) calculated for the solutions of Equation (8.2.1) is a nonincreasing time-function at V = V(t). Multiplying Equation (8.2.1) by μ and integrating over the space we obtain

$$\frac{dF}{dt} = -\int \Lambda(\nabla\mu)^2 dx \leq 0 \tag{8.2.16}$$

Expression (8.2.16) follows from the relationships

$$\int \frac{\delta F}{\delta\phi}\frac{\partial\phi}{\partial t}dx = \frac{dF}{dt},\ \int \frac{\delta F}{\delta\phi}\frac{\partial\phi}{\partial x}dx = 0 \tag{8.2.17}$$

which are proved using the $\phi(x,t)$ spatial periodicity. The property (8.2.16) also holds in the multidimensional case for periodic conditions, and for the impermeability conditions (8.2.4) and (8.2.5) at $V = 0$. If $V \neq 0$, the contribution in the free energy change connected with the influence of boundaries can be of arbitrary sign, generally. However, in the infinite region, Equation (8.2.16) holds for conditions (8.2.4) and (8.2.5) and $V = V(t)$. In this case the above transformations lead to the equality

$$\frac{1}{L}\frac{dF}{dt} = -\frac{F_o(L) - F_o(0)}{L} V - \frac{1}{L}\int_0^L \Lambda \left(\frac{\partial \mu}{\partial x}\right)^2 dx$$

which goes over to Equation (8.2.16) at $L \to \infty$.

The free energy decrease rate dF/dt vanishes at the solutions of Equation (8.2.11) only. Thus, the free energy is a Lyapunov function of the problem considered.

Stability of Stationary Solutions

The points of stability of the stationary solutions of Equation (8.2.1) have been discussed in the literature.[21,22,27] The quite popular scheme of the dynamic systems stability study was used under these considerations; namely, after constructing the Lyapunov function one finds out which stationary solutions provide its local minimums, and which are unstable.[29] Let us consider first the case of zero convection. Define the functional

$$J = \int G dx = \int (F_o + K\left(\frac{\partial \phi}{\partial x}\right)^2 - C\phi)\, dx \qquad (8.2.18)$$

whose extremals are steady-state solutions of Equation (8.2.11). The positivity condition of the second variation of functional (8.2.18) calculated at the periodic steady-state solutions can be not satisfied. This is clear from the following argument.[21] Let us consider a bunch of close extremals containing the steady-state solution under consideration. The extremals from the bunch are periodic stationary solutions of Equation (8.2.1) with close values of the parameter h. Let the extremals issue from the same point with the value $\phi = \phi_3$, and L is the multiple of the period of the steady-state solution considered. Then the extremals from the bunch should intersect (Figure 55), what is obvious from a geometrical viewpoint, since the steady-state solutions with a larger period have larger magnitude. As is known, the necessary and sufficient condition of the local minimum of a functional at a extremal is holding to the Legendre and Jacobi conditions.[30] The first of them holds that ($\partial^2G/\partial\gamma^2 = 2K > 0$), but the existence of the intersection points of the extremals (the points conjugated to $x = 0$) shows that the Jacobi condition does not hold and the local minimum of the functional considered is absent.

The above consideration did not use the fact that from the nonstationary problem one follows the restriction for its solutions

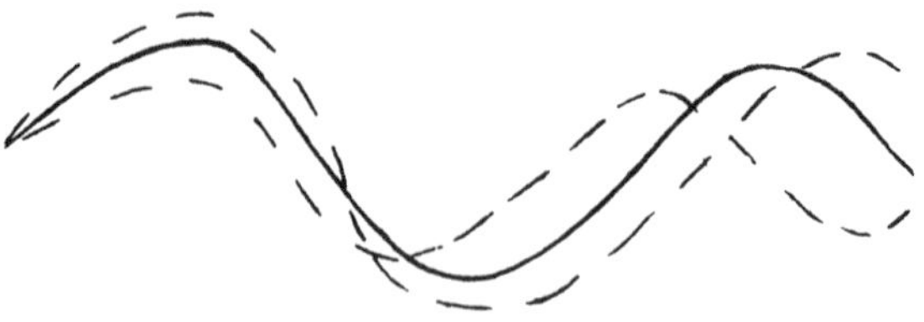

FIGURE 55. The bunch of extremals containing the stationary solution in which stability is studied.

$$\frac{1}{L}\int_0^L \phi dx = \phi_o \tag{8.2.19}$$

Accurately speaking, we could not use the fact that the larger the period of steady-state solution the larger amplitude it has, since one has to compare the steady-state solutions from several phase planes (i.e., with several C). However, our consideration can be modified by taking into account restriction (8.2.19). Let condition (8.2.19) hold at each extremal from the bunch in Figure 55. Owing to space periodicity the bunch vertex can be chosen at the point with the value $\phi = \phi_o$ which is reached compulsorily at every appropriate stationary solution. Then, if the extremal and one close by do not intersect, the latter lies either above or below the first. Therefore we should have

$$\int_0^L \phi_I (x)\, dx \neq \int_0^L \phi_{II} (x)\, dx$$

and one of these two extremals cannot satisfy condition (8.2.19). This contradiction proves the fact that the steady-state solution, whose periods are divisors of L are not the local minimum points of functional (8.2.18). This means that in each vicinity of such a steady-state solution there is a ϕ-distribution whose free energy is less than at the solution. But the system reaching the ϕ-distribution will not return to the steady-state solution because it would be in contradiction with the free energy decrease. Thus, such a steady-state solution is unstable.

As has been proved,[22] restriction (8.2.19) provides the stability of the stationary solution with period L (see also Novick-Cohen[27]). The only stable structure of problem (8.2.1) with the boundary conditions (8.2.4) and (8.2.5) is the wall-like solution which is exactly half of the periodic solution with the period $2L$. Notice that the stability study of steady-state space-inhomogeneous solutions of the equation "linear diffusion + nonlinear reaction" leads to the problem of finding the points of local minimum of the functional of the type in Equation (8.2.18). It has been proved that **all** the steady-state solutions of the problem with periodic and impermeable boundary conditions are unstable.[29,31] Equation (8.2.1) at K = const and Λ = const differs from the equation "diffusion + reaction" only by an additional Laplace operator in the right side, and this circumstance leads to the additional restriction (8.2.19) providing the stability of the steady-state solution with maximum period.

If V = V(t) ≠ 0, then according to the transformation (8.2.10) the unstable steady-state solutions of the problem without convection go over to the unstable solutions moving with the rate V. At V = const the linear analysis shows that the transfer through the spinodal corresponds to the Hopf bifurcation of the solutions of Equation (8.2.1).[32] The appearing periodic orbits are defined by equation $\mu(x - Vt) = C$. All the solutions having the time-period less than L/V are unstable. However, the solution with the time-period L/V appearing at

$$\left.\frac{\partial^2 F_o}{\partial \phi^2}\right|_{\phi = \phi_o} = -2K\frac{4\pi^2}{L^2}$$

and having the form of harmonic waves with small magnitude is stable. According to the Hopf theorem the orbit is the attractor of Equation (8.2.1) and can be continued to the region behind the spinodal by changing the bifurcation parameter (temperature or mean composition).

Description of Numerical Experiments

Thus, on the one hand one has the fluctuation growth with the characteristic scale ξ_M in the thermodynamically unstable region of phase diagram, on the other hand the system possesses a set of nontrivial steady-state solutions which are all unstable except the thermodynamically advantageous one. Notice that in some publications concerned with numerical solving of the spinodal decomposition equation the visually stable inhomogeneous structure with a period of the order of ξ_M, was observed forming right after ending the exponential growth stage, and the structure was close to a steady-state one. Owing to the above consideration this should mean that one observed process retardation at an unstable and long-lived structure considered as having reached a stable steady state. Particularly, such "pseudo-steady-state" structures were obtained in calculations[2] and they have been called kinetically stable structures (KSS).

The numerical modeling of the process over sufficiently long periods helps to understand what has happened. Here we will discuss such a calculation that was previously published in the literature.[21] The initial condition was accepted in the form

$$\phi(x,0) = 0.5 + 0.03\left(1 - \cos\frac{2\pi x}{L}\right), \; L = 40q_c^{-1}$$

(the length of the solution region is of the order of four periods of the maximum growth mode). The other initial data, and concrete form of the free energy and Onsager coefficient are taken from the literature;[21] convection was not considered (V = 0).

Below we describe the results of the calculations. First, the initial distribution transformed into the distribution with four "humps", which is in accordance with the linear analysis. However, it had a short lifetime, and the first long retardation happened at a KSS with three humps. The lifetime of the KSS was approximately 20 times longer than the time in the linear stage. Then one observed a rapid joining of two neighboring humps, and the profile became two-humped. This state also had a long lifetime, and then it transformed into the most thermodynamically advantageous one-humped one.

Owing to the symmetry of $\phi(x,0)$ in this calculation and, correspondingly, $\phi(x,t)$, regarding the point $x = L/2$, condition (8.2.5) holds at the boundaries. It is easily seen that, taking into account Equation (8.2.5), condition (8.2.4) can be reduced to the form $\phi_{xxx} = 0$ which also holds at the boundaries. To prove this, for example, for $x = 0$, let us consider the function $E(\Delta) = \phi(\Delta) - \phi(-\Delta)$. Clearly, $d^nE^n/d\Delta^n = 0$. But

$$\frac{dE}{d\Delta} = \left.\frac{\partial\phi}{\partial x}\right|_{x=\Delta} + \left.\frac{\partial\phi}{\partial x}\right|_{x=-\Delta}, \qquad \frac{d^3E}{d\Delta^3} = \left.\frac{\partial^3\phi}{\partial x^3}\right|_{x=\Delta} + \left.\frac{\partial^3\phi}{\partial x^3}\right|_{x=-\Delta}$$

Trending Δ to zero we obtain $dE/d\Delta = 0$ and $d^3E/d\Delta^3 = 0$, and consequently Equations (8.2.4) and (8.2.5) hold. However, the periodicity condition prohibits further evolution from the one-humped profile to the more thermodynamically advantageous monotonic steady-state wall-like solution which should be expected from conditions (8.2.4) and (8.2.5).

Figure 56 shows the evolution of some parameters. Curve 1 corresponds to the dependence $\phi(L/2, t)$; curve 2 corresponds to the dependence $\phi(0,t)$; curve 3 is mean free energy F per a monomer, and curve 4 is mean quantity $F_1 = K(\partial\phi/\partial x)^2$ per a monomer. The energy parameters in Figure 56 are shown in units of k_BT, where k_B is the Bolzmann constant, and T is the temperature. The unit of dimensionless time corresponds to the quantity $2 \cdot 10^7 \cdot \tau_{micro}$, where τ_{micro} is the time of diffusion motion of a monomer on the distance of the order of a characteristic monomer diameter. The time t_1 corresponds to the maximum of curve 1 (forming the four-humped profile), (t_2,t_3) and (t_4,t_5) are the KSS rearrangement intervals.

The free energy of the system is practically invariable at the retardation stages and decreases sharply at the KSS changing stages. The quantity $\langle F_1\rangle$ corresponds to the additional energy of the microphase transient layers and decreases while $\langle F\rangle$ is invariable, i.e., one observes the energy transfer from transient layers inside microphases during the KSS lifetime. The $\langle F_1\rangle$ decrease changes to an increase at the end of every rearrangement. As was noticed,[21] the increase in sections of the time-dependence of $\langle F_1\rangle$ are connected with the necessity to form transient layers with sufficiently high energy. This leads to temporary stabilization of the KSS formed and is probably a self-defense of the system from unavoidable KSS breakdown.

In Figure 57 we show the ϕ profile obtained in the calculation with the same initial data as that in the above case, except $L = 100\ q_c^{-1}$. The first KSS formed after 35 units of t was presented. One can see that the maximums and minimums of the profile are disposed according to the initial ϕ distribution. This means

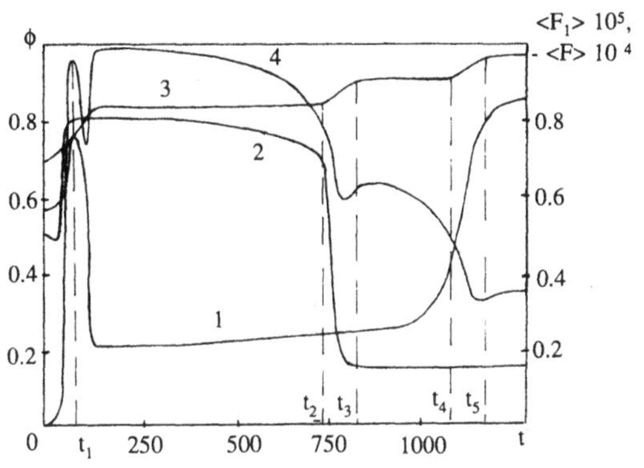

FIGURE 56. Evolution of the system in the process of spinodal decomposition: forming KSS.

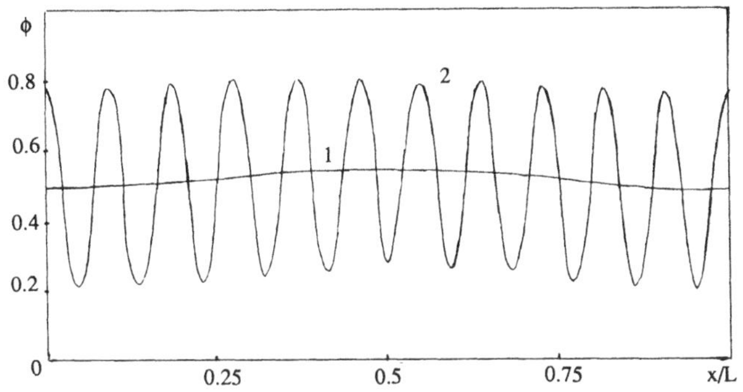

FIGURE 57. KSS (curve 2) saves the "memory" about the initial composition distribution (curve 1).

that the decomposition in several sections of the sufficiently large region of solution brings about several mean compositions which can differ from the total mean ϕ_o.

At $L < L_1 = \xi_c \cdot \sqrt{5}$ the wanderings between several KSS are absent, and the system just goes over into the thermodynamically advantageous state.

One can see that the decomposition ends at the steady state whose period is equal to the initial ϕ distribution period. Therefrom it follows that imposing the periodic boundary conditions is quite artificial;[2] for example, if the period of initial distribution is two times less than L, the final distribution with period L cannot materialize. However, "turning on" even a small noise (which always exists in a physical process) by modeling, eliminates this restriction on the way toward complete phase separation, and the final form of the ϕ profile should coincide with the real scale of the system.

Thus, the nonlinearity of Equation (8.2.1) leads to the interesting mechanism of uniting the decomposition structures. The structures are close to some basic ones which are the steady-state solutions of Equation (8.2.1). This is inherent in the behavior of many nonlinear systems.[1] The instability of the initial state leading to the growth of the fluctuations with a definite space scale plays an important role. The phenomenon of coalescence of the single structures discovered allows us to consider Equation (8.2.1) as the alternative to the most popular description of the later stage of phase separation based on the theory,[19] since using Equation (8.2.1) one naturally takes into account the interaction of neighboring embryos. Moreover, adding the random field in Equation (8.2.1) also enables one to model the fluctuational separation mechanism, when the point "mean composition – temperature" at the phase diagram is between spinodal and binodal, absolute instability is absent, and the system should overcome an energy barrier in a fluctuational manner on the way to phase separation.

The phenomenon of process stabilization at the unstable long-lived structures was also observed in a plasma thermodiffusion problem,[33] and by explaining some nonlinear features of glass transition.[34] Another area where such a phenomenon is discussed within the framework of a equation similar to Equation (8.2.1) is in the physics of thin films (dewetting process and the nucleation of dry regions for an unstable liquid film in contact with a solid[35,36]). The mathematical theory which would allow estimation of the retardation times at KSS is absent. Apparently, slowing down the stability loss in a vicinity of a bifurcation point, observed in the relaxation oscillation theory,[37] should be connected with the retardation at KSS.

"Turning on" the convection leads to uniting the decomposition structures against a background of the convective flow, i.e., wandering between unstable running wave solutions. The question of the behavior of the system at $V = V(x,t)$ is open. Particularly, it is not known how two single structures running with different speeds should interact in problem (8.2.1).

KSS and Dependence of the Coherent Equilibria of Binary Solid Solutions on Overall Composition

Below we will try to apply the results considered for explaining some questions which take place by description of the phase equilibria and decomposition structures of solid solutions of rock minerals.

As is known, the description of phase separation in a crystal lattice is complicated by the existence of elastic stresses in the process of components redistribution. Cahn[16] suggested the model of changing the boundary of thermodynamic instability due to the influence of elastic stresses, adding into the free energy functional (8.2.2) the term

$$\int P_{el} (\phi - \phi_o)^2 d\mathbf{r} \tag{8.2.20}$$

where P_{el} is a combination of the elastic constants of the crystal (it is assumed that the elastic constants are the same for both components and are defined by the parameters of a common crystal lattice). This leads to a shear of the "chemical" spinodal, and the instability of the one-phase state already has to be at

$$\partial^2 F_o/\partial\phi^2 = -2P_{el} < 0$$

The calculations of the "coherent" spinodal and binodal are carried out nowadays using Cahn's approach. A detailed survey of the geochemical methods is presented in a book by Khisina.[14]

However, using correction (8.2.20) one obtains some contradictions with experiments. The free energy of the homogeneous state has the form

$$G_o = F_o + P_{el}(\phi - \phi_o)^2 \qquad (8.2.21)$$

Though G_o depends on the overall composition ϕ_o, the abscissas of the contact points of the common tangent of the curve $G_c(\phi)$, which define the binodal values, do not depend on ϕ_o. This peculiarity of Cahn's theory has been studied,[38] and this actually does not correspond to many experiments.[14]

The existing methods of composition calculations by the coherent equilibrium are based on considering the stresses in an elementary cell of the crystal, and then using the correlations between the geometrical cell parameters and its composition. To obtain the correlations one supposes the composition, stresses, and strains to be homogeneous inside each lamella. However, the transient layer between the lamellae of coexisting phases can be significant; in some cases the actual lamella deformation is concentrated near its boundary and falls down toward to its center.

The discussion of Cahn's approach has been presented.[38,39] In this discussion one is guided to a conclusion that a new thermodynamic formulation of the coherent equilibrium in solid solutions is assumed to be necessary.

Let us show that both the inhomogeneity of the composition inside a lamella and the dependence of the lamellae compositions on the overall composition might be described using the free energy functional in form (8.2.2). Let the elastic contribution (8.2.20) be inserted into F_o. Then the energy functional (8.2.2) changes on the quantity $\int P_{el}\phi^2 dx$ and in fact does not depend on ϕ_o: $\int P_{el}(\phi_o^2 - 2\phi\phi_o)dx = -\int P_{el}\phi^2 dx$ and changing the free energy on a constant does not matter for our consideration. Thus, we have to again deal with the free energy functional of the kind in (8.2.2). The Ginzburg-Landau functional (8.2.2) is in some sense the basic one for the thermodynamic description of the systems with one order parameter in the long-wave approximation.

As was mentioned above, the one-dimensional steady state of Equation (8.2.1) with a given period ξ and a overall composition ϕ_o is defined from relationships (8.2.14) and (8.2.15) between ξ, ϕ_o, and C, h. Since in fact complete phase separation cannot be reached in the process of solid solution decomposition, the coherent equilibrium data are obtained experimentally by the analysis of the decomposition structures with spatial scales in the order of 100 to 10,000 Å, which is incomparably less than the macroscopic scale. Supposing that the structures are just the nonequilibrium long-lived KSS which are close to the stationary solutions of Equation (8.2.1) then one can obtain the dependence of the lamellae compositions on the overall composition, solving together Equations (8.2.14) and (8.2.15), and the equation

$$F_o(\phi) - C\phi - h = 0 \tag{8.2.22}$$

The larger root of Equation (8.2.22) determines the composition of the lamellae which are rich in the first component, the smaller root corresponds to the composition of the lamellae which are rich in the second component.

The ratio of the volume fractions of the phases, which are both poor and rich in the first component, correspondingly, is

$$\nu = \int_{\phi^{I}(C,h)}^{\phi_M(C,h)} \left(\frac{K}{F_o - C\phi - h}\right)^{1/2} d\phi \Bigg/ \int_{\phi_M(C,h)}^{\phi^{II}(C,h)} \left(\frac{K}{F_o - C\phi - h}\right)^{1/2} d\phi \tag{8.2.23}$$

In Equation (8.2.23), $\phi^{I}(C,h)$ and $\phi^{II}(C,h)$ are the smaller and the larger roots of Equation (8.2.22); $\phi_M(C,h)$ is the root of the following equation

$$\frac{\partial \gamma}{\partial \phi} = \frac{\partial}{\partial \phi}\left(\frac{F_o - C\phi - h}{K}\right) = 0 \tag{8.2.24}$$

At K = const Equation (8.2.24) goes over to $\mu_o = C$, and $\phi_M(C,h)$ goes over to ϕ_3.

From Equations (8.2.23) and (8.2.24) it follows that the ratio of the phase fractions is close to 1 at small magnitudes of the decomposition structure. When the magnitude increases the ratio deviates from 1 and the period of the structure becomes larger. In the framework of the theory under consideration one can predict the ν dependence on the overall composition at a given temperature and at a given period ξ. Let us do it in the case of the symmetric form of $F_o(\phi)$ (Figure 58b) and K = const. There is a certain interval of overall compositions corresponding to the region of one-phase state instability, inside which one can observe the structures with a given ξ. The boundaries of the interval $\phi_{d,i}$ are defined by the condition

$$\xi = \xi_c(\phi_{d,i}), \; i = 1,2$$

where $\xi_c = 2\pi/q_c$, and q_c is defined by formula (8.2.7). Outside the interval $(\phi_{d,1}, \phi_{d,2})$ all possible KSS have periods, which exceed ξ_c. At the ends of the interval $\nu = 1$; for the symmetric phase plane, for which the overall composition is exactly $\phi_3 = 1/2$, ν also equals 1. For the overall compositions within the

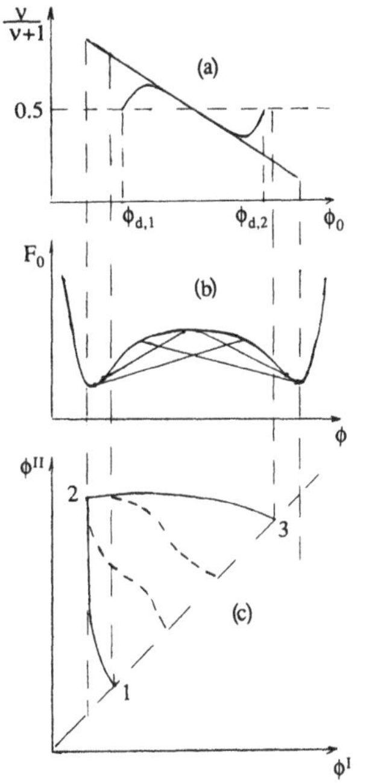

FIGURE 58. a: dependence of a fraction of the phase enriched by a second component on the overall composition for the nonequilibrium KSS with a given spatial period and for the equilibrium situation; b: construction of the diagram of the composition distribution in lamellae using the concentration dependence of free energy; c: distribution of the first component in lamellae at large spatial scales of decomposition structures.

interval $(\phi_{d,1}, 1/2)$ we have $\nu > 1$ (predominance of the phase which is poor in the first component), and $\nu < 1$ within the interval $(1/2, \phi_{d,2})$. Consequently, ν reaches its maximum and its minimum within the intervals $(\phi_{d,1}, 1/2)$ and $(1/2, \phi_{d,2})$, correspondingly. Thus, the qualitative behavior of ν is the following: while the overall composition changes from $\phi_{d,1}$ to $\phi_{d,2}$, quantity ν first increases from 1 to a certain maximum, then decreases reaching the value 1 at the overall composition 0.5, reaches a minimum, and then increases again till 1 at $\phi_{d,2}$. Taking into account the possibility of F_o asymmetry leads only to a displacement of the points of the dependence where $\nu = 1$. The $\nu/\nu + 1$ behavior is shown in Figure 58a. The straight line shows the traditional form of $\nu/\nu + 1$ dependence, i.e., when the phase compositions exactly correspond to the binodal values.

Using the theory proposed, one can predict the character of distribution of the lamellae compositions for several overall compositions and sufficiently large lamellae scales (i.e., essentially larger than the minimal ξ_c value at the given temperature). From the constructing the steady-state solutions it follows that at large ξ a pair of values (ϕ^{I}, ξ^{II}) of the first component fraction in the lamellae rich in it (upper index I) and poor of it (upper index II) is the abscissas of the point of intersection of a $F_o(\phi)$ tangent and $F_o(\phi)$ itself and the abscissas of the corresponding point of $F_o(\phi)$ tangency (Figure 58b). In Figure 58c we show the dependence of ϕ^{II} on ϕ^{I}. Points 1 and 3 correspond to the spinodal, and point 2 corresponds to the binodal. One can see that the dependence is not single-valued. The form of $\phi^{II}(\phi^{I})$ near point 1 can be determined from the equality

$$F(\phi_s + \Delta\phi^{II}) = F(\phi_s - \Delta\phi^{I}) + F'(\phi_s - \Delta\phi^{I}) \cdot (\Delta\phi^{II} + \Delta\phi^{I}) \quad (8.2.25)$$

Here $\phi_s + \Delta\phi^{II} = \phi^{II}$, $\phi_s - \Delta\phi^{I} = \phi^{I}$, and ϕ_s is the spinodal ϕ value at the point 1 of Fgure 58c. Expanding Equation (8.2.25) in the Taylor series near the value ϕ_s and retaining the terms up to $(\Delta\phi)^3$ leads to a cubic form which is transformed to the equation

$$\left(\frac{d\phi^{II}}{d\phi^{I}}\right)^3 - 3\frac{d\phi^{II}}{d\phi^{I}} + 2 = 0$$

This equation has two roots equal to 1, and one equals to -2. The appropriate root is only the last one because, according to the geometry of the problem, ϕ^{II} increases near point 1 in Figure 58c, while ϕ^{I} decreases. By analogy with this case, one obtains near point 3 in Figure 58c the equation

$$\left(\frac{d\phi^{I}}{d\phi^{II}}\right)^3 - 3\frac{d\phi^{I}}{d\phi^{II}} + 2 = 0$$

and now $d\phi^{II}/d\phi^{I} = -1/2$. Similar consideration near point 2 shows that by approaching it from point 1 the $\phi^{II}(\phi^{I})$ graph has a vertical tangent, and it has a horizontal tangent approaching it from point 3. Changing the graph convexity at sections 1–2 and 3–2 depends on the relationship of the slope of the straight line between points 1 and 2 and of the slope of the tangent at point 1 (correspondingly, on the relationship of the slope of the straight line between points 3 and 2 and of the slope of the tangent at point 3). For $F_o(\phi)$ dependence presented in Figure 58b, the convexities of the sections 1–2 and 2–3 conserve. If the ratio of the composition interval between a given spinodal branch and the closer binodal branch to the composition interval between the spinodal branch and the farther binodal branch exceeds 1/2 (i.e., the metastable composition region scale is of order of the absolute unstable composition region scale), the corresponding section of the graph in Figure 58c will change the convexity (dashed line).

It also follows from the construction of steady-state solutions that all the possible pairs (ϕ^{I}, ϕ^{II}) corresponding to the extrema of steady-state solutions of (8.2.1) and designating the lamellae compositions are collected inside the region bounded by the curve 1-2-3 and straight line $\phi^{II} = \phi^{I}$. However, the points corresponding to KSS should be concentrated near curve 1-2-3. This was demonstrated by all the spinodal decomposition calculations carried out by us: actually, the KSS rearrangements almost did not change the extreme ϕ values, though the characteristic scale increased.

Notice that there is a mechanism in the process of spinodal decomposition whose action is able to provide the real KSS stabilization, besides the nonlinear retardation considered above. As it is known, the growth of the lamellae leads to an increase of elastic energy, which is connected with the growth of ϕ gradients by coarsening the decomposition structure. This leads to a coherency loss at the lamellae boundaries, which slows down sharply the possibility of further diffusion redistribution between the lamellae. The lamellae formed before the coherency loss should be stabilized regarding their parameters (scale and composition) near a certain KSS.

Verification of the above theory can be shown by comparing the lamellae compositions obtained from Equations (8.2.14), (8.2.15), (8.2.22), and (8.2.23) and the corresponding experimental data for a given overall composition. One can estimate the lamellae composition data collected on a plane of the kind in Figure 58c. The closeness of the experimental data to a curve of the 1-2-3 type in Figure 58c, which could be obtained using $F_o(\phi)$ dependence for a mineral, also would be a theory corroboration. Similar problems can be considered in the framework of model (8.2.1) in order to describe the multidimensional decomposition structures, solving the equation numerically with partial derivatives in the form $\mu = C$ where μ is given by Equation (8.2.3). This could enable us to calculate the relationship between the parameters of lamellae for a given overall composition.

Thus, one has been shown that the nonlinear equation (8.2.1) could describe both the initial and later stages of phase separation. We have described a new type of spatially inhomogeneous structures which are unstable but long-lived. The "step" evolution of free energy, which is a consequence of the relaxation peculiarities of a system on the way to equilibrium, takes place. It has been shown that a possible explanation of the coherent equilibria dependence on the overall composition, observed experimentally in minerals, could be based on the nonequilibrium KSS. The relationship between lamellae parameters and overall composition, as well as the character of their deviation from equilibrium values, can be predicted. Generalizing the Cahn equation by taking convection into account leads to a wandering in the system between the unstable running wave solutions in the phase separation process.

Finally, notice that averaging the free energy decrease law (8.2.16) over the steady states leads to the evolution equations for the steady-state parameters C and h which are defined in Equation (8.2.12) and are supposed to be slow variables.[40] Besides that, the literature[41] discusses the mathematical statement of the problem about the contact of two pure components with limited mutual miscibility which leads to the formation and extension of a two-phase region. The process is described by average equations within the two-phase (decomposition) zone; in the one-phase zone one can apply the usual diffusion equation without using highest space derivatives, which are essential only near the phase transition point. The moving boundary of a two-phase region is determined from the condition $\partial^2 F_o/\partial\phi^2 = 0$, and one intertwines the equations there. A similar approach was applied in describing the oscillation zone in the theory of solitons.[42]

A complete in depth study of Equation (8.2.1) still has not been carried out analytically (probably, it is not possible). As is known, the Lifshitz-Slyozov theory,[19] which uses an embryo of the new phase as the main object of consideration and analyzes the phase separation process in terms of the embryo size distribution, gives for the mean size ξ the asymptotic expression $\xi \sim t^{1/3}$. Since the coalescence description in the theory of spinodal decomposition differs principally from the embryo approach,[19] nowadays one discusses actively the question of $\xi(t)$ asymptotical behavior for the Cahn equation. To simulate the decomposition in multidimensional systems with a statistically representative number of embryos, one uses supercomputers. The results of numerical modeling of isotropic two-dimensional decomposition have been reported by Elder and Rogers.[43] It is shown that the exponent in the structures growth law is close to 1/3. The later stage of spinodal decomposition was considered analytically by the author,[44] based on the so-called "mean field approximation" of Equation (8.2.1).[45] As a result, the scaling dependences $\xi(t)$ are obtained over long periods. The exponent in the structures growth law depends on the dimensionality of the problem: in the case of layered structures (dimensionality equals 1) it equals 1/2; for the two- and three-dimensional decomposition it equals 1/4. The approach to the later stage description is generalized for anisotropic systems,[46] among which are many solid solutions. Anisotropy is taken into account by considering the quadratic form

$$(\mathbf{K}\nabla\phi, \nabla\phi)$$

and the matrix of Onsager coefficients Λ_{ij} in Equations (8.2.1) and (8.2.2) instead of the isotropic gradient term $K(\Delta\phi)^2$ and one Onsager coefficient Λ. Over long periods the $\xi(t)$ behavior is defined both by the dimension of the problem and by the degeneracy of minimal positive eigenvalue of the matrix $\mathbf{K}$. Particularly, in the general case of the three-dimensional anisotropic decomposition (degeneracy is 1) the exponent in the structures growth law equals 1/2; for the isotropic case (degeneracy is 3) it is 1/4. A detailed description of these questions is provided in Chapter 9.

8.3. KINETICALLY STABLE STRUCTURES IN THE PHASE TRANSITION "VISCOUS GAS–VISCOUS LIQUID"

In the previous section we mentioned the retardation phenomenon near space-inhomogeneous unstable steady-state structures observed in publications.[34,47] The theory presented in those papers, in fact, was proposed for polymer glasses and was designed to explain the appearance of density inhomogeneities with the spatial scale essentially exceeding the typical intermolecular distance (more exactly, in the order of 10^3 to 10^4 Å; see also the discussion of this point by others in Reference 48). To obtain the deduced system of equations of fluctuation relaxation for a one-component compressible polymer system a quite cumbersome space averaging technique was used.[34] Below we will consider a special case when the theory can be expressed more simply. At the same time, this case corresponds to the situation of phase transition "viscous liquid–viscous gas", which is essentially more propagated in nature compared to the polymer systems. Besides that, supposing the viscosity and density of the physical system considered are large quantities one can reduce the complicated coupled system of mass and momentum transfer relationships to one nonlinear equation with partial derivatives having second order with respect to time; then it is possible to study the density relaxation process already for this basic equation. In this section the results of such analysis will be described.[49]

Main Equations

Let us consider the lattice model of a compressible one-component substance. Namely, we assume that at each site of a cubic lattice with parameter "a" there is either a monomer (atom) or a hole which corresponds to the absence of monomer at a site. The thermodynamics of such a system can be described by the Flory potential (evaluated per one site)

$$F_o = \frac{\phi \ln \phi}{N_m} + (1 - \phi) \ln (1 - \phi) - \chi \phi^2 \tag{8.3.1}$$

where ϕ is the fraction of occupied lattice sites; χ is the positive interaction parameter which describes Van der Waals attraction in the system; energy is written down in units of k_BT (k_B is the Boltzmann constant, T is absolute temperature). The first two terms in formula (8.3.1) describe the contribution of the configurational entropy of a mixture of monomers and holes, while the third term corresponds to interaction energy. The multiplier $1/N_m$ in the first term describes decreasing the entropy due to complexity of molecular structure (for instance, our system can be a mixture of N_m - mers and holes, and in this case the parameter N_m has the sense of the polymerization degree).

In the region of phase transition the expression for free energy must be completed by a gradient term to take into account the fluctuational inhomogeneities. In such a case we have for full free energy

$$F = \int (F_o + a^2 K(\phi)(\nabla \phi^2) \frac{d\mathbf{r}}{a^3} \tag{8.3.2}$$

Expression for K consists in general of two components. The first component may be found by transition from the discrete Ising model to the continual description; in a three-dimensional case it equals $\chi/6$. The second one takes into account the complexity of the space arrangement of one molecule among similar molecules; in the case of chain molecules it equals $1/36\phi$, then $K = \chi/6 + 1/36\phi$.

Hydrodynamics of our system is described by two equations:

$$\frac{\partial \phi}{\partial t} + \nabla (\phi \mathbf{V}) = 0 \tag{8.3.3}$$

$$\rho \left[\frac{\partial \mathbf{V}}{\partial t} + \mathbf{V} \nabla \mathbf{V} \right] = \nabla \boldsymbol{\sigma} \tag{8.3.4}$$

where

$$\sigma_{ij} = \eta \left(\frac{\partial v_i}{\partial x_j} + \frac{\partial v_j}{\partial x_i} - \frac{2}{3} \delta_{ij} \operatorname{div} \mathbf{V} \right) - \zeta \delta_{ij} \operatorname{div} \mathbf{V} - P \delta_{ij} \tag{8.3.5}$$

$\mathbf{V}$ is velocity; $\rho = \rho_o \phi = M\phi_o/a^3$ is density, $\boldsymbol{\sigma}$ is the stress tensor, η and ξ are dynamic shear and bulk viscosities, M is the mass of a monomer,

$$\eta = \eta_o \phi \tag{8.3.6}$$

$$P = k_B T \phi^2 \frac{\delta F^*}{\delta \phi}, \ F^* = \int \left(\frac{F_o}{\phi} + \frac{Ka^2}{\phi} (\nabla \phi)^2 \right) \frac{d\mathbf{r}}{a^3} \tag{8.3.7}$$

The form of function $\eta(\phi)$ takes into account the dependence of viscosity on density for sufficiently dense substances. Expression (8.3.7) generalizes the local dependence $P(\phi)$ in the lattice model[50]

$$P = - \frac{k_B T}{a^3} \frac{\partial (F_o/\phi)}{\partial (1/\phi)}$$

in the case of functional (nonlocal) representation of F.

Exclusion of the Hydrodynamic Mode

System (8.3.1) to (8.3.7) is closed and describes completely the relaxation of density and velocity. Under certain conditions it may be simplified. First, we shall consider a sufficiently viscous and sufficiently dense system. For the hydrodynamic processes with sufficiently small Reinholds numbers it is possible to neglect the convective term $\mathbf{V}\nabla\mathbf{V}$ in Equation (8.3.4). Secondly, the influence of bulk viscosity ζ is essential for fast processes only in systems with sufficiently low density, and we can neglect it also. Lastly, the order parameter in our system is in the density mode ϕ. This means that the characteristic scale of its change is essentially larger than the corresponding scale for the velocity mode $\mathbf{V}$. Actually, the relaxation rate of the hydrodynamic mode is defined by the coefficient η before $\nabla^2\mathbf{V}$ in the right side of Equation (8.3.4), and by the above assumption the viscosity is quite large. From the other side, the rate of density relaxation in fact is described by value

$$\frac{\partial^2 F_o}{\partial \phi^2} = \frac{1}{N_m \phi} + \frac{1}{1 - \phi} - 2\chi$$

Near the boundary of thermodynamic stability (spinodal) this value is either a small positive (slow ϕ - relaxation) or negative (spontaneous growth of fluctuations). For such nonequilibrium physical systems with many essentially different space and time scales of change, the adiabatic approximation including the "killing" of fast modes is applicable.[1] In this approximation the difference between $\Delta\phi\mathbf{V}$ and $\phi\Delta\mathbf{V}$ is negligible (a similar approach was used by Haken in the study of a coherent emission model[1]). As a result, Equation (8.3.4) may be rewritten in the form

$$\rho_o \frac{\partial \mathbf{V}\phi}{\partial t} = \eta_o \nabla^2(\mathbf{V}\phi) + \frac{\eta_o}{3} \operatorname{grad} \operatorname{div}(\mathbf{V}\phi) - \nabla P \qquad (8.3.8)$$

After differentiation of Equation (8.3.3) by time we get

$$\frac{\partial^2 \phi}{\partial t^2} + \frac{\partial}{\partial t} \operatorname{div}(\phi\mathbf{V}) = 0 \qquad (8.3.9)$$

In order to exclude the hydrodynamic mode we apply the divergence operator to Equation (8.3.8)

$$\rho_o \frac{\partial}{\partial t} \operatorname{div}(\phi\mathbf{V}) = \eta_o \operatorname{div} \nabla^2(\phi\mathbf{V}) + \frac{\eta_o}{3} \operatorname{div} \operatorname{grad} \operatorname{div}(\phi\mathbf{V}) - \nabla^2 P \qquad (8.3.10)$$

Combining Equations (8.3.9) and (8.3.10) we get

$$\rho_o \frac{\partial^2 \phi}{\partial t^2} + \eta_o \operatorname{div} \nabla^2(\phi\mathbf{V}) + \frac{\eta_o}{3} \operatorname{div} \operatorname{grad} \operatorname{div}(\phi\mathbf{V}) - \nabla^2 P = 0 \qquad (8.3.11)$$

Then, using the identities

$$\nabla^2 \mathbf{f} = \operatorname{grad} \operatorname{div} \mathbf{f} - \operatorname{rot} \operatorname{rot} \mathbf{f}\,, \quad \operatorname{div} \operatorname{rot} \mathbf{f} = 0$$

we obtain from Equation (8.3.11)

$$\rho_o \frac{\partial^2 \phi}{\partial t^2} + \frac{4}{3} \eta_o \operatorname{div} \operatorname{grad} \operatorname{div}(\phi\mathbf{V}) - \nabla^2 P = 0 \qquad (8.3.12)$$

Lastly, according to Equation (8.3.3)

$$\operatorname{div}(\phi\mathbf{V}) = -\frac{\partial \phi}{\partial t}$$

Therefore Equation (8.3.12) may be finally rewritten in the form

$$\rho_o \frac{\partial^2 \phi}{\partial t^2} = \nabla^2 \left(\frac{4}{3} \eta_o \frac{\partial \phi}{\partial t} + P \right) \tag{8.3.13}$$

Thus, by accounting for the momentum transfer processes, the relaxation of density is described by an equation of the second order with respect to time, while the nonlinear effects are determined by dependence of P on ϕ. In this form Equation (8.3.13) resembles an acoustic equation with dissipation.

Let us write Equation (8.3.13) in the dimensionless form. By changing the variables

$$t \longrightarrow \frac{t}{C_1}, \ \mathbf{r} \longrightarrow \frac{\mathbf{r}}{a}, \ C_1 = \frac{3\rho_o a^2}{4\eta_o}$$

$$P^* = \frac{a^3}{k_B T} P = \phi \left[\frac{\partial F_o}{\partial \phi} - 2K\nabla^2\phi - K_\phi (\nabla\phi)^2 \right] - F_o - K(\nabla\phi)^2$$

we obtain

$$\frac{\partial^2 \phi}{\partial t^2} = \nabla^2 \left(\frac{\partial \phi}{\partial t} + CP^* \right), \ C = \frac{9\, k_B T \rho_o}{16\, a\eta_o^2} \tag{8.3.14}$$

For sufficiently viscous systems the value C is a small parameter. For instance, it was shown that $C \sim N_m^{-6} \sim \eta^{-2}$ for chain molecules,[34] i.e., even for small integer N_m the value C is quite small. The simplest nonlinear equation which possesses the main properties of Equation (8.3.14) can be obtained by the use of the power series for $P(\phi)$ up to terms of the third degree. After some transformations of dependent and independent variables it takes the form

$$\frac{\partial^2 \phi}{\partial t^2} = \nabla^2 \left(\frac{\partial \phi}{\partial t} + u_b\phi + \phi^3 - \nabla^2\phi \right)$$

Alternation of the sign of the only external parameter u_b (from plus to minus) corresponds here to transition through the boundary of thermodynamic stability.

Linear Approximation

Let us study the dynamics of fluctuations ϕ in the linear approximation. Linearizing Equation (8.3.14) and then considering the solution of the problem in the form of superposition of elementary waves of the type

$$\exp(\omega t)\exp(j\mathbf{q}\mathbf{r}), \ j^2 = -1$$

we obtain the following dispersion relation

$$\omega^2 + q^2\omega + \epsilon_1 q^2(k_c^2 + q^2) = 0 \tag{8.3.15}$$

where

$$k_c^2 = \frac{1}{2K}\frac{\partial^2 F_o}{\partial \phi^2}, \quad \epsilon_1 = 2C\phi K$$

From Equation (8.3.15) it follows that

$$\omega_{1,2} = \frac{-q^2 \pm (q^4 - 4\epsilon_1 q^2(q^2 + k_c^2))^{1/2}}{2} \tag{8.3.16}$$

First let us study the most interesting case: $k_c^2 < 0$. Here ω_1 turns to zero at $q^2 = 0$ and $q^2 = -k_c^2$, being positive within the interval $(0, -k_c^2)$. The maximum value of ω_1 within this interval is reached at

$$q_M^2 = \frac{2k_c^2\epsilon_1^2 + |k_c^2|\epsilon_1^{3/2}}{\epsilon_1(1 - 4\epsilon_1)}$$

so that at $C \ll 1$ or $\epsilon_1 \ll 1$ we have

$$q_M^2 = -k_c^2\sqrt{\epsilon_1} \tag{8.3.17}$$

Thus, for viscous systems the maximum of the amplification coefficient is shifted mainly toward longer waves ($\sqrt{\epsilon_1} \sim \eta^{-1}$).

The quantity ω_2 corresponds to a damped mode. It determines the relaxation due to viscous forces without any acoustic effects. At small C and $k_c^2 < 0$, the discriminant of Equation (8.3.15) is positive at all q^2, i.e., there can be no temporal oscillations. At $q^2 > -k_c^2$ the value ω_1 is always negative.

By substituting Equation (8.3.17) into Equation (8.3.16) we get the maximum growth rate of the fluctuations

$$\omega_M = -k_c^2 \cdot \epsilon_1 \tag{8.3.18}$$

Let us note that in the case of usual spinodal decomposition of binary mixtures we have $q_M^2 = -1/2\, k_c^2$ and $\omega_M \sim k_c^4$.[15]

At $k_c^2 > 0$ (above the phase transition point), as one can see from Equation (8.3.16), the fluctuations are always damped. If $C \ll 1$ ($\epsilon_1 \ll 1$), then at sufficiently small **q** ($q^2 < 4\epsilon_1 k_c^2/(1 - 4\epsilon_1)$), we have oscillations with a frequency that turns to zero at $q^2 = 0$ and at $q^2 = 4\epsilon_1 k_c^2/(1 - 4\epsilon_1)$. The oscillation frequency is maximum at $q^2 = 2\epsilon_1 k_c^2/(1 - 4\epsilon_1)$, where it approximately equals $4\epsilon_1^{3/2}k_c^2$. Within the range

$$0 < q^2 < \frac{4\epsilon_1 k_c^2}{1 - 4\epsilon_1}$$

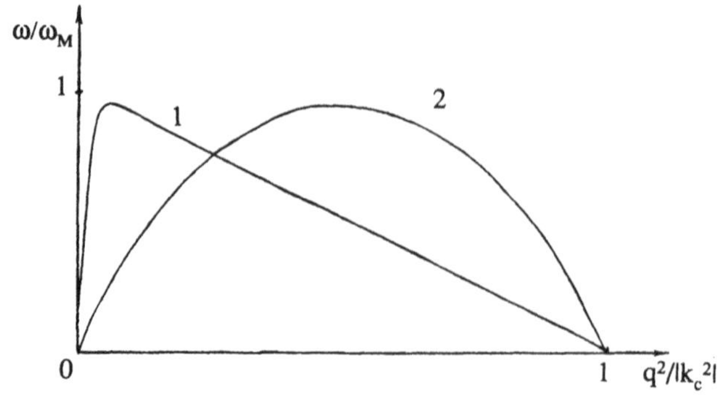

FIGURE 59. Form of the fluctuation amplification coefficient for the phase transition "viscous gas - viscous liquid" (curve 1) and for the phase separation of a binary mixture (curve 2).

the relaxation rate is $q^2/2$. At $q^2 > 4\epsilon_1 k_c^2/(1 - 4\epsilon_1)$, $\omega_{1,2}$ are real numbers, and in the case of a large q^2 one can distinguish the slow (ω_1) and the fast (ω_2) modes. At $q^2 \gg 4\epsilon_1 k_c$ the relaxation rate is

$$\omega_1 = -\epsilon_1(k_c^2 + q^2)$$

Let us note that in the case of spinodal decomposition of binary mixtures there is no such range of q-values where the fluctuations would die out with oscillations.[15]

Figure 59 shows the dependence of the amplification coefficient normalized to its maximum value, on the value $q^2/|k_c^2|$ at $k_c^2 < 0$ (curve 1). The form of the amplification coefficient corresponds to a C value equal to 0.009; the decrease of C practically does not change the dimensionless $\omega_1(q^2)$ dependence. One observes the quantitative similarity in the case of spinodal decomposition of a binary mixture (curve 2). Though the dependence of ω_1 on $q^2/|k_c^2|$ is still of extreme character, in the case under consideration the maximum is strongly shifted toward the long-wavelength region. Furthermore, at $C \ll 1$ for almost all the wave numbers corresponding to growing fluctuations, the amplification coefficient can be approximated by the following linear dependence

$$\frac{\omega_1}{\omega_M} = 1 - \frac{q^2}{|k_c^2|} \tag{8.3.19}$$

or, taking in account Equation (8.3.18),

$$\omega_1 = \epsilon_1 \, (- k_c^2 - q^2)$$

Since in addition the ϵ_1 value is believed to be small, the maximum growth mode is much less distinguished than in the case of spinodal decomposition of binary mixtures.

Thus, the model proposed predicts the formation of the density microstructure, which is connected with the extremum presence of the ω_1 dependence on q^2 at $k_c^2 < 0$. We considered above only the early stage of density fluctuation development. The similarity between the case under consideration and the spinodal decomposition gives one reason to expect observing the same qualitative regularities in the latter (nonlinear) relaxation stage as those recorded in the literature.[21,41] The main theme here is that there are kinetically stable structures formed with longlife times near unstable stationary states. One can expect that the system will go through the consequence of the KSS in the process of its evolution to the total thermodynamic equilibrium. The closeness of the growth rates of the modes with quite different wavelength values, as in Equation (8.3.19), would cause stronger dependence of the nonlinear dynamics on the initial conditions. Below we will discuss the nonlinear properties of Equation (8.3.14).

Stationary Solutions

For time-independent solutions of Equation (8.3.14) we have

$$\nabla^2 P^* = 0 \tag{8.3.20}$$

Below, the case of one spatial variable will be considered. Because of the P* definition (8.3.7), in the case of the problem with periodic boundary conditions or with the attenuation conditions at the infinity one can write instead Equation (8.3.20)

$$\frac{d}{dx}\left(\phi^2 \frac{\delta F^*}{\delta \phi}\right) = 0$$

and then

$$\phi^2 \frac{\delta F^*}{\delta \phi} = -C_2 \tag{8.3.21}$$

where C_2 is the integration constant. According to Equation (8.3.7), relationship (8.3.21) can be presented in the form

$$\frac{\partial(F_o/\phi)}{\partial \phi} - \frac{2K}{\phi}\frac{d^2\phi}{dx^2} - \frac{\partial(K/\phi)}{\partial \phi}\left(\frac{d\phi}{dx}\right)^2 = -\frac{C_2}{\phi^2} \tag{8.3.22}$$

After integrating Equation (8.3.22) over ϕ we get

$$F_o - K\gamma^2 = h\phi + C_2\,,\ \gamma = \frac{d\phi}{dx} \tag{8.3.23}$$

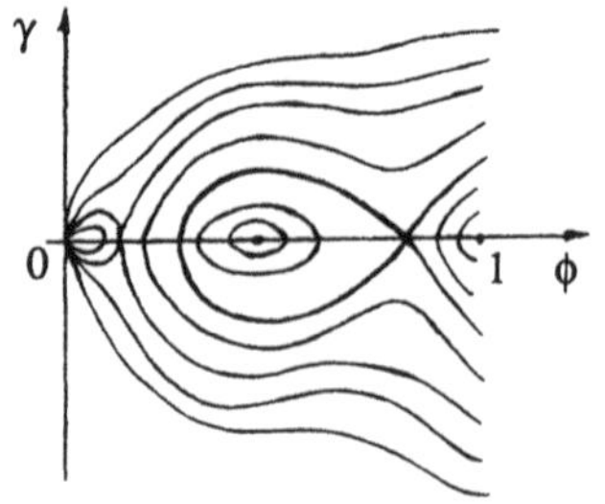

FIGURE 60. Phase plane of the one-dimensional stationary problem of phase transition "viscous gas - viscous liquid".

By using solution (8.3.23), the stationary structures can be studied the same way as in the case of spinodal decomposition (see the previous section). The singular points of Equation (8.3.22) are determined from the equation

$$\frac{1}{K\phi}(F_o - C_2 - \mu_o\phi) = 0 \ , \ \mu_o = \frac{\partial F}{\partial \phi}$$

This gives us

$$F_o - C_2 = \phi \frac{\partial(F_o - C_2)}{\partial \phi} \tag{8.3.24}$$

If $\chi > \chi_{cr}$ (for model (8.3.1), $\chi_{cr} = 1/2\,(N_m^{-1/2} + 1)^2$), then function $F_o(\phi)$ has two points of inflection. It is clear on purely geometrical grounds that in this case Equation (8.3.24) has three non-zero roots; the largest and the smallest ones correspond to the saddle points, while the one with the intermediate value corresponds to the vortex point. The phase plane is featured in Figure 60. The case of function $K(\phi)$ which is regular inside the whole interval (0, 1) and tends to infinity at $\phi \to 0$ (for example, because of the contribution into K in the order of ϕ^{-1}) is shown.

The spatially periodic solutions exist in the region limited by the separatrix (the thick line), and their periods become greater the closer they are to the separatrix. Let us note that, depending on the parameters of the problem, the separatrix bounding the region of periodic solutions can also run through the left saddle point in Figure 60.

The minimum period is found by solving the stationary problem linearized near the vortex point of the phase plane

$$\xi_c = -\frac{4\pi K}{\partial^2 F_o/\partial \phi^2} = \frac{4\pi K}{2\chi - \dfrac{1}{1-\phi} - \dfrac{1}{N_m \phi}}$$

and it tends to infinity as the system approaches the spinodal.

At $\chi < \chi_{cr}$ the only stationary solution of the problem with periodic boundary conditions is the constant function.

The form of relationship (8.3.23) shows that a general solution of the stationary problem (8.3.20) coincides with the solution of the problem for the equation

$$\frac{\partial}{\partial x}\frac{\delta F}{\delta \phi} = 0 \tag{8.3.25}$$

Let us notice now that in the case of the nonstationary problem with periodic boundary conditions the space-averaged value of ϕ must be constant: otherwise from Equation (8.3.14) it would follow that the mean value of ϕ should grow linearly with time. Problem (8.3.25) is equivalent to the variational problem of finding the extremum of the functional

$$J = \int (F_o + K(\nabla\phi)^2 - C_2\phi)\, dx$$

within the interval $(0, L)$ with the periodic boundary conditions for ϕ and an additional condition

$$\int \phi dx = \text{const}$$

As has been shown,[21,22] this variational problem has one stable solution which has the period L and corresponds to the most thermodynamically advantageous state.

The fact that a stationary solution of period L is stable within the interval $(0, L)$ is essentially related to the specific class of variations of functional J we consider here ($\int \delta\phi dx = 0$). If no such restriction is imposed upon the class of variations, none of the extremes of a Ginzburg-Landau type of functional (in the case of one spatial variable) will be its minimum point.[31,51] The constraint imposed upon the variations when one searches for a stationary solution of the problem is a consequence of the divergent form of the right-hand side of the nonstationary Equation (8.3.14) describing the nonlinear process of relaxation. The latter process consists in the evolution of the density distribution from the one formed at the initial (linear) stage to the equilibrium (i.e., thermodynamically advantageous) state.

Results of Calculations

Let us illustrate one of the stages of the nonlinear process of relaxation with the results of the numerical solution of Equation (8.3.14) by the Runge-Kutta method (with preliminary discretization of the spatial variable). At the end points of the interval $(0, L)$ the periodicity conditions are imposed. The initial condition was taken in the form

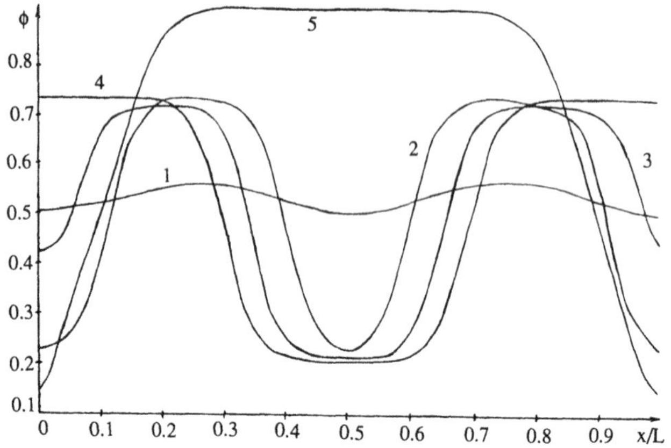

FIGURE 61. Profiles φ forming in the process of phase transition.

$$\phi(0,t) = \phi_o + 2A_1 \sin\frac{\pi x n}{L} + A_2 \frac{x}{L}$$

The values of the parameters were: $\chi = 3$, $L = 40\ \Delta x$, $n = 1$, $N_m = 2$, $C = 0.009$, $\phi_o = 0.73$, $A_1 = 0.02$, and $A_2 = 0$; the spatial step of the finite-difference method was $\Delta x = 0.06\ q_M^{-1}$, where q_M is given by formula (8.3.17). At these values of the parameters the spatially homogeneous solution $\phi = 0.75$ is unstable.

Figure 61 shows the stationary profile of φ (curve 5). In the course of calculation the maximum and minimum values of $\phi(x,t)$ were varied monotonically. The stationary structure has been formed at $t \approx 1600$. One can see the profile is very different from a sinusoidal one and is characterized by an almost constant value of φ within most of the period and its sharp variation within a narrow zone.

Besides the density profile, we also calculated the mean free energy of the system per one lattice site:

$$\langle F \rangle = \frac{1}{L}\int (F_o + K\,(\nabla\phi)^2)\,dx$$

and the mean value of its part $\langle F_1 \rangle$ that depends on $\nabla\phi$ and thus can be interpreted as the "surface" energy of transient layers between the high-density and low-density regions. It turned out that the free energy of the system monotonically becomes smaller with evolution of the density distribution, and the surface energy monotonically increases. This is a consequence of the fact that in this case the minimum value of the wave number exceeds q_M, the mode with the maximum amplification coefficient has maximum wave length, and spatial structures of period L/n with $n > 1$ are not formed. However, with an increase of L one should observe, as in the case of spinodal decomposition, the transformation of

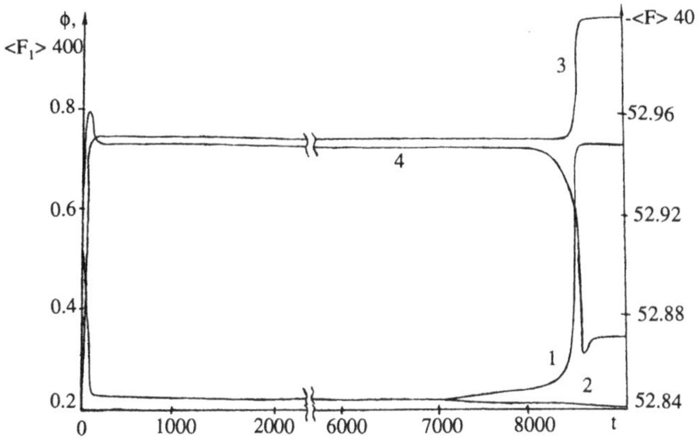

FIGURE 62. Evolution of parameters of the system: phenomenon of anomalous retardation at the "two-humped" structure.

the structure with a subsequent increase in the length. Each of the intermediate structures has a very long lifetime (compared to the structure formation time), i.e., they are kinetically stable.

In Figures 61 and 62 we present the calculation results for relaxation at the following values of parameters: $\chi = 2.2$, $N_m = 1$, $L = 40\ \Delta x$, $n = 2$, $\phi_o = 0.5$, $\Delta x = 0.5\ q_M^{-1}$, $C = 0.9$, $A_1 = 0.03$, and $A_2 = 0.01$. The constant solution $\phi = 0.535$ is unstable. The initial two-humped profile (curve 1 in Figure 61) was selected to initiate the growth of the fastest mode. The initial distribution was made slightly asymmetric ($A_2 \neq 0$) to ensure subsequent relaxation of the system after the end of the stage of exponential fluctuation growth and formation of a two-humped profile (curve 2 in Figure 61) close in form to the periodic stationary solution (at $L = 20\Delta x$ the process would end at this point).

As one can see in Figure 62, the stage of exponential fluctuation growth lasts 40 to 50 units of time. The structure that forms at this stage lives an extraordinarily long time. The process abruptly slows down, and though the relaxation still goes on, within the next 6000 units of time it is approximately 10 times slower than at the initial stage. Between the moments $L \approx 6000$ and $t \approx 8200$ the value of $\phi(0,t)$ increases on 0.02. After that the process is abruptly accelerated. The transition from a two-humped profile to the one-humped profile (curve 4 in Figure 61), which is the most advantageous one thermodynamically, takes about 60 units of time. After that the process ends.

The mean free energy of the system per site $\langle F \rangle$ (curve 3 in Figure 62) decreases with time in a nonuniform manner. It is practically constant (up to the fifth significant digit) between $t = 50$ and $t = 8400$. After that $\langle F \rangle$ rapidly decreases to its minimum value. The quantity $\langle F_1 \rangle$ that corresponds to the additional energy in the transient layers, increases at the linear stage of the process; then, so far as $\langle F \rangle$ is invariant, decreases starting at about $t = 7500$ (curve 4 in Figure 62). This means that during the lifetime of the kinetically stable state the energy is transferred from the transient layers inside the microphases (the quantity $\langle F_o \rangle = \langle F \rangle - \langle F_1 \rangle$ can be interpreted as the fraction of free energy corresponding to the bulk phases). At the end of the transition to the most thermodynamically advantageous profile the decline in $\langle F_1 \rangle$ is replaced by its increase (t = 8500). All these regularities are similar to those discovered in the studies of the spinodal decomposition equation.

The maximum time step, admissible for numerical stability of the calculations obtained, was found according to the formula $\Delta t = A_3(\Delta x)^2$, where A_3 was taken to be 0.05 in the first calculation and 0.5 in the second. In the first case the calculations took 20 h of processor time on an ES-1055 computer; in the second case the calculations lasted 6 h. Nevertheless, in the first case the slow stage was absent, while the transition to the stationary profile took about the same time as the exponential fluctuation growth time. The corresponding stage in the second case took only about 2 min of processor time. This indicates that the increase of the viscosity of the substance and, correspondingly, the reduction of parameter c, leads to a sharp increase of the process duration and of the calculation time.

Thus, the transition from a dense viscous gas to a liquid state can be regarded as a decay of a nonstable spatially homogeneous density distribution and formation of KSS. In a real physical system the number of rearrangements approximately equals $m_r = L\, q_M$ (for the conditional one-dimensional case), and the growth of microstructure size proceeds as follows:

$$q_M^{-1} \sim L_{m_r} = \frac{L}{m_r} \rightarrow L_{m_r-1} = \frac{L}{m_r - 1} \rightarrow L_{m_r-2} = \frac{L}{m_r - 2} \rightarrow \ldots \; L_1 = L$$

The value of m_r for real systems makes one expect extraordinarily long times to reach the thermodynamic equilibrium.

The observed similarity between the transition under consideration and spinodal decomposition should be studied and specified further. First of all, one should calculate the characteristic rearrangement time of KSS. The specific feature of this problem is that owing to the concrete form of the amplification coefficient at the early stage of the process many modes in a wide range of wave numbers have close growth rates. This results in the nonlinear effects beginning to manifest themselves sooner than they do in the case of spinodal decomposition; the evolution of the system under consideration depends on the initial conditions more heavily. Actually, calculations were also made for the same initial data as in the last calculation discussed, but n = 1. Correspondingly, the initial distribution had one hump (curve 1 in Figure 63). It was expected that the calculation time would be extensive because the linear theory predicted the fastest growth of the mode with space period *L*/2. However, it has been found that the process is arranged much more simply, the distinguishing of the fastest mode was not observed, and ϕ distribution went on to the thermodynamically advantageous state in about 200 units of time (curve 3 in Figure 63) changing time-monoton-

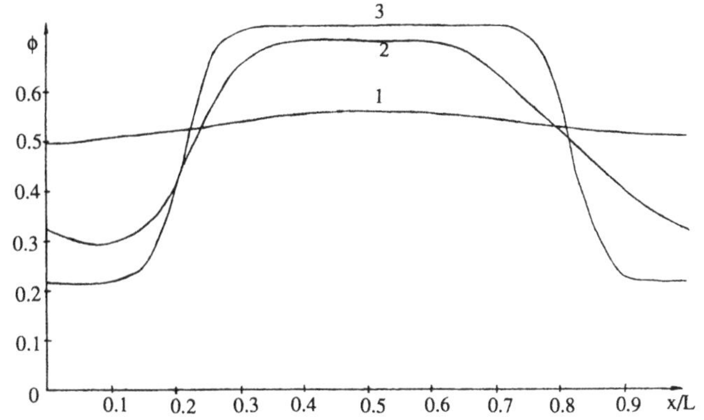

FIGURE 63. The evolution of ϕ distribution for the "one-humped" initial distribution. In this case KSS does not form (compare with Figures 61 and 62).

ically in every point of space and avoiding the retardation near KSS. The second surprising fact in the calculation was that the final stationary profile did not coincide with curve 4 in Figure 61 and was shifted in regard to it on half of the spatial period, so that minimum the value achieved in the points $x = 0$ and $x = L$, and maximum value was in the point $x = L/2$. Thus, in the case of absence of the two-humped initial profile the nonlinearity acted quite strongly, excluding the linear stage of the process and turning the stationary profile bottom up.

The above calculations were made for the case of one spatial variable. In the multi-dimensional case the situation is much more complicated. For a problem with periodic boundary conditions in an d-dimensional parallelepiped with a size vector $(L_1, \ldots, L_d)$ one can formulate conditions which would distinguish such regions in the size space that, at the stage of exponential fluctuation growth, each of them would have its own maximum growth rate mode with a definite $\mathbf{q}$ that leads to the appearance of the corresponding KSS. By analogy,[2] with variation of the parallelepiped sizes, the bifurcation surfaces separating the regions in the size space are determined for each number i of size vector component L_i by the relationships

$$\omega_1(\mathbf{q}) = \omega_1(\mathbf{q} + \boldsymbol{\Delta}_i\mathbf{q}) \tag{8.3.26}$$

$$\omega_1(\mathbf{q} + \boldsymbol{\Delta}_i\mathbf{q}) \geq \omega_1(\mathbf{q} + \boldsymbol{\Delta}_j\mathbf{q}), \; j = 1, \ldots d \tag{8.3.27}$$

where

$$\mathbf{q} = \left(\frac{2\pi n_i}{L_1}, \ldots, \frac{2\pi n_d}{L_d}\right)$$

and $\boldsymbol{\Delta}_j\mathbf{q}$ is the vector with all zero components except j-th, which equals $2\pi/L_j$.

If inequality (8.3.27) does not hold for a j, vector $\mathbf{q} + \mathbf{\Delta}_j\mathbf{q}$ corresponds to a higher growth rate than $\mathbf{q}$ and $\mathbf{q} + \mathbf{\Delta}_i\mathbf{q}$. Requirements (8.3.26) and (8.3.27) are general conditions for all problems with nonmonotonic q^2-dependence of the fluctuation amplification coefficient. For each i they determine the bifurcation surfaces, each of which is labeled by d natural numbers $n_1, \ldots, n_d$. Particularly, for the kinetics of the spinodal decomposition equation considered in the previous section these surfaces are sections of d-dimensional pseudoellipsoids.[2]

The analysis of conditions (8.3.26) and (8.3.27) shows that a small variation of the size vector may result in that the system will intersect a bifurcation surface in the size space and the corresponding KSS will acquire a qualitatively different form (for instance, at d = 3 in addition to the possibility of existence of spherical structures one may get a foliated structure or a cylindrical one[2,47]). This point can be easily explained by the following example. Let us suppose that at d = 2, $L_1 = 100.5$, and $L_2 = 50$, where $\xi_M = 2\pi/q_M = 1$. In this case the KSS forming after linear stage will be close to homogeneous along side 1 of the parallelepiped (L_2 is multiple to ξ_M). However, by small changes in the sizes (compared to themselves) the whole picture could change: for $L_1 = 100$ and $L_2 = 50.5$, the advantageous KSS will be close to homogeneous along side 2 of the parallelepiped (now L_1 is multiple to ξ_M). This means that the transition of a real physical system to a thermodynamically unstable state ($\partial^2 F_o/\partial\phi^2 < 0$) will lead to formation of a system of gaseous-like (low density) and liquid-like (high density) clusters. Owing to the arbitrariness of the initial ϕ distribution, the parallelepipeds, already having distorted sides, will randomly adjoin one another, and the formed clusters will interrupt and, in general, have a finite size. Drawings of similar clusters obtained by numerical solution of the spinodal decomposition equation have been published.[52] The authors also observed the changing of character of the order of the parameter spatial distribution from a spherical-like KSS to a reptile-like one by changing the mean value of the order parameter and, consequently, increasing q_M. The nonlinear stage for the multidimensional Equation (8.3.14) still has not been studied.

The results discussed in this chapter have common features. Namely, there is always one certain basic state of the system considered which becomes unstable by changing an external parameter (in Section 8.1 such a parameter is the flow rate, in Section 8.2, the temperature or the composition, and in Section 8.3, the temperature or the density). The breakdown of the basic state occurs, and is accompanied by the retardation near an inhomogeneous KSS (in Section 8.1 that is the discontinuous structure of the condensate bank, in Sections 8.2 and 8.3, the stationary solutions). The typical property of KSS is the irreversibility, which was discussed at the end of Section 8.1 in connection with the reservoir flow theory. Concerned to KSS irreversibility in the phase transition theory, such effects can be observed by an instant change of temperature during the KSS rearrangement and the retransference of the system to the thermodynamically stable state. In this case the relaxation will happen essentially more quickly en route to damping ϕ fluctuations.

A system on the way to equilibrium can be represented as a ball rolling down a mountain. The mountain could be smooth, and then the ball would easily roll down to the bottom. There could be natural chasms and barriers: the ball could reach one of them and stop. This situation corresponds to the dissipative structures. However, a mountain can have quite a smooth surface, but some of the paths from its top to the bottom can possess at some points almost zero slope. At these points the ball would greatly slow down its movement, but would not stop at all, and this situation corresponds to KSS. The kinetically stable structures differ principally by their nature from dissipative ones, and thus are of general theoretical interest.

8.4. DEWETTING OF A SOLID SURFACE: ANALOGY WITH SPINODAL DECOMPOSITION

In this section we would like to consider an analogy between the mathematical descriptions of two nonlinear phenomena. These are the model of dewetting of a solid surface covered by a film of a fluid[35,36,53-56] and the Cahn model of spinodal decomposition described in Section 8.2. Our main goal is to show that the nonlinear equation of the film dynamics proposed in the literature[35,36,53-56] is able to describe the dewetting process in the entire time range, including the initial stage of film rupture as well as the later stage of the nucleation and growth of dry regions. Due to the formal identity of the models of spinodal decomposition and dewetting, all the results obtained for the Cahn equation can be applied directly to the dewetting description.

The nonlinear equation of the film dynamics can be written in the form[36]

$$\frac{\partial H}{\partial t} = \frac{\partial}{\partial x}\left(\frac{H^3}{3\eta}\frac{\partial}{\partial x}\left(-\sigma\frac{\partial^2 H}{\partial x^2} + \rho g H - \Pi(H)\right)\right) \tag{8.4.1}$$

where H is the film thickness, η is the viscosity of the film, σ is the surface tension between the film phase and the above bulk fluid phase, ρ is the density of the film phase, g is the gravitation force acceleration, $\Pi(H)$ is the disjoining pressure (the difference between the pressures in the film and bulk phases[57]).

The stability condition of a film with the thickness H_o is given by the following inequality,

$$\frac{\partial \Pi}{\partial H}(H_o) < \rho g \tag{8.4.2}$$

There are two main forms of $\Pi(H)$ considered in the literature in connection with the problem of film stability.[58] First, Π can be a monotonically decreasing function of H, which tends to infinity at $H \to 0$ and to zero at $H \to \infty$ (curve 1 in Figure 64). Secondly, $\Pi(H)$ can drop from the infinity to a certain negative value, and then increases to zero at $H \to \infty$ (curve 2 in Figure 64). According to the literature,[36] the first case corresponds to the action of disjoining forces

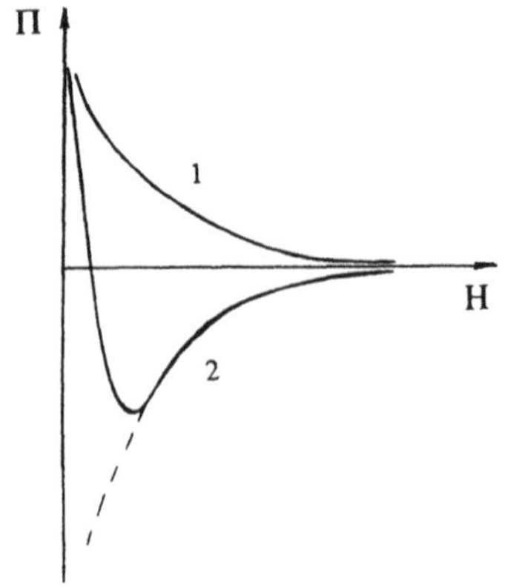

FIGURE 64. Dependences of disjoining pressure on thickness for a wetting film (curve 1) and a nonwetting film (curve 2).

stabilizing the film, the second one derives the action of conjoining forces which tend to destabilize the film and which should lead to its rupture.[36]

For the case of action of disjoining forces, the derivative $\partial\Pi/\partial H$ is negative in the entire range of H, and condition (8.4.2) is always satisfied. For the case of conjoining forces, there can be an interval (H_s^-, H_s^+) in the H range, such that the films with the thicknesses from the interval are unstable. The values H_s^- and H_s^+ are equivalent to the spinodal values of concentration in the theory of spinodal decomposition. Below we will consider the behavior of solutions of Equations (8.4.1) when the conjoining forces are acting.

In the existing studies of the film rupture model (8.4.1) the decreasing branch of the nonmonotonic dependence $\Pi(H)$ was apparently not used (though the qualitative curves analogous with curve 2 in Figure 64 were presented.[35,36] For the charge-neutralized fluids, the standard form of $\Pi(H)$ used in calculations of Equation (8.4.1) was

$$\Pi(H) = \frac{A}{H^3}, \ A < 0 \tag{8.4.3}$$

i.e., curve 2 in Figure 64 was replaced in the range of low h to the dependence shown qualitatively by a dashed line. Such an approximation imposed a strict limitation of the time interval within which Equation (8.4.1) was applicable. Namely, after the film thickness reached zero value at a point in the solution region, model (8.4.1) with closure (8.4.3) became inapplicable. As a result, only the initial stage of the proces could be described. The physical mechanisms of the later (nucleation and growth of dry regions) stage were considered by Kheshgi and Scriven.[36]

Actually, the descending branch of curve 2 in Figure 64, describing the adsorbed layer of fluid on the solid, corresponds to the narrow H range (~1 nm) which is essentially less than the characteristic film thickness of interest. However, as will be shown, it plays an important role for regularizing model (8.4.1) and its applicability in the entire time range.[59]

Let us rewrite Equation (8.4.1) in the dimensionless form

$$\frac{\partial H}{\partial t} = \frac{\partial}{\partial x}\left(H^3 \frac{\partial}{\partial x}\left(-2K\frac{\partial^2 H}{\partial x^2} + H - \Pi^*(H)\right)\right) \tag{8.4.4}$$

where

$$T \longrightarrow t\,\frac{\rho g H_o}{3\eta},\ H \longrightarrow \frac{H}{H_o},\ x \longrightarrow \frac{x}{H_o};\ K = \frac{\sigma}{2H_o^2\rho g},\ \Pi^* = \frac{\Pi}{\rho g H_o}$$

Here H_o is a characteristic thickness scale.

Now we present Equation (8.4.4) in the standard form,

$$\frac{\partial H}{\partial t} = \frac{\partial}{\partial x}\left(\Lambda \frac{\partial}{\partial x}\frac{\delta F}{\delta H}\right) \tag{8.4.5}$$

where

$$\Lambda = H^3,\ F = \int\left[f(H) + K\left(\frac{\partial H}{\partial x}\right)^2\right]dx,\ f(H) = \frac{H^2}{2} + G^*(H),\ G^*(H) = \int_H^\infty \Pi^*(H')\,dH' \tag{8.4.6}$$

Here F is the free energy of the film,[60] f(H) is the free energy density of a film with invariant thickness, H;$\delta F/\delta H$ is the variational derivative of F;$G^*(H)$ is the so-called disjoining potential.[57] Obviously, the form of $G^*(H)$ quantitatively repeats the form of $\Pi^*(H)$.

If $\Pi^*(H)$ looks like curve 2 in Figure 64, then f(H) can have either none or two inflection points (Figure 65). This corresponds to the existence of a two-phase region on the phase diagram of a binary mixture in the theory of spinodal decomposition, and H in Equation (8.4.5) corresponds to the fraction of a component of binary mixture in Section 8.2. The boundary of the film stability is given by the condition $\partial^2 f/\partial H^2 = 0$ at $H = H_s^{\pm}$. Two equilibrium values of thickness, $H_b^{\pm}$, which present the final result of film decomposition, are defined by the double tangent of f(H) (binodal values in the theory of spinodal decomposition). Lastly f(H) tends to infinity when H tends either to zero or to infinity.

Since Equation (8.4.5) looks exactly like the basic equation of Section 8.2, all the results obtained there can be directly applied to the dewetting model. Namely, if $H_s^- < H < H_s^+$ then the infinitesimal thickness perturbations will grow, and there is the fastest-growing mode corresponding to the wave number

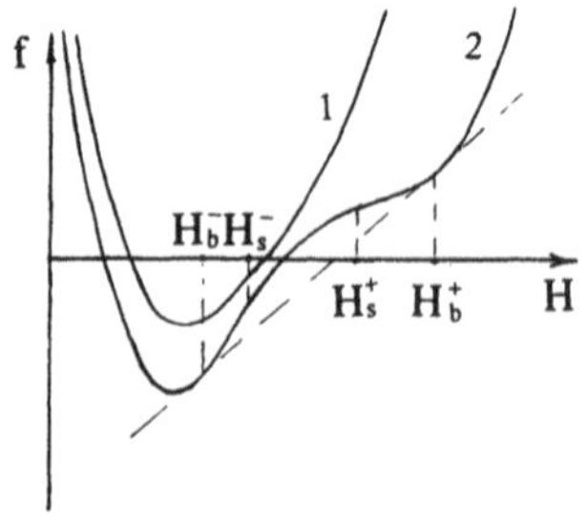

FIGURE 65. Dependences of the free energy density of a homogeneous film on thickness. When the contribution of the gravitation force in f(H) is large enough, the film is stable (curve 1). In the opposite case, film rupture is possible for some thicknesses (curve 2).

$q_M = q_c/\sqrt{2}$, where q_c is given by Equation (8.2.7). The evolution in the later stage looks like a wandering between the unstable stationary solutions of Equation (8.4.5), accompanied by an increase of the characteristic space scale of inhomogeneities in the system. At the initial stage of decomposition, H(x) has nearly sinusoidal form; at the late stages, a typical H(x) distribution looks like a sequence of plateaux with the narrow zones of high H gradients between. The H values at the plateaux are close to the binodal values, H_b^- and H_b^+. The plateaux with $H \approx H_b^-$ correspond to zones of adsorbed layers of fluid on solid (some authors[55,56] contend they correspond to dry zones where $H \equiv 0$). The plateaux with $H \approx H_b^+$ correspond to the fluid film zones.

The most important parameter of the nucleation dynamics is the mean scale of decomposition structures, $\xi(t)$. In the theory of phase transitions where model (8.4.5) is widely applied, ξ is derived as the reciprocal of the wave number giving the maximim value to the structure factor at a given moment of time (see Section 9.1.). The behavior of $\xi(t)$ depends on the concrete structure of functions f, Λ, and K, but over long periods (substantially exceeding the duration of the linear stage) $\xi(t)$ is a power function with a positive exponent. At early stages of nucleation, when the mean-field approximation[45] for model (8.4.5) is applicable, it was found that $\xi(t) \sim t^{1/2}$.[44]

We still considered the process with a one-space variable. If one permits more real film dynamics characterized by two-space variables, one obtains Equations (8.4.5) and (8.4.6) where all the space derivatives are replaced to the gradient operators. The nucleation of dry regions is characterized by a power law $\xi(t)$, whose exponent will be different from the one-dimensional case. Particularly, at early stages of nucleation, $\xi(t)$ increases as $t^{1/4}$.[44,46]

The characteristic ratio of areas occupied by the adsorbed layer and film, $\xi_a(t)$ and $\xi_f(t)$ — $\xi_a(t) + \xi_f(t) = \xi(t)$ — can be estimated over long periods of time from the equation

$$\frac{\xi_a(t)}{\xi_f(t)} = \frac{H_b^+ - H_o}{H_o - H_b^-}$$

since the decomposition process consists over large time scales mainly in uniting the decomposition structures whose maximum and minimum thicknesses are characterized by almost established values of $H \approx H_b^{\pm}$.

In the infinite system, with the conditions of the space derivatives attenuation at $x \to \pm \infty$, the final result of the decomposition process is the formation of a kink-like solution with $H = H_b^{\pm}$ at $x \to \pm \infty$ (one half-space covered by the adsorbed layer and another covered by film). Of course, one should remember that the assumptions on which the model (8.4.1) is based can be violated over long periods of time. Particularly, Equation (8.4.1) itself is obtained by a perturbation method from a more complex model with the supposition that $\partial H/\partial x$ is small enough (formally, essentially less than 1).[35]

Within the framework of the dewetting scenario presented, one can suggest a method of $\Pi^*(H)$ evaluation using the measurements of thicknesses of the adsorbed layer and film, which should be about H_b^- and H_b^+ when the film decomposition process is visually completed. Actually, f(H) is connected with H_b^- and H_b^+ through the relationships

$$f(H_b^+) - f(H_b^-) = f'(H_b^-)(H_b^+ - H_b^-),\ f'(H_b^-) = f'(H_b^+) \qquad (8.4.7)$$

In this particular case, when the dispersion forces provide the main contribution to the disjoining pressure,

$$\Pi(H) = \frac{B}{H^9} + \frac{A}{H^3}\ ,\ A < 0\ ,\ B > 0 \qquad (8.4.8)$$

if one is to assume that the fluid-solid interactions are described by the 6-12 Lennard-Jones potential. In the dimensionless variables derived above,

$$\Pi^*(H) = \frac{B_1}{H^9} + \frac{A_1}{H^3}\ ,\ B_1 = \frac{B}{H_o^{10}\rho g}\ ,\ A_1 = \frac{A}{H_o^4\rho g}$$

Using Equations (8.4.7) and (8.4.8) we obtain the system of linear equations for deriving A_1 and B_1

$$A_1\left[\frac{1}{(H_b^-)^3} - \frac{1}{(H_b^+)^3}\right] + B_1\left[\frac{1}{(H_b^-)^9} - \frac{1}{(H_b^+)^9}\right] = H_b^+ - H_b^-\ ,$$

$$A_1\left[\frac{1}{2(H_b^-)^2} - \frac{1}{2(H_b^+)^2} + \frac{H_b^- - H_b^+}{(H_b^-)^3}\right] + \qquad (8.4.9)$$

$$B_1\left[\frac{1}{8(H_b^-)^8} - \frac{1}{8(H_b^+)^8} + \frac{H_b^- - H_b^+}{(H_b^-)^9}\right] = \frac{1}{2}(H_b^+ - H_b^-)$$

Solving system (8.4.9) in regard to A_1 and B_1 one can find the concrete form of disjoining pressure.

Lastly, we will discuss the behavior of rupture time, t_{rup}. This quantity was defined for models (8.4.1) and (8.4.3) as the time when H becomes zero at a point of the solution region.[53-56] The rupture time for the linearized models (8.4.1) and (8.4.3) was found to be essentially larger than that in the nonlinear models (8.4.1) and (8.4.3), i.e., the nonlinearity accelerated the rupture process. This result is correct if one uses closure (8.4.3) for model (8.4.1), since in that case the effective 'diffusion coefficient' tends to minus infinity as H tends to zero, and at small H one sees the sharp acceleration of the rupture.

If one uses the form presented by curve 2 in Figure 64 (say, Equation (8.4.8) for the disjoining pressure, with the accurate $\Pi(H)$ asymptotic at $H \to 0$), the nonlinearity stabilizes the solution behavior at small H. Moreover, since f(H) $\to \infty$ at $H \to 0$, the possibility of approaching H zero would be in contradiction with F temporal decrease, which was established in Section 8.2 in Equation (8.2.16). Thus, the zero value of H just cannot be reached.

The qualitative behavior of t_{rup} in the regularized model of dewetting can be considered in the one-mode approximation. This approximation cannot describe the nucleation process connected with the interaction of modes with different wavelengths, but it enables us to consider the role of the nonlinear retardation at the initial stage. If $H_b^+ - H_b^-$ is small enough, one can approximate f(H) by a polynomial of the fourth order. For the dimensionless amplitude of a growing mode, a, after simple transformation of the dependent and independent variables in Equation (8.4.5), one obtains the equation

$$\frac{da}{dt} = a - a^3 \tag{8.4.10}$$

which is well known in synergetics.[1] Solving Equation (8.4.10) we find the time between an initial state with magnitude a_o and a reference state with magnitude a_r ($1 > a_r > a_o > 0$),

$$t_{rup} = \ln\left(\frac{a_r}{a_o}\right) + \frac{1}{2}\ln\left(\frac{1 - a_o^2}{1 - a_r^2}\right) \tag{8.4.11}$$

The last term on the right-hand side of Equation (8.4.11) shows the role of nonlinearity in the t_{rup} increase. Therefore, the nonlinearity can lead to the t_{rup} decrease, as well as to its increase, depending on the concrete form of the disjoining pressure used in calculations.

Thus, it is shown that the nonlinear Cahn equation appears by the description of one more class of hydrodynamical phenomena — dewetting of a solid surface. It is important to note that the dewetting model which is applicable at the nucleation stage of process should take into account the attraction between molecules of the solid and fluid at large H, the repulsion between the molecules of the solid and fluid at small H, the capillary pressure, and the gravitation force, simultaneously. In fact, all these effects have been put forward,[35,36,53-56] and we just showed that the explicit consideration of the repulsion effect at small H should alleviate some troubles with the applicability of the dewetting model over long time periods.

The scenario of the film decomposition considered above is a direct consequence of the absolute instability of a film with thickness from the interval, $H_s^- < H < H_s^+$. As is known in the theory of phase transitions, the breakdown also is possible in the metastable states, $H_b^- < H < H_s^-$ and $H_s^+ < H < H_b^+$ (the nucleation and growth mechanism[18,19]). To produce such a mechanism within the framework of the dewetting model, it is necessary to add a random noise (influence of thermal fluctuations which give rise to the local H fluctuations) into the right-hand side of Equation (8.4.1). The numerical studies of the two-dimensional Cahn equation with thermal noise show that it does describe the nucleation process in the entire range $H_b^- < H < H_b^+$, and the morphologies of the decomposition structures are different for the metastable (circular domains) and unstable (percolating domains) states.[52]

In order to estimate the typical values of H_s^+, H_s^-, H_b^+, and H_b^-, we set $A = -10^{-21}$ J, $\rho g = 10^4$ kg/m^{2}s^2, and suppose that the minimum value of $\Pi(H)$ in Equation (8.4.8) is reached at $H = 1$ nm.[36] Deriving B and solving Equations (8.4.7) one can find that $H_s^- = (1 + 5.5 \cdot 10^{-13})$ nm, $H_s^+ = 740.09$ nm, $H_b^- = 0.83$ nm, $H_l^+ = 328891$ nm. The value of H_s^- is in qualitative accordance with the estimation of 1000 nm given by Kheshgi and Scriven.[36] It is noticeable that the thickness region (H_s^+, H_b^+) is much larger than (H_b^-, H_s^+). This means that film rupture can happen due to thermal fluctuations even though the film thickness is essentially greater than the upper boundary of the absolute instability. This is confirmed by the experiments which were carried out by Padday[61] in order to measure the critical rupture thicknesses for different kinds of fluids. The range of the critical rupture thickness was found to be 0.01 to 0.05 cm, which is in accordance with our estimation of H_b^+.

Thus, the film rupture scenario described above enables us to explain quantitatively the unusually large values of the critical rupture thicknesses observed in the Padday's experiments. It seems to be an significant argument in favor of the model presented.

When $\Pi(H)$ is taken in form (8.4.8), some explicit expressions for H_s^+, H_s^-, H_b^+, and H_b^- have been reported.[59] It turns out that, for a typical fluid, the parameter

$$\epsilon = -\frac{\rho g H_{ex}^4}{2A} > 0 \qquad (8.4.12)$$

characterizing the ratio of 'gravitational' and 'disjoining' contributions in f(H), is a small one. In Equation (8.4.12) H_{ex} is the thickness value corresponding to the $\Pi(H)$ minimum. Using a perturbation method,[59] one can show that, at small ϵ,

$$H_b^+ = \frac{\vartheta_1 H_{ex}}{\epsilon^{1/2}}, \; H_s^+ = \frac{\vartheta_2 H_{ex}}{\epsilon^{1/4}}, \; H_b^- = \vartheta_3 H_{ex}, \; H_s^- = H_{ex} \qquad (8.4.13)$$

where

$$\vartheta_1 = \frac{3^{2/3}}{2^{3/2}}, \ \vartheta_2 = \frac{3^{1/4}}{2^{1/4}}, \ \vartheta_3 = \frac{1}{3^{1/6}}$$

For concrete ρ, H_{ex}, and A values presented above, the small parameter, ϵ, equals $0.5 \cdot 10^{-11}$. One can ensure that, in this case, the asymptotic expressions (8.4.13) provide at least five correct digits for H_b^+, H_s^+, H_b^-, and H_s^-.

REFERENCES

1. **Haken, G.,** *Advanced Synergetics. Instability Hierarchies of Self-Organizing Systems and Devices,* Springer-Verlag, Berlin, 1983.
2. **Mitlin, V. S., Manevich, L. I., and Erukhimovich, I. Ya.,** *Zh. Eksp. Teor. Fiz.,* 88(2), 495, 1985; *Sov. Phys. JETP,* 61(2), 290, 1985.
3. **Mitlin, V. S.,** *Izv. Akad. Nauk SSSR, Mekh. Zhid. Gaza,* 6, 130, 1988; *Fluid Dyn. (USSR),* 23(6), 909, 1988.
4. **Guzhov, N. A. and Mitlin, V. S.,** *Izv. Akad. Nauk SSSR, Mekh. Zhid. Gaza,* 4, 83, 1986; *Fluid Dyn. (USSR),* 21(4), 576, 1986.
5. **Rozhdestvenskii, B. L. and Yanenko, N. N.,** *Systems of Quasilinear Equations and Their Applications to Gas Dynamics,* Nauka, Moscow, 1978.
6. **Panfilov, M. B.,** *Izv. Akad. Nauk SSSR, Mekh. Zhid. Gaza,* 4, 94, 1985; *Fluid Dyn. (USSR),* 20(4), 574, 1985.
7. **Nikolaevskii, V. N., Bondarev, E. A., Mirkin, M. I., et al.,** *Motion of Hydrocarbon Mixtures in Porous Media,* Nedra, Moscow, 1968.
8. **Mitlin, V. S.,** New Methods of Calculation of Enriched Gas Action Upon a Gas-Condensate Seam, Ph.D. thesis, VNIIGAS, Moscow, 1987 (in Russian).
9. **Khristianovich, S. A.,** *Prikl. Mat. Mekh.,* 5(2), 277, 1941 (in Russian).
10. **Whitham, G. B.,** *Linear and Nonlinear Waves,* Wiley-Interscience, New York, 1974.
11. **Amiks, J. W., Bass, D. M., and Whiting, J. R.,** *Petroleum Reservoir Engineering. Physical Properties,* McGraw-Hill, New York, 1960.
12. **Mitlin, V. S.,** *Izv. Akad. Nauk SSSR, Fiz. Zemli,* 1, 46, 1990; *Physics of the Solid Earth,* 27(1), 31, 1990.
13. **Putnis, A. and McConnel, J. D. C.,** *Principles of Mineral Behavior,* Elsevier, New York, 1980.
14. **Khisina, N. R.,** *Subsolidus Transformations of Solid Solutions of Rock-Formed Minerals,* Nauka, Moscow, 1987.
15. **Cahn, J. W.,** *Acta Metall.,* 9, 795, 1961.
16. **Cahn, J. W.,** *Acta Metall.,* 10, 179, 1962.
17. **Cahn, J. W.,** *Trans. Metall. Soc. AIME,* 242, 166, 1968.
18. **Scripov, V. V. and Scripov, A. B.,** *Usp. Fiz. Nauk,* 128, 193, 1979; *Sov. Phys. Usp.,* 22, 389, 1979.
19. **Lifshitz, I. M. and Slyozov, V. V.,** *Zh. Eksp. Teor. Fiz.,* 35(2), 479, 1958; *Sov. Phys. JETP,* 8, 331, 1958.
20. **De Gennes, P. G.,** *J. Chem. Phys.,* 72(9), 4756, 1980.
21. **Mitlin, V. S. and Manevich, L. I.,** *Vysokomol. Soedin.,* 30A(1), 9, 1988; *Polymer Sci. USSR,* 30(1), 8, 1988.
22. **Givental, A. B.,** *Teor. Mat. Fiz.,* 76(3), 462, 1988; *Theor. Math. Phy. (USSR),* 76(3), 989, 1988.

23. **Manevich, L. I., Mitlin, V. S., and Shaginian, S. A.,** *Khim. Fiz.*, 3(2), 283, 1984 (in Russian).
24. **Mitlin, V. S. and Manevich, L. I.,** *Vysokomol. Soedin.*, 27B(6), 409, 1985; *Int. Polymer. Sci. Tech.*, 12(11), 98, 1985.
25. **Landau, L. D. and Lifshitz, E. M.,** *Hydrodynamics*, Nauka, Moscow, 1986 (in Russian).
26. **Nikolaevskii, V. N.,** *Mechanics of Porous and Fractured Media*, Singapore World Publ., 1990.
27. **Novick-Cohen A.,** *J. Stat. Phys.*, 38(3/4), 707, 1985.
28. **Andronov, A. A., Vitt, A. A., and Khaikin, C. E.,** *Theory of Oscillations*, Nauka, Moscow, 1981 (in Russian).
29. **Henry, D.,** *Geometrical Theory of Half-Linear Parabolic Equations*, Mir, Moscow, 1985 (in Russian).
30. **Gelfand, I. M. and Fomin, S. V.,** *Calculus of Variation*, Nauka, Moscow, 1961 (in Russian).
31. **Belintsev, B. I.,** *Usp. Fiz. Nauk*, 141(1), 51, 1983; *Sov. Phys. Usp.*, 29(9), 775, 1983.
32. **Marsden, J. E. and McCracken, M.,** *The Hopf Bifurcation and its Applications*, Springer-Verlag, New York, 1976.
33. **Bunkin, F. B., Kirichenko, N. A., and Morosov Yu. Yu.,** *Pis'ma Zh. Eksp. Teor. Fiz.*, 41(9), 378, 1985.
34. **Mitlin, V. S. and Manevich, L. I.,** *Vysokomol. Soedin.*, 31A(5), 1020, 1989; *Polymer Sci. USSR*, 31(5), 1989.
35. **Teletzke, G. F., Scriven, L. E., and Davis, H. T.,** *Chem. Eng. Commun.*, 55, 41, 1987.
36. **Kheshgi, H. S. and Scriven., L. E.,** *Chem. Eng. Sci.*, 46, 519, 1991.
37. **Anon.,** Modern Problems of Mathematics. Vol. 5, III., *Theory of Bifurcations*, VINITI, Moscow, 1986 (in Russian).
38. **Robin, P.-Y.,** *Am. Mineral.*, 59, 1299, 1974.
39. **Tullis, I. and Yund, R. A.,** *Am. Mineral.*, 64(9/10), 1063, 1979.
40. **Mitlin, V. S. and Manevich, L. I.,** *Vysocomol. Soedin.*, 30B(8), 597, 1988 (in Russian).
41. **Mitlin, V. S. and Manevich, L. I.,** *J. Polymer Sci.*, 28B(1), 1, 1990.
42. **Zakharov, V. E., Manakov, C. V., Novikov, C. P., et al.,** *Theory of Solitons. Method of Inverse Problem*, Nauka, Moscow, 1980 (in Russian).
43. **Elder, K. R., Rogers, T. M., and Desai, R. C.,** *Phys. Rev. Sect. B.*, 38(7), 4725, 1988.
44. **Mitlin, V. S.,** *Zh. Eksp. Teor. Fiz.*, 95(5), 1826, 1989; *Sov. Phys. JETP*, 68(5), 1056, 1989.
45. **Langer, J. S.,** *Ann. Phys.*, 65, 53, 1971.
46. **Mitlin, V. S.,** *Zh. Eksp. Teor. Fiz.*, 98(2), 554, 1990.
47. **Mitlin, V. S. and Manevich, L. I.,** *Vysokomol. Soedin.*, 31A(8), 1674, 1989; *Polymer Sci. USSR*, 31, 8, 1989.
48. **Rostiashvili, V. G., Irzhak, B. I., and Rosenberg, B. A.,** *Glass Transition of Polymers*, Khimiya, Leningrad, 1987 (in Russian).
49. **Mitlin, V. S. and Manevich, L. I.,** *Int. J. Eng. Sci.*, 1992, v. 30, No. 2, p. 237.
50. **de Gennes, P. G.,** *Scaling Concepts in Polymer Physics*, Cornell University Press, Ithaca, NY, 1979.
51. **Ma, S.,** *Modern Theory of Critical Phenomena*, W. A., Benjamin, New York, 1976.
52. **Rogers, T. M. and Desai, R. C.,** *Phys. Rev. Sect. B*, 39(16), 11956, 1989.
53. **Sheludko, A.,** *Adv. Colloid Interface Sci.*, 1, 391, 1967.
54. **Ruckenstein, E. and Jain, R. K.,** *Chem. Soc. Faraday Trans. 2*, 70(2), 132, 1974.
55. **Williams, M. B. and Davis, S. H.,** *J. Colloid Interface Sci.*, 90, 220, 1982.
56. **Sharma, A. and Ruckenstein, E.,** *J. Colloid Interface Sci.*, 113, 456, 1986.
57. **Deryagin, B.,** *Kolloidn. Zh.*, 17, 205, 1955.
58. **Dzyaloshinskii, I. E., Lifshitz, E. M., and Pitaevskii, L. P.,** *Sov. Phys. JETP*, 37, 161, 1960.
59. **Mitlin, V. S.,** *J. Colloid Interface Sci.*, (accepted for publication).
60. **de Gennes, P. G.,** *Rev. Mod. Phys.*, 57, 827, 1985.
61. **Padday, J. F.,** in *Thin Liquid Films and Boundary Layers, (Spec. Discuss. Faraday Soc.)*, No. 1, 64; Academic Press, New York, 1970.

9

Applications of the Asymptotical Methods in Gradient Theory

9.1. GROWTH KINETICS OF STRUCTURES IN THE NONLINEAR THEORY OF PHASE SEPARATION: ISOTROPIC CASE

The coalescence occurring in solutions in the process of phase separation is often described using the classical theory of Lifshitz and Slyozov,[1] dealing with the time evolution of the size of a spherical nucleus and predicting a dependence $\xi \sim t^{1/3}$ after a long period of time. The Lifshitz-Slyozov theory is based on the assumption that the interaction between nuclei is weak and the ratio of their size to the average distance between them is small. The asymptotic behavior of $\xi(t)$ predicted by the theory applies to a wide range of physical systems, as confirmed by numerous experiments.

The initial distribution of the nuclei, when they are far from one another, is typical of the formation of a new phase by the mechanism of nucleation and growth when the parameters of the system in question lie in the range of metastable states and the frequency of spontaneous nucleation is not too high at the stage of formation of a nucleus of critical size.[2] If a system can evolve rapidly beyond the spinodal without phase separation in accordance with the nucleation and growth mechanism in the range of metastable states, the next stage is spinodal decomposition of a solution (see Section 8.2). In this process there is no need for overcoming any energy barrier, and nuclei of the new phase may appear in the immediate vicinity of one another. In other words, during spinodal growth we can have situations when the interaction between nuclei cannot initially be regarded as weak.

Figure 57 in Section 8.2 shows the profile of the distribution of the fraction of one of the components of a binary mixture obtained by solving a one-dimensional spinodal decomposition equation and corresponding to the moment of termination of the stage of exponential growth of fluctuations. We can see that large-amplitude fluctuations, which form a modulated structure and can be regarded as nuclei of a new phase, are separated by distances equal to their spatial size. The continuum theory of spinodal decomposition was proposed by Cahn.[3] The results discussed in Section 8.2 suggest that a nonlinear Cahn equation can provide a unified description of the process of spinodal decomposition at the initial and later stages. The nonlinear terms make it possible to effectively allow for the interaction of nuclei; they limit the exponential growth of fluctuations in a thermodynamically instable region, and give rise to "coalescence" of a modulated structure, i.e., they result in modifications accompanied by an increase in the spatial scale. It would therefore be of interest to consider how the average size in a structure varies in accordance with the nonlinear theory of spinodal decomposition.

Notice that the complete analytical study of this question within the framework of the Cahn equation was not carried out. We shall try to provide such a description on the basis of the spinodal decomposition equation "truncated" in a certain manner, and we shall estimate the validity of this truncation. We shall employ asymptotic methods to derive an equation describing the change in the characteristic spatial size of a structure. An analysis considered in the limit of long time periods yields simple power dependences $\xi(t)$.[4] We shall discuss the isotropic case in this section; the asymptotic theory of anisotropic decomposition, which is the generalization of the isotropic case and is directed mainly to describe solid solutions, will be proposed in Section 9.2.

Derivation of the Truncated Equation

As in Section 8.2, the spinodal decomposition is described by the equations[3,5,6]

$$\frac{\partial \phi}{\partial t} = \nabla(\Lambda(\phi)\nabla\mu) \tag{9.1.1}$$

$$F = \int \rho[F_o(\phi) + K(\phi)(\nabla\phi)^2]\, d\mathbf{r} \tag{9.1.2}$$

$$\mu = \frac{1}{\rho}\frac{\delta F}{\delta \phi} = \mu_o(\phi) - 2K(\phi)\nabla^2\phi - K_\phi(\nabla\phi)^2, \quad \mu_o = \frac{\partial F_o}{\partial \phi} \tag{9.1.3}$$

The following notation is used in Equations (9.1.1) to (9.1.3): ϕ is the fraction of one of the components of a binary mixture; Λ is the Onsager coefficient; ρ is the number of molecules of the mixture per unit volume; F is the functional of the free energy of the system in units of k_BT; μ is the chemical potential expressed in terms of the variational variable F. We shall consider the problem with impermeable boundaries in a rectangular parallelepiped of linear dimensions $L_1, \ldots, L_d$.

We shall transform Equation (9.1.1) as follows. We shall consider fluctuations $\delta\phi$ relative to the average value ϕ_o over the whole region, when the functions in Equation (9.1.1) nonlinear in ϕ can be expanded as the Taylor series. We shall then introduce Fourier components $\delta\phi_{\mathbf{q}}$; it follows from the condition of impermeability that $\delta\phi_{\mathbf{q}} = \delta\phi_{-\mathbf{q}}$. We shall multiply the resultant equation by $\exp(-j\mathbf{qr})$, $j^2 = 1$, and integrate it over the space so that Equation

(9.1.1) reduces to a system of equations for $\delta\phi_{\mathbf{q}}$. On the right-side of this system there are sums of the type

$$\sum_{\mathbf{q}_1, \ldots, \mathbf{q}_n} \delta\phi_{\mathbf{q}_1} \cdot \ldots \delta\phi_{\mathbf{q}_n}, \ \mathbf{q}_1 + \ldots + \mathbf{q}_n = \mathbf{q}$$

which are transformed into convolution-type integrals in accordance with the rule

$$\sum_{\mathbf{q}_1, \ldots, \mathbf{q}_n} \delta\phi_{\mathbf{q}_1} \ldots \delta\phi_{\mathbf{q}_n} = u^{n-1} \int \delta\phi(\mathbf{q} - \mathbf{q}_2 - \ldots - \mathbf{q}_n) \cdot$$

$$\delta\phi(\mathbf{q}_2) \ldots \delta\phi(\mathbf{q}_n) d\mathbf{q}_2 \ldots d\mathbf{q}_n, \ u = (2\pi)^{-d} \cdot \prod_{j=1}^{d} L_j \tag{9.1.4}$$

The spatial derivatives in Equation (9.1.1) transform in the usual way on adoption of the Fourier components. Expanding now the integrand $\delta\phi(\mathbf{q} - \mathbf{q}_2 - \ldots - \mathbf{q}_n)$ near $\mathbf{q}$, we obtain a system of partial differential equations with respect to $\mathbf{q}$ ($\mathbf{q}$ is already considered as a continuous quantity). Our truncation of this system results in retention of only the "bulk" terms and dropping all the terms with derivatives with respect to $\mathbf{q}$. The use of such a system is permissible if the processes resulting in the smearing out of the spectrum by nonlocal interaction between modes are weak compared with the separation mechanism. Essentially this is equivalent to neglect of the diffusion terms in the "diffusion + reaction" equation. The problem of applicability of this approximation will be considered separately.

We thus find that

$$\frac{\partial \delta y \mathbf{q}}{\partial t} = -q^2 \Lambda(\phi_o + z) \left[\frac{\mu_o(\phi_o + z) - \mu_o(\phi_o)}{z} + 2q^2 K(\phi_o + z) \right] \delta\phi_{\mathbf{q}} \tag{9.1.5}$$

which corresponds to the equation

$$\frac{\partial \phi}{\partial t} = \Lambda(\phi_o + z) \nabla^2 \left[\frac{\mu_o(\phi_o + z) - \mu_o(\phi_o)}{z} \phi - 2K(\phi_o + z) \nabla^2 \phi \right] \tag{9.1.6}$$

In Equations (9.1.5) and (9.1.6) we have

$$z = u \int \delta\phi(\mathbf{q}) \, d\mathbf{q} = \delta\phi(\mathbf{r}) \big|_{\mathbf{r}=0}$$

It should be noted that Equations (9.1.5) and (9.1.6) do not include the terms of Equation (9.1.1) proportional to $K_\phi (\nabla\phi)^2$ and $\nabla\Lambda \cdot \nabla\mu$. This is due to the fact that in the adopted approximation they are expressed in terms of integrals $\int q_j \delta\phi(\mathbf{q}) d\mathbf{q}$, where $j = 1, \ldots, d$, which vanish because $\delta\phi(\mathbf{q})$ is an even function.

The structure of the truncated Equation (9.1.5) obtained above represents generalization, to the case when Λ and K depend on ϕ, of the expression derived by Langer starting from the functional equation of continuity of the density of the statistical distribution of configurations.[7] Langer predicted the correct tendency of behavior of the solutions after a long period of time, namely that the maximum of the Fourier spectrum of fluctuations would shift toward longer wavelengths. It is shown above that one can derive equations of this kind[7] issuing directly from Equation (9.1.1).

Having solved Equation (9.1.5), we obtain

$$\delta\phi(\mathbf{q}) = C(\mathbf{q}) \exp\left\{ - q^2 \int_0^t \Lambda(\phi_o + z) \left[\frac{\mu_o(\phi_o + z) - \mu_o(z)}{z} + 2K(\phi_o + z)\, q^2\right] dt\right\} \tag{9.1.7}$$

whereas integration of Equation (9.1.7) with respect to **9** gives

$$z = u \int \delta\phi(\mathbf{q})\, d\mathbf{q}\ , \ C(\mathbf{q}) = \delta\phi(\mathbf{q})\big|_{t=0} \tag{9.1.8}$$

Equation (9.1.5) in the Limit of Long Time Periods

We shall be interested in the solutions of Equation (9.1.5) after a long time period, i.e., sufficiently after the linear stage of the process. Inserting Equation (9.1.7) into Equation (9.1.8), we shall rewrite Equation (9.1.8) in the form

$$z = u \int C(\mathbf{q}) \exp[-(Aq^2 + Bq^4)t]\, d\mathbf{q} \tag{9.1.9}$$

where

$$A = \frac{1}{t}\int_0^t \Lambda(\phi_o + z)\, \frac{\mu_o(\phi_o + z) - \mu_o(\phi_o)}{z}\, dt$$

$$B = \frac{2}{t}\int_0^t \Lambda(\phi_o + z)\, K(\phi_o + z)\, dt \tag{9.1.10}$$

Since all the integrands in Equation (9.1.10) are bounded, it follows that 0 and 0 are bounded functions of time. We shall introduce polar coordinates q, $\beta_1, \ldots,$ β_{d-1}, in Equation (9.1.9); here, β_j are the angular variables. Then, Equation (9.1.9) becomes

$$z = u \int C_*(q) S(q) \exp[-(Aq^2 + Bq^4)\, t]\, dq \tag{9.1.11}$$

where $C_*(q)$ is the average (over a sphere of radius q) initial distribution; $S(q)$ is the surface area of this sphere. In the one-dimensional case we have $S(q) = 2$, since we are going over from an integral over the whole space to an integral in the region $q > 0$, whereas for $d = 2$ we find that $S = 2\pi q$ and $S = 4\pi q^2$ if $d = 3$.

The quantity B is always positive, since $\Lambda > 0$ and $K > 0$. The quantity A can generally be of any sign. We can easily see that as long as ϕ_o and $\phi_o + z$ correspond to the region below the spinodal, the integrand in the expression for A is negative. It follows from formal considerations that there are two types of asymptotic behavior of Equation (9.1.11).

We shall now consider the case when $A < 0$, which at any rate will be first to appear. Then, the function $I_e = -(Aq^2 + Bq^4)$ has a maximum at

$$q_*^2 = -A/2B \tag{9.1.12}$$

Applying the Laplace method to Equation (9.1.11), we obtain

$$z = uC_*(q_*)S(q_*)\exp\left(\frac{A^2t}{4B}\right)\left(-\frac{\pi}{2At}\right)^{1/2} \tag{9.1.13}$$

If $A > 0$, a maximum of I is located at the boundary of the integration domain and, moreover, we have $S(0) = 0$ when $d > 1$. We then find that $I_e'(0) = 0$ and $I_e''(0) < 0$. Let us assume that $d = 1$ (one-dimensional case) and that $C_*(q)$ is a smooth function of q^2. We then obtain[8]

$$z = uC_*(0)\,(\pi/At)^{1/2} \tag{9.1.14}$$

If $d = 2$, we can rewrite the integral in Equation (9.1.11) in the form

$$\int C_*(q^2)\pi\exp(I_e t)\,d(q^2) = \frac{\pi C_*(0)}{At},\quad z = \frac{u\pi C_*(0)}{At} \tag{9.1.15}$$

If $d = 3$, we can transform the integral in Equation (9.1.11) as follows[8]

$$\int C_*(q)\,4\pi q^2\exp(I_e t) = \int \frac{4\pi q^2 C_*(q)}{I_e' t}\,d\,(\exp(I_e t)) =$$

$$-\int \frac{\partial}{\partial q}\left[\frac{4\pi q^2 C_*(q)}{I_e' t}\right]\exp(I_e t)dq = \left(\frac{\pi}{At}\right)^{1/2}\frac{2\pi C_*(q)}{t(2A + 4Bq^2)}\bigg|_{q=0}, \tag{9.1.16}$$

$$z = uC_*(0)\left(\frac{\pi}{At}\right)^{3/2}$$

Equations (9.1.14) to (9.1.16) can now be combined:

$$z = uC_*(0)\left(\frac{\pi}{At}\right)^{d/2} \tag{9.1.17}$$

However, in the case of a formal change in A it is found that the type of saddle point is modified (a singularity merges with a boundary). Therefore, in going over from $A > 0$ to $A < 0$ the asymptotes of Equation (9.1.13) and (9.1.17) should be refined in accordance with analysis provided by Fedoryuk.[8]

Equations (9.1.13) and (9.1.17) include integrals of type (9.1.10). The next transformation of Equation (9.1.13) gives rise to a pair of ordinary differential equations for A and B. We introduce

$$A_*(z) = \Lambda(\phi_o + z)\,[\mu_o(\phi_o + z) - \mu_o(\phi_o)]/z$$

$$B_*(z) = 2\Lambda(\phi_o + z)K(\phi_o + z)$$

$$R_-(A,B,t) = uC_*(q_*)S(q_*)\exp(A^2t/4B)(-\pi/2At)^{1/2} \tag{9.1.18}$$

We apply to Equation (9.1.13) the operators A_* and B_*:

$$A_*(z) = A_*(R_-)\ ,\ B_*(z) = B_*(R_-)$$

Equations (9.1.10) and (9.1.18) yield the required pair of equations:

$$\dot{A} = [A_*(R_-(A,B,t)) - A]/t$$

$$\dot{B} = [B_*(R_-(A,B,t)) - B]/t \tag{9.1.19}$$

Similarly, introducing

$$R_+(A,t) = uC_*(0)(\pi/At)^{d/2}$$

we derive from Equation (9.1.10) one equation

$$\dot{A} = [A_*(R_+(A,t)) - A]/t \tag{9.1.20}$$

We shall be interested in the future in the solutions of Equations (9.1.13) and (9.1.19), because these equations correspond to the motion of a maximum of the spectrum of fluctuations toward longer wavelength.

The system can be closed by defining the function $C_*(q)$. It is known that at the initial stage of spinodal decomposition the short-wavelength part of the spectrum, corresponding to the condition $q > q_c$, where $q_c^2 = -\partial^2F_o/\partial\phi^2/2K$ is suppressed.[3] The process is concentrated at long wavelengths, which is simplest to allow for with the aid of C_*, which is constant in the range $q < q_c$ and is equal to zero for $q > q_c$. A characteristic value of C_* can be found from the normalization condition

$$z|_{t=0} = z_o = u \int C(\mathbf{q})d\mathbf{q}$$

which yields

$$C_* = z_o/uq_c^2\gamma_d \tag{9.1.21}$$

Here $\gamma_d = 2, \pi, 4\pi/3$ applies when $d = 1$, 2, and 3, respectively.

Asymptotics of the Growth of Decomposition Structures

We shall consider the model of spinodal decomposition in the case when $K = \text{const}$ and $\Lambda = \text{const}$ and use the Ginzburg-Landau potential in the form

$$F = \int \rho\,[a\phi^2 + b\phi^4 + K(\nabla\phi)^2]\,d\mathbf{r} \tag{9.1.22}$$

The form (9.1.22) is obtained by expanding Equation (9.1.2) as a Taylor series in $(\delta\phi)^4$ and by a shift by a suitable constant; the remaining term of the $\text{const}_1\,\phi + \text{const}_2$ type does not affect the growth kinetics. Since a fourth-degree polynomial is usually sufficient to satisfactorily approximate the concentration dependence of the free energy when, at a given temperature, there is only one interval of compositions corresponding to thermodynamically unstable states, and usually only the order of magnitude of K and Λ are known, it follows that an analysis of such a model is a very important task.

We shall investigate the asymptotic change in the wave number corresponding to a maximum of the fluctuation spectrum. It follows from Equations (9.1.10), (9.1.12), and (9.1.22) that

$$q_*^2 = q_M^2 - \frac{3b\phi_o}{Kt}\int_0^t z dt - \frac{b}{Kt}\int_0^t z^2 dt\ , \quad q_M^2 = \frac{q_c^2}{2}$$

Using Equation (9.1.13), we obtain

$$\frac{d(q_*^2 t)}{dt} = q_M^2 - \frac{3b\phi_o}{K}\,uC_*(q_*)S(q_*)\exp(Bq_*^4 t)\left(-\frac{\pi}{2At}\right)^{1/2}$$
$$- \frac{b}{K}\left[uC_*(q_*)S(q_*,\ \exp(Bq_*^4 t)\left(-\frac{\pi}{2At}\right)^{1/2}\right]^2 \tag{9.1.23}$$

We shall now adopt dimensionless variables

$$v = \frac{q_*^2}{q_M^2}, \qquad \tau = tq_*^4 B$$

(in other words, unit of τ corresponds to the temporal duration of the initial stage of the process). It then follows from Equation (9.1.21) that

$$\frac{dv}{d\tau} = [G(v,\tau) - v]/\tau$$

where

$$G(v,\tau) = 1 - \frac{A_1 \exp(v^2\tau)v^{(d-2)/2}}{\tau^{1/2}} - \frac{A_2 \exp(2v^2\tau)v^{d-2}}{\tau} \qquad (9.1.24)$$

$$A_1 = \frac{3b\phi_o}{Kq_M^2} \cdot \frac{\pi^{1/2} \cdot d \cdot z_o}{2^{(d-2)/2}}, \quad A_2 = \frac{b}{Kq_M^2} \frac{\pi d^2 z_o^2}{2^{d-2}} \qquad (9.1.25)$$

Equation (9.1.23) plays the same role in the present analysis as the main equation in the Lifshitz-Slyozov theory,[1] by relating the supersaturation of the system to the radius of a nucleus.

Let us suppose for the sake of discussion that $C_* > 0$; the case $C_* < 0$ can then be considered in an equivalent manner. Solving Equation (9.1.23) as a quadratic equation, we find that the expression in brackets becomes

$$uC_*(q_*)S(q_*) \exp(Bq_*^4 t)\left(-\frac{\pi}{2At}\right)^{1/2} =$$

$$\left\{\left[\left(\frac{3b\phi_o}{K}\right)^2 + \frac{4b}{K}\left(q_M^2 - \frac{d(q_*^2 t)}{dt}\right)\right]^{1/2} - \frac{3b\phi_o}{K}\right\} \Big/ \frac{2b}{K} \qquad (9.1.26)$$

If in Equation (9.1.26) $d(q_*^2t)/dt$ the quantity is a rising function of time, then there is no solution for long periods. If this quantity tends to a constant non-zero value, then $q_*^2 t \sim t$, and $q_*^2 \to$ const in the limit $t \to \infty$, which again is physically meaningless. The solution (9.1.26) exists only if $d(q_*^2t)/dt$ decreases after a long time, which is equivalent to an asymptotic fall of q_*^2. But then the asymptotic behavior of the solution is governed by an equation which does not contain time derivatives:

$$uC_*(q_*)S(q_*, \exp(Bq_*^4 t)\left(-\frac{\pi}{2At}\right)^{1/2}$$

$$= \left\{\left[\left(\frac{3b\phi_o}{K}\right)^2 + \frac{4bq_M^2}{K}\right]^{1/2} - \frac{3b\phi_o}{K}\right\} \Big/ \frac{2b}{K} \qquad (9.1.27)$$

In the case of the adopted initial condition (9.1.21), Equation (9.1.27) is equivalent to

$$G(v,\tau) = 0$$

which yields to the equation

$$\frac{\exp(v^2\tau)\, v^{(d-2)/2}}{\tau^{1/2}} = [(A_1^2 + 4A_2)^{1/2} - A_1]/2A_2 = y \qquad (9.1.28)$$

The form of Equation (9.1.28) depends on the dimensionality d of the problem.

If d = 1, we can represent Equation (9.1.28) in the form

$$\frac{4}{y^4\tau} = \frac{g}{\exp g}, \; g = 4v^2\tau \qquad (9.1.29)$$

Then, for a sufficiently large value of τ the solution of Equation (9.1.29) is a Burmann-Lagrange series[8]

$$g = \sum_{n=1}^{\infty} \left(\frac{4}{y^4\tau}\right)^n \frac{n^{n-1}}{n!} \qquad (9.1.30)$$

Using the Stirling formula, we can transform the n-th term of the series (9.1.30) at large n values into

$$(4/y^4\tau)^n e^n/(2\pi n^3)^{1/2}$$

which shows that Equation (9.1.30) converges when

$$\tau > 4e/y^4 \qquad (9.1.31)$$

For these values of τ, we find that

$$g = \frac{4}{y^4\tau} + O\left(\frac{1}{\tau^2}\right) \qquad (9.1.32)$$

It follows from Equation (9.1.32) that at high values of τ the spatial size ξ of a structure increases in accordance with the law

$$\xi = \xi_M y\tau^{1/2} + O(\tau^{-1/2}), \; \xi_M = \frac{2\pi}{q_M} \qquad (9.1.33)$$

If d = 2, the solution of Equation (9.1.28) is exact:

$$v^2 = \ln(\tau y^2)/2\tau \qquad (9.1.34)$$

which exists for $\tau > y^{-2}$. If $\tau \gg y^{-2}$, the characteristic size of the structure is

$$\xi = \xi_M \cdot \left(\frac{2\tau}{\ln(\tau y^2)}\right)^{1/4} \cdot \left[1 + O\left(\frac{1}{\ln\tau}\right)\right] \qquad (9.1.35)$$

In logarithmic coordinates we have, instead of Equation (9.1.35),

$$\ln\frac{\xi}{\xi_M} = \frac{1}{4}\ln(2\tau)\left[1 + O\left(\frac{\ln\ln\tau}{\ln\tau}\right)\right]$$

If $d = 3$, we can represent Equation (9.1.28) in the form

$$g \exp g = 4y^4\tau^3 \qquad (9.1.36)$$

where g is defined by Equation (9.1.29). At high values of τ the solution (9.1.36) is[8]

$$g = \ln(4y^4\tau^3) - \ln\ln(4y^4\tau^3) + O\left(\frac{\ln\ln(4y^4\tau^3)}{\ln(4y^4\tau^3)}\right) \qquad (9.1.37)$$

The solution (9.1.37) is defined if $\tau \cdot (4y^4)^{1/3} > 1$. However, if $\tau \gg (4y^4)^{-1/3}$ we have

$$\xi = \xi_M \cdot \left[\frac{4\tau}{\ln(4y^4\tau^3)}\right]^{1/4} \cdot \left[1 + O\left(\frac{1}{\ln\tau}\right)\right] \qquad (9.1.38)$$

Using logarithmic coordinates, we find that Equation (9.1.38) becomes

$$\ln\frac{\xi}{\xi_M} = \frac{1}{4}\ln(4\tau)\left[1 + O\left(\frac{\ln\ln\tau}{\ln\tau}\right)\right]$$

Therefore, when an analysis is made using logarithmic coordinates, after a long time period the dependence of $\ln \xi$ on $\ln \tau$ approaches a straight line with a slope 1/2 if $d = 1$ and 1/4 if $d = 2$ or $d = 3$, and this slope is independent of the initial distribution of ϕ. The difference between the power exponent in the case of different dimensionalities d apparently reflects the fact that the boundaries of growing decomposition structures are planar if $d = 1$, but are curved if $d > 1$. If $d = 1$, we obtain a larger power exponent than others[1], whereas for $d > 1$ we obtain a smaller exponent.

It is clear from Equation (9.1.33) that the rate of growth of spinodal structures is different for $\phi = \phi_o$ and $\phi = -\phi_o$. This asymmetry of the solution is related to the fact that in both cases we use in Equation (9.1.25) the same quantity z_o from Equation (9.1.21) so that the corresponding initial distributions $\phi(\mathbf{r})$ are

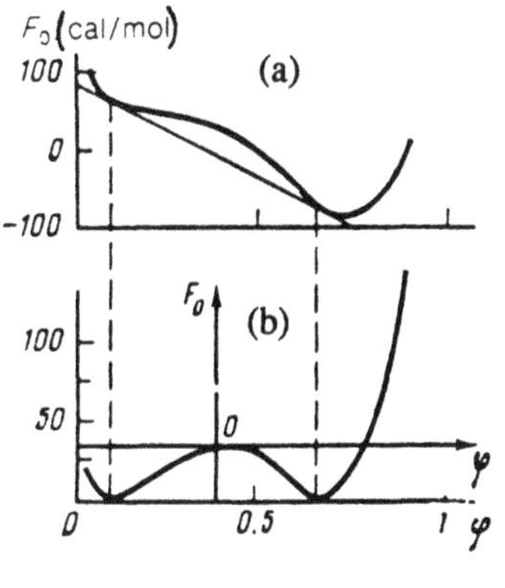

FIGURE 66. a: initial dependence $F_o(\phi)$; b: replotted dependence $F_o(\phi)$ representing along the ordinate the discrepancy between $F_o(\phi)$ and the general tangent to the curve in Figure 66a. The transition to this symmetric form occurs as a result of a shift of the axes in Figure 66b. This gives a function $a\phi^2 + b\phi^4$ which has minima at the points of minima of F_o and passes through the origin of the coordinates.

asymmetric relative to $\phi = 0$. The asymmetry disappears if opposite signs are attributed to z_o in Equation (9.1.25) for $\phi = \phi_o$ and $\phi = -\phi_o$.

These expressions can now be used to consider the problem of the influence of the rate of cooling of a solution on the kinetics of growth of decomposition structures. For such a problem we should bear in mind that A_1 and A_2 are known functions of temperature and the fact that temperature depends on time.[9]

Allowance for the asymmetry of $F_o(\phi)$ leads us to the problem of finding the roots of a polynomial of form (9.1.23) and at a degree higher than second. This simply alters the constant y in the resultant expressions. In general (if $\Lambda \neq$ const and $K \neq$ const) we should study equations of the (9.1.19) and (9.1.20) type.

Example: Decomposition in Feldspar

We shall now use the expressions obtained above to estimate the parameters of the kinetics of spinodal decomposition of alkali feldspars belonging to a high albite-sanidine series (solid solutions with the formula $NaAlSi_3O_8$- $KAlSi_3O_8$). The dependence of the free energy on the fraction of sanidine at T = 774 K at a pressure of 1 kbar is known,[10] and can be reduced to the symmetric Ginzburg-Landau form, as shown in Figure 66. The corresponding energy $a\phi^2 + b\phi^4$ converted to a value per a molecule of the mixture and normalized to k_BT is characterized by $a = -0.58$ and $b = 3.68$. The scale L, representing the size of a monomineral grain, is $\sim 10^4$ to 10^7 Å. The effective volume of a position of a molecule in the crystal lattice (volume of a unit cell) in feldspar is of the order of 700 Å at 774 K,[3] and the distance between the neighboring lattice sites is 7 to 12 Å.[10] The size ξ_M of the spinodal structures at the end of the linear stage of the process was determined experimentally earlier for the high albite-sanidine series and amounted to 75 ± 10 Å over a wide range of temperatures.[10]

The value of K, which in our derivation is independent of ϕ, can be estimated from the equality

$$\frac{4\pi^2}{\xi_M^2} = -\frac{a}{2K}$$

which gives 40 Å.[2] If we consider the case when $\phi_o = 0$ (transition to an unstable region via a critical point) we find that $y = A_2^{1/2}$. The value of z_o (representing the maximum deviation from the average composition at $\tau = 0$) is assumed to be 0.1; we also have $q_M = 2\pi/\xi_M$. We shall define the times at which the expressions obtained above are valid:

$$\nu_*(1) = \frac{4e}{y^4}, \ \nu_*(2) = \frac{e}{y^2}, \ \nu_*(3) = \frac{e}{(4y^4)^{1/3}} \tag{9.1.39}$$

We shall consider decomposition in a d-dimensional region with linear dimensions L, along each direction. The case $d = 1$ corresponds to coherent decomposition structures (cryptoperthites in the case of feldspars[10]) in the form of platelets; $d = 2$ corresponds to cylindrical structures and $d = 3$ represents three-dimensional, nearly spherical structures. These types of structure are encountered in crystalline solid solutions during precipitation at the postcrystallization stage. Precipitation of platelet, cylindrical, or spherical structures is due to the presence of certain specific most favorable crystallographic directions in the host lattice, and it is along these directions that the process of perturbation is developing[5] (also see the next section). Substitution of the data for feldspars gives the following values of y: 4.44 for $d = 1$, 3.15 for $d = 2$, and 2.97 for $d = 3$.

The expressions of ξ can be rewritten subject to Equation (9.1.39)

$$\xi = \xi_M \tau^{1/2}\left(\frac{4e}{\nu_*}\right)^{1/4}, \ d = 1; \ \xi = \xi_M\left(\frac{2\tau}{\ln(e\tau/\nu_*)}\right)^{1/4}, \ d = 2;$$
$$\xi = \xi_M\left(\frac{4\tau}{3\ln(e\tau/\nu_*)}\right)^{1/4}, \ d = 3 \tag{9.1.40}$$

Substituting the values of y, we find from Equation (9.1.40) that

$$\xi = \xi_M \cdot 4.44\,\tau^{1/2}, \ d = 1; \ \xi = \xi_M\left(\frac{2\tau}{2.3 + \ln\tau}\right)^{1/4}, \ d = 2;$$
$$\xi = \xi_M \cdot \left(\frac{4\tau}{3\ln\tau + 5.74}\right)^{1/4}, \ d = 3 \tag{9.1.41}$$

We must bear in mind that if $\tau \sim 1$, the integrand in Equation (9.1.11) does not yet have a sufficiently sharp maximum, so that expressions of the type given by Equation (9.1.41) can be employed only for values of τ substantially larger than 1.

Evolution of the Structure Factor of the System

Using the results of the above consideration, we shall obtain the asymptote of the change in the square of the Fourier component $\delta\phi_{\mathbf{q}}$, because this quantity is proportional to a so-called structure factor, which is measured in X-ray and neutron scattering experiments. Using Equation (9.1.7), we find that

$$\Gamma(\tilde{q}^2,\tau) = (\delta\phi_{\mathbf{q}})^2/C^2 = \exp[2\tau\tilde{q}^2(2v - \tilde{q}^2)],\ \tilde{q}^2 = \frac{q^2}{q_M^2} \qquad (9.1.42)$$

Substituting the expressions (9.1.33), (9.1.35), (9.1.38), and (9.1.39) in Equation (9.1.42), we obtain

$$\ln\Gamma = 2\tilde{q}^2[(\nu_*/e)^{1/2} - \tilde{q}^2\tau]\ ,\ d = 1$$

$$\ln\Gamma = 2\tilde{q}^2\ \{[2\tau\ln(\tau e/\nu_*)]^{1/2} - \tau\tilde{q}^2\}\ ,\ d = 2 \qquad (9.1.43)$$

$$\ln\Gamma = 2\tilde{q}^2\cdot\left\{\left[\tau\ln\left(\frac{\tau^3e^3/\nu_*^3}{\ln(\tau^3e^3/\nu_*^3)}\right)\right]^{1/2} - \tau\tilde{q}^2\right\}\ ,\ d = 3$$

The maximum value of $\ln\Gamma$ is equal to $2v^2\tau$ and it is attained at $\tilde{q}^2 = v$. We can see that it shifts with time toward longer wavelengths. It also follows from Equation (9.1.43) that if $d = 1$, the value of $\ln\Gamma$ at the maximum decreases as τ^{-1}, whereas for $d = 2$ and $d = 3$ it rises as $\ln\tau$. If $\tilde{q}^2 = 2v$, the value of Γ falls from its maximum to 1. The effective width of the spectrum is thus $(2v)^{1/2}$. If $d = 1$, it decreases as $(\ln\tau/\tau)^{1/4}$, whereas for $d = 3$ it decreases as $[\tau^{-1}\ln(\tau/\ln\tau)]^{1/4}$. The qualitative behavior of $\Gamma(\tilde{q}^2)$ for a fixed time is shown in Figure 67.

We shall compare the results obtained with one of the best-known theories of spinodal decomposition, proposed by Langer,[7] as well as with the results of mathematical modeling of spinodal decomposition in three-dimensional space by the methods of molecular dynamics. As pointed out in the beginning of this section, Langer generalized Cahn's theory and obtained an equation for the density of the statistical distribution over all possible configurations.[7] Certain consequences follow from this equation for various moments of the distribution function. In particular the "mean-field approximation" is obtained for the structure factor of the system $G_2(\mathbf{q},t)$ on the assumption that the density of the distribution is of Gaussian form with zero average. In this approximation the relevant equations are as follows:

$$\partial G_2(\mathbf{q},t)/\partial t = 2\Omega_L(\mathbf{q},t)G_2(\mathbf{q},t) \qquad (9.1.44)$$

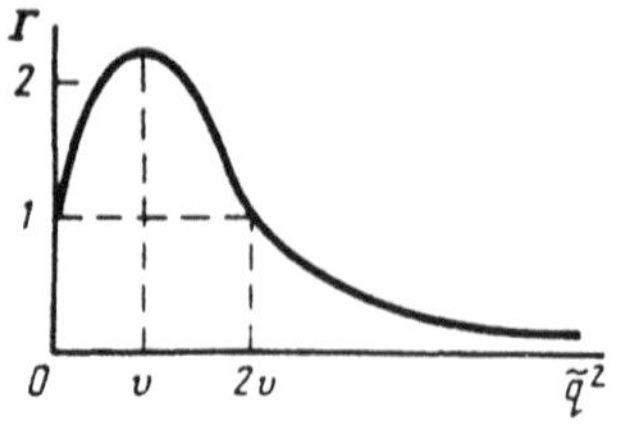

FIGURE 67. Dependence of the structure factor on the square on the dimensionless wave number.

$$\Omega_L(\mathbf{q},t) = -\Lambda q^2\left(\frac{\partial\mu_o}{\partial\phi}\bigg|_{\phi=\phi_o} + \frac{1}{2}\frac{\partial^3\mu_o}{\partial\phi^3}\bigg|_{\phi=\phi_o}\cdot\langle U^2\rangle + 2Kq^2\right) \quad (9.1.45)$$

$$\langle U^2\rangle = \frac{1}{(2\pi)^3}\int G_2(\mathbf{q},t)\,d\mathbf{q} \quad (9.1.46)$$

We can easily see that Equations (9.1.44) to (9.1.46) for G_2 represent a particular case of Equation (9.1.5) for $\delta\phi(\mathbf{q},t)$. The whole asymptotic analysis of this investigation can be repeated verbatim for Equations (9.1.44) to (9.1.46): the same results are obtained, apart from the constant y.

Allowance for the asymmetry of the fluctuations and for the influence of the thermal noise results in the generalization of Equation (9.1.44) to (9.1.46).[11] However, the asymptotic of the growth of structures is once again governed by an exponent close to 1/4. Figure 68a shows variation of the spatial size corresponding to a maximum of $G_2(\mathbf{q})$[11] (see also the review by Ustinovshchicov[12]). An analysis of the results can be carried out using the coordinates $\ln\xi$ and 1/4 $\ln t$. We can see that the slope of the line governing the change in the size of the structures is close to a straight line with a slope of 45°. This means that according to the Langer theory the structures grow with an exponent close to 1/4. Some deviation from the exact 1/4 law may occur because of a logarithmic correction in Equation (9.1.38).

Allowance for the thermal fluctuations in the Langer theory gives rise on the right-hand side of Equation (9.1.44) to the additional term

$$2\Lambda k_B T q^2 \quad (9.1.47)$$

As follows from the numerical analysis,[11] the deviation of the gain factor $(q^2G_2)^{-1}\partial G_2/\partial t$ from a linear dependence occurs mainly in the range of large values for q where the main contribution into the G_2 change comes from the thermal fluctuations and the gain factor dependence on q^2 reaches a horizontal asymptote. On the other hand, in the long-wavelength range the dominant term in the right side of Equation (9.1.44) is $2\Omega_L G_2$, describing the intrinsic nonlinear kinetics of the process. In the absence of the expression (9.1.47) in the theory described by Equations (9.1.44) to (9.1.46), the dependence of the gain factor on q^2 is linear, as expected on the basis of the above analysis (see Equation

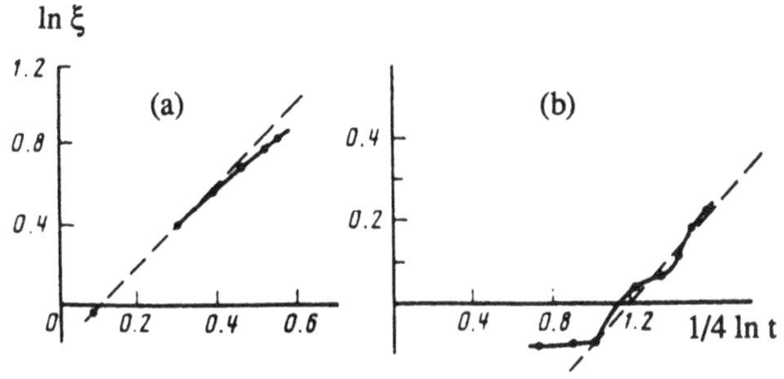

FIGURE 68. Comparison of the resultant asymptote (dashed line) with the results of calculations of the decomposition process obtained using the Langer theory (a); and with the results of modeling of the process by the methods of molecular dynamics (b).

[9.1.42]). In the two-dimensional calculations of the Cahn Equation (9.1.1) one also obtains the q^2-dependence of the gain factor consisting of two sections: linear decreasing in the long-wave region and constant in the short-range region.[13]

The logarithm of the maximum of the structure factor in the Langer theory increases as $\ln t$, which is also in agreement with our analysis of the multidimensional case.

To model spinodal decomposition in a binary alloy and to study the behavior of the structure factor, the Monte Carlo method has been used.[14] Figure 68b shows, using the coordinates $\ln\xi$ and $1/4 \ln t$, the evolution of the spatial size corresponding to the maximum of the structure factor. The first three points represent the linear stage of the process when the characteristic spatial size remains constant. This size then increases: it is clear from Figure 68b that the points, representing the calculations of Marro, et al.,[14] are grouped near a straight line with a slope of 45°. This implies the growth of structures with an exponent close to 1/4.

Calculations made by using another familiar theory of precipitation, proposed by Binder,[15] also predict growth of structures characterized by an exponent smaller than 1/3.

Thus, a comparison with the Langer theory and with the results of mathematical modeling of the process demonstrates a good agreement between the calculated asymptotes and the numerical results published earlier.

Validity of the Truncated Equation

We shall now consider in greater detail the validity of our truncation of Equation (9.1.1), corresponding to the approximation of point equations in the theory of autowave processes.[16] For simplicity, we shall consider the case when Λ = const, K = const. Then, in estimating the contribution of the diffusion of the modes to the intermode interaction we have to compare quantities

$$\delta\phi(q) \int \delta\phi(q) S(q)\, dq\, , \quad \frac{\partial^2 \delta\phi(q)}{\partial q^2} \int q^2 S(q)\, \delta\phi(q)\, dq \qquad (9.1.48)$$

The second quantity in Equation (9.1.48) corresponds to the first significant term dropped by transfer from Equation (9.1.1) to Equation (9.1.5). Substituting Equation (9.1.42) into Equation (9.1.48) and differentiating, we can see that at high values of τ we obtain

$$\frac{\partial^2 \delta\phi(q)}{\partial q^2} \int q^2 S(q)\, \delta\phi(q)\, dq / \delta\phi(q) \int \delta\phi(q) S(q)\, dq \sim v^2\tau \qquad (9.1.49)$$

If $d = 1$, we have $v^2\tau \sim \tau^{-1}$. Thus, in the one-dimensional case we can say that the calculated asymptote accurately describes the process after a long period of time. If $d > 1$, we obtain from Equation (9.1.49) that $v^2\tau \sim \ln\tau$. This means that in the multidimensional case the correction terms are generally not small compared with those already allowed for. It is quite clear that since the process in question is related to the collapse of the spectrum at long wavelengths, the diffusion of modes in q-space, controlled by the second term of Equation (9.1.48), should be manifested most strongly in the multidimensional case. Nevertheless, the study demonstrates a good agreement between the analytical results obtained above and those of the numerical calculations carried out using the familiar theories of spinodal decomposition. This may mean that mutual compensation of the dropped terms of higher orders is possible. On the other hand, one claims that the exponent 1/4 can be observed at the intermediate stages of structures growth, and one should expect a transfer to exponent 1/3 over longer time periods.[13] We will discuss this question further in the next section. Notice only that the analytic method proposed above provides asymptotically exact solutions of the "mean field approximation" in the theory of spinodal decomposition. It can also be used to analyze nonlinear integrodifferential equations of the type (9.1.5) in other areas of physics.

9.2. THEORY OF DIFFUSION-CONTROLLED PHASE TRANSITIONS IN AN ANISOTROPIC MEDIUM

The $d = 3$ case in the consideration of Section 9.1 corresponds to several physical systems (liquid mixtures, melts, and solid solutions) in the phase separation process. The $d < 3$ cases, discussed there, describe extremely anisotropic states of a solid solution, because all the characteristics depend only on the modulus of the wave vector. If $d = 1$, this should mean that decomposition along a certain straight line is preferred from the point of view of minimization of the elastic energy of a crystal rather than decomposition on a plane perpendicular to this straight line.[17] If $d = 2$, then energy considerations make it likely that a solid solution becomes stratified along two spatial directions. In general, solid solutions are anisotropic systems, to describe which it is necessary to generalize the expression (9.1.2). Namely, the anisotropy of decomposition should be manifested by an anisotropy of the scattering of particles passing through the crystal lattice of a solid solution. The corresponding generalization of the theory yields a free energy functional of the form[18,19]

$$F = \int \rho \left[F_o(\phi) + \sum_{i,l=1}^{d} K_{il} \frac{\partial \phi}{\partial r_i} \frac{\partial \phi}{\partial r_l} \right] dr \tag{9.2.1}$$

where

$$K_{il} = \frac{1}{4} \frac{\partial^2 (1/G_2)}{\partial q_i q_l} \tag{9.2.2}$$

whereas the quantity

$$G_2(\mathbf{q}) = \rho \int \mathbf{dr} \exp(-j\mathbf{qr}) \langle \delta\phi(\mathbf{r}) \delta\phi(0) \rangle \tag{9.2.3}$$

is proportional to the scattering cross section of particles transferring a momentum in a medium.[20,21] Based on its physical significance the quadratic form $(\mathbf{Kx},\mathbf{x})$ is positive-definite

$$(\mathbf{Kx},\mathbf{x}) \geq 0 \tag{9.2.4}$$

and the spectrum of the symmetric matrix $\mathbf{K}$ consists of positive real eigenvalues.

The necessity of an anisotropic correction in order to describe the decomposition of solid solutions stems primarily from the fact that the most general functional of one variable ϕ is expressed in the long-wavelength approximation in form (9.2.1), if we assume that F is an analytic function of ϕ and its spatial derivatives.[18,19] The possibility of describing the process with the help of only one variable ϕ (the order parameter) in turn indicates that the other parameters of the system (strains in the crystal, the components of the stress tensor, etc.), vary over much shorter spatial scales and with much shorter relaxation times. For this reason they can be expressed in terms of the leading mode (in the case under consideration, the concentration). This principle for excluding fast modes is widely employed in the theory of self-organization of dynamic systems,[22] and Cahn used it implicitly to construct a theory with one order parameter.[17]

An asymptotic analysis of Equation (9.1.1) with the potential (9.2.1), generalizing that described in Section 9.1, will be carried out in this section using a multidimensional version of the Laplace method.[23]

Linear Analysis: Stationary Points of the Amplification Factor

We study the initial stage of decomposition with the help of the linearized equation (9.1.1). Transforming in Equation (9.1.1), after linearization, to the Fourier components of the fluctuations about the average value ϕ_o, and taking into account Equations (9.2.1) to (9.2.4), we obtain

$$\frac{d\delta\phi(\mathbf{q})}{dt} = R(\mathbf{q})\delta\phi(\mathbf{q}) \ , \ R(\mathbf{q}) = -\Lambda q^2[h_s + 2(\mathbf{Kq},\mathbf{q})] \tag{9.2.5}$$

Here $\delta\phi(\mathbf{q})$ is the Fourier component of the fluctuation, $\mathbf{q}$ is the wave vector, and $h_s = \partial^2 F_o/\partial\phi$. For $h_s > 0$ the Fourier components $\delta\phi(\mathbf{q})$ decay for all $\mathbf{q}$. For $h_s < 0$ the Fourier components corresponding to the condition

$$2(\mathbf{K}\mathbf{q},\mathbf{q}) < -h_s \qquad (9.2.6)$$

grow; modes for which $\mathbf{q}$ lies outside the ellipsoid (9.2.6) decay. The mode of maximum growth is determined by the condition

$$\frac{\partial R}{\partial q_\kappa} = 0\,, \qquad \kappa = 1, \ldots, d \qquad (9.2.7)$$

The symmetry of $\mathbf{K}$ makes it possible to write the condition (9.2.7) in the form

$$2q^2\mathbf{K}\mathbf{q} = -[h_s + 2(\mathbf{K}\mathbf{q},\mathbf{q})]\mathbf{q} \qquad (9.2.8)$$

From here we obtain the solution $\mathbf{q} = 0$. If $\mathbf{q} \neq 0$, then, multiplying Equation (9.2.8) by $\mathbf{b}_\kappa$ — the eigenvector of the matrix K corresponding to the eigenvalue λ_κ — we obtain

$$[2q^2\lambda_\kappa + h_s + 2(\mathbf{K}\mathbf{q},\mathbf{q})](\mathbf{q},\mathbf{b}_\kappa) = 0\,, \qquad \kappa = 1, \ldots, d \qquad (9.2.9)$$

The case of the general position corresponds to the case in which all eigenvalues of the matrix $\mathbf{K}$ nondegenerate. In this case there are several possibilities for solving Equation (9.2.9) for $\mathbf{q}$. The relationship

$$(\mathbf{q},\mathbf{b}_\kappa) = 0 \qquad (9.2.10)$$

cannot hold for all κ if $\mathbf{q} \neq 0$, since the vectors $\mathbf{b}_\kappa$ form an orthogonal basis.[24] If in the condition (9.2.10) is violated for only one value of κ, then because $\mathbf{b}_\kappa$ and $\mathbf{b}_l$ are orthogonal for $K \neq l$ the eigenvector $\mathbf{b}_l$, determined to within a constant factor, will be a solution of the system (9.1.9). The length of $\mathbf{b}_l$ is given by the relationship

$$2q^2\lambda_l + h_s + 2(\mathbf{K}\mathbf{q},\mathbf{q}) = 0 \qquad (9.2.11)$$

Since $(\mathbf{K}\mathbf{q},\mathbf{q}) = \lambda_l q^2$, Equation (9.2.11) can be rewritten in the form

$$q^2 = -h_s/4\lambda_l \qquad (9.2.12)$$

Finally, if the vector $\mathbf{q}$ is orthogonal to less than $d - 1$ eigenvectors $\mathbf{b}_\kappa$, then at least two of the d relationships (9.2.9) can be written in the form (9.2.12). By assumption, all λ_κ are nondegenerate, so that the last possibility cannot materialize.

Thus, except at $\mathbf{q} = 0$, the function $R(\mathbf{q})$ has another d stationary points corresponding to the eigenvectors of the matrix $\mathbf{K}$ with the normalization Equation (9.2.12). The value of the amplification factor at these points is

$$R(\mathbf{q}) = 2\Lambda\lambda_l q^4 = \Lambda h_s^2/8\lambda_l, \quad \lambda_1 < \lambda_2 < \ldots \tag{9.2.13}$$

i.e., smaller values of $R(\mathbf{q})$ correspond to larger values of λ. Comparing Equations (9.2.6) and (9.2.13) one can see that all stationary points lie inside the region $\{\mathbf{q}: R(\mathbf{q}) \geq 0\}$.

If the eigenvalues of the matrix $\mathbf{K}$ are degenerate, the wave vector space must be represented as a direct sum of the invariant subspaces Q_l of the eigenvectors of the matrix $\mathbf{K}$ corresponding to different values of Λ_l. As is well known, vectors from the different subspaces Q_l are orthogonal.[24] It can be shown that the stationary points of the function $R(\mathbf{q})$, except $\mathbf{q} = 0$, lie in the intersection of Q_l and the sphere Z_l determined by the relationship (9.2.12). For stationary points $\mathbf{q} \in Q_l \cap Z_l$, $R(\mathbf{q})$ is constant and given by Equation (9.2.13).

In the case of the eigenvalue with degeneracy d any vector is an eigenvector of $\mathbf{K}$ and the appropriate vectors lie on the sphere (9.2.12). This is not the general case for anisotropic systems, where $R(\mathbf{q})$ has only d stationary points.

Kinetically Favored Directions of Decomposition

To determine which solutions of Equation (9.1.7) give the function $R(q)$ an extremum we shall study the quadratic form

$$(\mathbf{B}(\mathbf{q})\mathbf{x}, \mathbf{x}) = \sum_{\kappa,l=1}^{d} B_{\kappa l}(\mathbf{q}) x_\kappa x_l, \quad B_{\kappa l} = \frac{1}{\Lambda} \frac{\partial^2 R}{\partial q_\kappa \partial q_l} \tag{9.2.14}$$

Differentiating $R(\mathbf{q})$ we obtain from Equation (9.2.14)

$$\begin{aligned} B_{\kappa l} = &- 2h_s\delta_{\kappa l} - 4K_{\kappa l}q^2 - 4\delta_{\kappa l}(\mathbf{Kq}, \mathbf{q}) - 8q_\kappa \sum_{i=1}^{d} K_{li} q_i \\ &- 8q_l \sum_{i=1}^{d} K_{\kappa i} q_i \end{aligned} \tag{9.2.15}$$

where $\delta_{\kappa l}$ is the Kronecker symbol. One can see from Equation (9.2.15) that the amplification factor $R(\mathbf{q})$ is minimum when $\mathbf{q} = 0$, since

$$(\mathbf{B}(0)\mathbf{x}, \mathbf{x}) = -2h_s x^2 > 0, \qquad \text{if} \quad \mathbf{x} \neq 0$$

Let the stationary point being considered as an extremum lie in the intersection $Q_l \cap Z_l$. Substituting into Equation (9.2.15) the expression (9.2.11) for h, we obtain with the help of the properties of an eigenvector

$$\sum_{\kappa=1}^{d} K_{i\kappa} q_\kappa = \lambda_l q_i, \quad (\mathbf{Kq}, \mathbf{q}) = \lambda_l q^2 \tag{9.2.16}$$

the following expression for $(\mathbf{B}(\mathbf{q})\mathbf{x},\mathbf{x})$:

$$(\mathbf{B}(\mathbf{q})\mathbf{x},\mathbf{x}) = 4\lambda_l q^2 x^2 - 4q^2(\mathbf{K}\mathbf{x},\mathbf{x}) - 16\lambda_l(\mathbf{q},\mathbf{x})^2 \qquad (9.2.17)$$

In this circumstance we shall assume that the magnitude of the vector **x** is equal to 1. This is not important for the subsequent check, since the function (9.2.17) is a homogeneous function of **q** and **x**. We shall study first the case $\lambda_l = \lambda_1$, where λ_1 is the smallest eigenvalue in the spectrum. The fact that λ_1 is the minimum eigenvalue implies[24]

$$(\mathbf{K}\mathbf{x},\mathbf{x}) \geq \lambda_1$$

(e.g., see Gelfand[24]), whence we have

$$(\mathbf{B}(\mathbf{q})\mathbf{x},\mathbf{x}) \leq -16\lambda_1(\mathbf{q},\mathbf{x})^2$$

The quadratic form $(\mathbf{K}\mathbf{x},\mathbf{x})$ is minimum if, and only if, **x** is the eigenvector corresponding to the eigenvalue λ_1. Hence if λ_1 is not degenerate, the form (9.2.17) only takes on negative values. Indeed, if **x** is not an eigenvector corresponding to λ_1, then the sum of the first two terms on the right side of Equation (9.2.17) is strictly less than zero. If **x** is such an eigenvector, then the sum of these terms is equal to zero, but because if **q** and **x** are collinear the third term is less than zero. Hence if λ_1 is not degenerate, the function R(**q**) has an unconditional maximum at the corresponding stationary point, it is evident from Equation (9.2.13) that this maximum will be a global maximum. If λ_1 is degenerate, then a global maximum is reached at all points in $Q_1 \cap Z_1$ and it will be a conditional maximum.

The stationary points corresponding to λ_l with $l > 1$ can be studied in an analogous manner. Analyzing Equation (9.2.17) it can be shown that at these points R(**q**) does not have a local maximum. The directions of **x** in which the form (9.2.17) is positive are determined by the eigenvectors from Q_i with $i < l$; the directions in which the form (9.2.17) is negative are determined by the vectors from Q_i with $i > l$ and from the set $\mathbf{x} \in Q_l$, $(\mathbf{q},\mathbf{x}) \neq 0$.

Thus the amplification factor R(**q**) has one local minimum at $\mathbf{q} = 0$ and one global maximum in the set $Q_l \cap Z_l$; all remaining stationary points will be saddle points. This shows that in the anisotropic case of the general position both the kinetically favored spatial size of the structures and the direction of decomposition are determined at the initial stage (in contrast to the isotropic case when all modes whose wave vectors lie on the sphere (9.2.12) grow with the same amplification factor).

Nonlinear Stage of Decomposition

Following the consideration in previous section, we shall use a truncated form of Equation (9.1.1) to study the later stage of anisotropic decomposition (the coalescence stage, at which the structures grow). This equation is constructed by rewriting Equation (9.1.1) with the potential (9.2.1) in terms of the Fourier coefficients of the fluctuations ϕ, expanding in the Taylor series integrals of the type

$$\int \delta\phi(\mathbf{q} - \mathbf{q}_2 - \dots - \mathbf{q}_n)\delta\phi(\mathbf{q}_2) \dots \delta\phi(\mathbf{q}_n)d\mathbf{q}_2 \dots d\mathbf{q}_n$$

which arise in the nonlinear terms, and dropping all terms containing derivatives with respect to $\mathbf{q}$. In the previous section it was shown that going from Cahn's equation to the truncated form is equivalent to the "mean-field approximation" in the well-known nonlinear theory of spinodal decomposition;[7] the exact asymptotic solutions obtained agree with some earlier numerical results.

We study the decomposition process in a region in the shape of a rectangular parallelipiped of volume V with impermeable boundaries. The truncated equation has the form

$$\frac{d\delta\phi(\mathbf{q})}{dt} = -q^2\Lambda(\phi_o + z)\left[\frac{\mu_o(\phi_o + z) - \mu_o(\phi_o)}{z} + 2(\mathbf{K}(\phi_o + z)\mathbf{q},\mathbf{q})\right]\delta\phi(\mathbf{q}) \tag{9.2.18}$$

$$z = u\int \delta\phi(\mathbf{q})d\mathbf{q} = \delta\phi(\mathbf{r})\big|_{\mathbf{r}=0}\,,\quad u = V/(2\pi)^d \tag{9.2.19}$$

We represent Equation (9.2.18) in the form

$$\delta\phi(\mathbf{q}) = C(\mathbf{q})\cdot\exp\left\{-q^2\int_0^t \Lambda(\phi_o + z)\cdot\left[\frac{\mu_o(\phi_o + z) - \mu_o(\phi_o)}{z} + 2(K(\phi_o + z)\mathbf{q},\mathbf{q})\right]dt\right\}$$

where $C(\mathbf{q}) = \delta\phi(\mathbf{q})|_{t=0}$. Integrating over $\mathbf{q}$ we obtain by using Equation (9.2.19)

$$z = u\int C(\mathbf{q})\exp\left\{-q^2\int_0^t \Lambda(\phi_o + z)\cdot\left[\frac{\mu_o(\phi_o + z) - \mu_o(\phi_o)}{z} + 2(K(\phi_o + z)\mathbf{q},\mathbf{q})\right]dt\right\}d\mathbf{q} \tag{9.2.20}$$

Let us study Equation (9.2.20) at times much longer than the duration of the linear stage of the process. We write Equation (9.2.20) in the form

$$z = u\int C(\mathbf{q})\exp\{-[Aq^2 + q^2(\mathbf{D}\mathbf{q},\mathbf{q})]t\}d\mathbf{q} \tag{9.2.21}$$

where

$$A = \frac{1}{t}\int_0^t \Lambda(\phi_o + z)\,\frac{\mu_o(\phi_o + z) - \mu_o(\phi_o)}{z}\,dt$$

$$\mathbf{D} = \frac{2}{t}\int_0^t \Lambda(\phi_o + z)\mathbf{K}(\phi_o + z)\,dt \qquad (9.2.22)$$

The form $(\mathbf{Dq},\mathbf{q})$ is positive definite since $(\mathbf{Kq},\mathbf{q})$ is positive definite and Λ is positive; generally speaking, A can be positive or negative. Below, as in the previous section, we shall be interested in the solution of Equation (9.2.21) for $A < 0$, since this case is always seen first and corresponds to the motion of the maximum of the fluctuation spectrum in the region of small $\mathbf{q}$.

We note that the full Equation (9.1.1) also can be replaced by an equation in the form (9.2.21), but A will be determined in terms of expressions of the kind

$$\left[\int \delta\phi(\mathbf{q} - \mathbf{q}_2 - \ldots - \mathbf{q}_n)\delta\phi(\mathbf{q}_2)\ldots\delta\phi(\mathbf{q}_n)d\mathbf{q}_2\ldots d\mathbf{q}_n\right] \Big/ \delta\phi(\mathbf{q})$$

Our approximation (9.2.18) means that these $\mathbf{q}$- and t-dependent expressions are replaced by a function of t only.

We shall now apply to the integral on the right side of Equation (9.2.21) the multidimensional Laplace method.[8] In so doing we shall employ the results of the above linear analysis of anisotropic models (9.1.1) and (9.1.2), since the form of the function $R(\mathbf{q})$ is similar to that of the argument of the exponential function in Equation (9.2.21) and the time variable in Equation (9.2.21) is just an external parameter. In the case of the general position the spectrum of the matrix $\mathbf{D}$ consists of nondegenerate eigenvalues. The argument of the exponential in Equation (9.2.21) assumes a maximum value for the eigenvector $\mathbf{q}_*$ corresponding to the minimum eigenvalue λ_D in the spectrum of the matrix D and having the length $q_*^2 = -A/2\lambda_D$. The asymptotic value of the integral in Equation (9.2.21) is given by the expression

$$z = 2\,\frac{uC(\mathbf{q}_*)\,\exp(A^2t/4\lambda_D)(2\pi/t)^{d/2}}{|\det \boldsymbol{I}_{\mathbf{qq}}\,(\mathbf{q}_*)|^{1/2}} \qquad (9.2.23)$$

where

$$I(\mathbf{q}) = -\,q^2[A + (\mathbf{Dq},\mathbf{q})]$$

and $\boldsymbol{I}_{\mathbf{qq}}$ is the matrix of the second derivatives of the function I with respect to the components of the wave vector — the multiplier 2 appears in Equation (9.2.23) because there are two symmetric maxima, one at $\mathbf{q}_*$ and another at

$-\mathbf{q}_*$. Using Equations (9.2.15) and (9.2.16) we obtain

$$I_{qq}(\mathbf{q}_*) = 2q_*^2(-\mathbf{D} + \lambda_D\mathbf{E} - 4\lambda_D\mathbf{P}) , P_{\kappa l} = \frac{q_{*,\kappa} \cdot q_{*,l}}{q_*^2}$$

$$\det I_{qq}(\mathbf{q}_*) = 2^d q_*^{2d} \det(\lambda_D\mathbf{E} - \mathbf{D} - 4\lambda_D\mathbf{P}) \quad (9.2.24)$$

The remaining analysis is simplest for the case of ϕ-independent Λ and $\mathbf{K}$. In this case the elements of the matrix $\mathbf{P}$, expressed in terms of the direction cosines of the vector $\mathbf{q}_*$, are constant, just like the elements of the matrix $\mathbf{D}$ and the quantity Λ. Introducing the notation

$$\sigma_D = |\det(\lambda_D\mathbf{E} - \mathbf{D} - 4\lambda_D\mathbf{P})|$$

we obtain instead of Equation (9.2.23), by using Equation (9.2.24),

$$z = u_1C(\mathbf{q}_*) \exp(A^2t/4\lambda_D)(-\pi/At)^{d/2} \quad (9.2.25)$$

where $u_1 = 2u(2\lambda_D)^{d/2}/\sigma_D^{1/2}$. Particularly in the one-dimensional case we have $\det\mathbf{P} = 1$, $\sigma_D = 4\lambda_D$, $\lambda_D = 2\Lambda K$, $u_1 = u/2^{1/2}$ and Equation (9.2.25) goes over to Equation (9.1.13) from the previous section.

An equation for the squared wave vector $\mathbf{q}_\kappa$, corresponding to the maximum of the fluctuation spectrum of the anisotropic system, can be derived from Equation (9.2.25).

Asymptotic Growth of Structures

We again study the case $F_o = a\phi^2 + b\phi^4$, which corresponds to the Ginzburg-Landau form of the free energy functional extended to the anisotropic case:

$$F = \int \rho[a\phi^2 + b\phi^4 + (\mathbf{K}\nabla\phi,\nabla\phi)]\,d\mathbf{r} \quad (9.2.26)$$

The importance of considering this form was discussed in the previous section. From Equations (9.2.26) and (9.2.22) and the definition of q_*^2 we obtain

$$q_*^2 = q_M^2 - \frac{3b\phi_o}{\lambda_1 t}\int_0^t z\,dt - \frac{b}{\lambda_1 t}\int_0^t z^2 dt \quad (9.2.27)$$

Here λ_1, as above, is the minimum eigenvalue in the spectrum of the matrix $\mathbf{K}$, $\lambda_D = 2\Lambda\lambda_1$, and $q_M^2 = -h_s/4\lambda_1$. We rewrite Equation (9.2.27) in the form

$$\frac{d}{dt}(tq_*^2) = q_M^2 - \frac{3b\phi_o}{\lambda_1}z - \frac{b}{\lambda_1}z^2 \quad (9.2.28)$$

Solving Equation (9.2.28) for z, using Equation (9.2.25), we obtain

$$z = u_1 C(q_*) \exp\left(\frac{A^2 t}{8\lambda_1 \Lambda}\right)\left(-\frac{\pi}{At}\right)^{1/2} = \left\{-\frac{3b\phi_o}{\lambda_1}\right.$$

$$\left.\pm\left[\left(\frac{3b\phi_o}{\lambda_1}\right)^2 + \frac{4b}{\lambda_1}\left(q_M^2 - \frac{d}{dt}(q_*^2 t)\right)\right]^{1/2}\right\} \bigg/ \frac{2b}{\lambda_1} \qquad (9.2.29)$$

The "+" sign in front of the square root in Equation (9.2.29) corresponds to $C > 0$ and the "−" sign corresponds to $C < 0$. If in Equation (9.2.29) the quantity $d(q_*^2 t)/dt$ is an increasing function of time, then at long times a solution does not exist. $d(q_*^2 t)/dt$ approaches a constant non-zero value, then $q_*^2 t \sim t$ and in the limit $t \to \infty$ the quantity $q_*^2 \to \text{const} \neq 0$, which is also physically meaningless. The solution of Equation (9.2.29) exists when at long times $d(q_*^2 t)/dt$ decreases to zero. This is equivalent to q_*^2 decreasing. But then the asymptotic behavior of the solution is determined by an equation with no derivatives

$$u_1 C(q_*) \exp\left(\frac{A^2 t}{8\lambda_1 \Lambda}\right)\left(-\frac{\pi}{At}\right)^{d/2} = g_\pm$$

$$g_\pm = \left\{-\frac{3b\phi_o}{\lambda_1} \pm \left[\left(\frac{3b\phi_o}{\lambda_1}\right)^2 + \frac{4b}{\lambda_1} q_M^2\right]^{1/2}\right\} \bigg/ \frac{2b}{\lambda_1} \qquad (9.2.30)$$

As noted above, at the initial stage of decomposition the short-wavelength part of the spectrum of fluctuations is suppressed. The decomposition process is concentrated in the long-wavelength region; this can be taken into account with the leap of the initial condition, constant in the sphere $q^2 < 2q_M^2$ and equal to zero outside it. The quantity C is determined from the normalization condition $z|_{t=0} = z_o = u \int C d\mathbf{q}$, so that

$$C = z_o / u q_M^d 2^{d/2} \gamma_d \qquad (9.2.31)$$

where γ_d is equal to 2, π, and $4\pi/3$ for $d = 1, 2$, and 3, respectively. Let us discuss for an example the case $C > 0$.

Transferring to the dimensionless variables $v = q_*^2/q_M^2$ and $\tau = 2q_M^4 \Lambda \lambda_1 t$ (putting t into a dimensionless form in which the duration of the initial stage of the process corresponds to unity), we obtain

$$\frac{dv}{d\tau} = \frac{G(v,\tau) - v}{\tau}, \quad G = 1 - A_1 \frac{\exp(v^2\tau)}{(v\tau)^{d/2}} - A_2 \frac{\exp(2v^2\tau)}{(v\tau)^d}$$

$$A_1 = \frac{3b\phi_o}{\lambda_1 q_M^2} \cdot \frac{z_o \pi^{d/2} \lambda_1^{d/2}}{\gamma_d \nu_1^{1/2} 2^{(d-2)/2}}, \quad A_2 = \frac{b}{\lambda_1 q_M^2} \frac{z_o^2 \lambda_1^d \pi^d}{\nu_1 \gamma_d^2 2^{d-2}}$$

where

$$\nu_1 = |\det(\lambda_1\mathbf{E} - \mathbf{K} - 4\lambda_1\mathbf{P})|$$

The condition (9.2.30) is equivalent to the Equation $G(v,\tau) = 0$, whence

$$\exp(v^2\tau)/(v\tau)^{d/2} = \left[\left(\frac{1}{4}A_1^2 + A_2\right)^{1/2} - \frac{1}{2}A_1\right] \Big/ A_2 = y \quad (9.2.32)$$

Raising Equation (9.2.32) to the power 4/d, we write Equation (9.2.32) in the form

$$\frac{g}{\exp g} = \frac{4}{d \cdot y^{4/d}\tau}, \; g = \frac{4}{d}v^2\tau \quad (9.2.33)$$

For sufficiently large τ the solution of Equation (9.2.33) is given by the Burmann-Lagrange series[8]

$$g = \sum_{n=0}^{\infty} \left(\frac{4}{dy^{4/d}\tau}\right)^n \frac{n^{n-1}}{n!} \quad (9.2.34)$$

The series (9.2.34) converges for $\tau > 4e/dy^{4/d}$ and then

$$v^2 = y^{-4/d} \cdot \tau^{-2} + O(\tau^{-3})$$

The spatial size ξ of the decomposition structures grows as

$$\xi = \xi_M \tau^{1/2} y^{1/4} + O(\tau^{-1/2}), \; \xi_M = 2\pi/q_M \quad (9.2.35)$$

Thus for any d the growth exponent of the structures in an anisotropic system is equal to 1/2. However as the degeneracy of λ_1 increases the asymptotic behavior changes. If the degeneracy is equal to m, then the general form of Equation (9.2.32) will be

$$\frac{\exp(v^2\tau)\,v^{(m-1)/2}}{(v\tau)^{(d-m+1)/2}} = \frac{\exp(v^2\tau)}{v^{(d-2m+2)/2}\tau^{(d-m+1)/2}} = y \quad (9.2.36)$$

Indeed, if we choose in the wave space a basis consisting of the eigenvectors of the matrix **K**, transform in the subspace Q_1 to the coordinates

$$q_\lambda = \left(\sum_{j=1}^{m} q_j^2\right)^{1/2}, \; \beta_1, \ldots, \beta_{m-1}$$

(β_j are angular variables), and integrate over β_j, then the integral in Equation (9.2.21) is transformed into an integral over a space with dimension $d - m + 1$. The minimum eigenvalue of the matrix **K** in the new space is nondegenerate. Using the scheme described in this section we obtain Equation (9.2.36) instead Equation (9.2.32), and the factor $v^{(m-1)/2}$ appears in the numerator of Equation (9.2.36), because on transforming to a lower-dimensional space the surface area of an m-dimensional sphere with radius q_λ appears in the integrand. It can be shown that for $d - 2m + 2 > 0$ the quantity $\xi = 2\pi/q_*$ grows with the exponent

$$\frac{(d - m + 1)}{2d - 4m + 4}$$

For $d - 2m + 2 \le 0$ the quantity ξ grows as

$$(\tau/\ln\tau)^{1/4}$$

i.e., the structures grow with an exponent of 1/4. The last case is close to the isotropic ($m = d$) case studied in the previous section; we should recall that the following results were obtained there in the multidimensional case

$$\begin{aligned} \xi &= \xi_M[2\tau/\ln(\tau y^2)]^{1/4} \cdot [1 + O(1/\ln\tau)]\ ,\ d = 2 \\ \xi &= \xi_M[4\tau/\ln(4y^4\tau^3)]^{1/4} \cdot [1 + O(1/\ln\tau)]\ ,\ d = 3 \end{aligned} \qquad (9.2.37)$$

Indeed, let us show how to solve Equation (9.2.36). We write Equation (9.2.36) in the form

$$\frac{\exp(v^2\tau)}{(v^2\tau)^{(d-2m+2)/4}} = y\tau^{d/4}$$

Next, three cases are distinguished. For $d - 2m + 2 = 0$, Equation (9.2.36) can be solved exactly:

$$v^2 = \tau^{-1}\ln(y\tau^{d/4})$$

For $d - 2m + 2 > 0$, introducing $g = 4v^2\tau/(d - 2m + 2)$ we obtain instead of Equation (9.2.36)

$$\frac{g}{\exp g} = \frac{4}{d - 2m + 2} y^{-4/(d-2m+2)} \cdot \tau^{-d/(d-2m+2)}$$

This equation is solved by representing the solution in form of a Burmann-Lagrange series, as Equation (9.2.33); the leading term in such a asymptotic expansion is

$$v^2 = y^{-4/(d-2m+2)} \cdot \tau^{-(2d-2m+2)/(d-2m+2)}$$

For $d - 2m + 2 < 0$, introducing $g = 4v^2\tau/(2m - d - 2)$, we obtain from Equation (9.2.36)

$$g \exp g = \frac{4}{2m - d - 2} y^{4/(2m-d-2)} \cdot \tau^{d/(2m-d-2)} = f$$

This equation is equivalent to Equation (9.1.28) at $d = 3$, and its solution has the following leading term

$$v^2 = \frac{2m - d - 2}{4\tau} (\ln f + \ln \ln f)$$

One can see from the solutions presented that for $d = 3$, the exponent in the growth law is equal to 1/2 for $m = 1$, 1 for $m = 2$, and 1/4 for $m = 3$. Thus, as one goes from an anisotropic situation ($m = 1$) to an isotropic situation ($m = 3$) the exponent changes from 1/2 to 1/4 nonmonotonically. In other words, for every d there exists a critical dimension of the manifold of wave vectors that correspond to modes of maximum growth,

$$m_* = (d + 2)/2$$

which marks the boundary between decomposition with a "quasi-isotropic" asymptotic law ($m \geq m_*$, the exponent is equal to 1/4) and decomposition with an "anisotropic" asymptotic law ($m < m_*$, the exponent is greater than or equal to 1/2).

It follows from the results obtained above that the anisotropic decomposition in a multidimensional system proceeds generally as a quasi one-dimensional process and predominantly in the direct $\mathbf{q}_*$. We also note that isotropic decomposition is slower than anisotropic decomposition. This is connected with the quasi one-dimensionality of the anisotropic decomposition, in which there is virtually no competition between decomposition in the most favorable spatial direction and in other directions. This situation is different from the isotropic situation, in which modes with wave vectors of the same length are equally suitable irrespective of the direction of the vector.

Averaging Over the Initial Conditions

Equations (9.2.35) and (9.2.37) contain the quantity z_o, which is the maximum deviation of the composition in the sample (monomineral grain) from the average value at $t = 0$. Estimates of z_o may not be available or they may not be very accurate, so it is more convenient to represent the asymptotic expressions as an average over an ensemble of equilibrium initial distributions of ϕ. Assume that initially the system has a temperature T_o above the spinodal, and decomposition occurs by means of quite rapid cooling to the temperature T. We average over all possible distributions of the form (9.2.31). Using dimensionless variables defined above and taking into account Equation (9.2.31), we rewrite Equation (9.2.30) in the form

$$\frac{u_2^2}{g_\pm^2} C^2 = \frac{(v\tau)^d}{\exp(2v^2\tau)}, \quad u_2 = \frac{2\pi^{d/2} \cdot q_M^d \lambda_1^d u}{\nu_1^{1/2}} \tag{9.2.38}$$

The Gibbs distribution at the temperature T_o sufficiently far from the thermodynamic instability region has the form[19]

$$w_g \sim \exp\left\{ - \int \rho[(a_o + 6b_o\phi_o^2)(\delta\phi)^2 + (\mathbf{K}_o\nabla\delta\phi, \nabla\delta\phi)]\,d\mathbf{r} \right\}$$

which is obtained from Equation (9.2.26) (the fluctuations magnitude is small far from the spinodal, and we should hold in Equation (9.2.26) the terms up to the quadratic ones). The zero index here denotes the quantities a, b, and **K** at T_o. Substituting Equation (9.2.31), we obtain

$$w_g \sim \exp(\rho V u C^2 \gamma_d 2^{d/2} q_M^d \alpha_g)$$

$$\alpha_g = \int [a_o + 6b_o\phi_o^2 + (\mathbf{K}_o\mathbf{q}, \mathbf{q})]\,d\mathbf{q} \Big/ \int d\mathbf{q} \tag{9.2.39}$$

$$= a_o + 6b_o\phi_o^2 + 2q_M^2 \cdot \left(\sum_{j=1}^{d} \lambda_{j,o}\right) \Big/ (d + 2)$$

Averaging Equation (9.2.38) over the distribution (9.2.39), allowing for the fact that $g_\pm = g_+$, if $C > 0$ and $g_\pm = g_-$, if $C < 0$ we obtain after transformations the equation

$$\exp(v_*^2\tau)/(v_*\tau)^{d/2} = y_*$$

where v_* is the average of v obtained in the manner indicated,

$$y_* = \left(\frac{g_+^2 g_-^2}{g_+^2 + g_-^2} \frac{2^{(3d+2)/2} \cdot \rho\gamma_d\nu_1\alpha_g}{q_M^d \lambda_1^d} \right)^{1/2}$$

$$\frac{g_+^2 g_-^2}{g_+^2 + g_-^2} = \frac{2q_M^2}{(3b\phi_o/\lambda_1)^2 + 2bq_M^2/\lambda_1} \tag{9.2.40}$$

Correspondingly, Equation (9.2.35) assumes the form

$$\xi_* = \xi_M \tau^{1/2} y_*^{1/d}$$

If the determinant ν_1 in Equation (9.2.40) is estimated to be of the order of λ_1^d in order of magnitude and the initial state is not too close to the spinodal, then Equation (9.2.40) can be simplified as follows:

$$y_* = \left[\frac{g_+^2 g_-^2}{g_+^2 + g_-^2} \frac{2^{(3d+2)/2} \cdot (a_o + 6b_o\phi_o^2)\rho\gamma d}{q_M^d}\right]^{1/2}$$

$$q_M^2 = -\frac{a + 6b\phi_o^2}{2\lambda_1} \qquad (9.2.41)$$

After similar averaging in the isotropic case the quantity y in Equation (9.2.37) must be replaced by

$$y_* = \left(\frac{g_+^2 g_-^2}{g_+^2 + g_-^2} \frac{2^{(3d+6)/2} \cdot \pi^{d-1} \cdot \rho\alpha_g}{\gamma_d d^2 q_M^d}\right)^{1/2}$$

$$\alpha_g = a_o + 6b_o\phi_o^2 + 2dKq_M^2/(d + 2) \qquad (9.2.42)$$

and in the definition (9.2.40) of $g_\pm$ the quantity λ_1 must be replaced everywhere by K.

It is important to explain more precisely what the case $d < 3$ studied above means. For $d = 1$, the initial conditions for Equation (9.1.1) are such (in the three-dimensional space) that they do not depend on two spatial coordinates. For $d = 2$, the initial conditions remain constant along one coordinate axis. Accordingly, for $d = 1$ the quantity ρ in Equation (9.2.2) is the linear density of lattice sites along the coordinate axis along which the initial condition varies, and it has the dimension of inverse length. For $d = 2$, ρ has the dimension of inverse length squared. In principle, such states are possible for some specific character of the ordering of the solid solution at the temperature T_o, but it is clear that in themselves they are exceptional. The fundamental result of the consideration in this section is the following property of anisotropic decomposition: asymptotically the process is organized as a one-dimensional process.

By analogy with the previous section, we can consider the point of the validity of the asymptotics (9.2.35) and (9.2.37). Now we need to compare the terms, taken into account and omitted in Equation (9.2.18), in the following form

$$\delta\phi(\mathbf{q}) \int \delta\phi(\mathbf{q})d\mathbf{q}\ , \quad \frac{\partial^2\delta\phi(\mathbf{q})}{\partial q_\kappa \partial q_l} \int q_\kappa q_l \delta\phi(\mathbf{q})d\mathbf{q}$$

The ratio of the first term to the second one estimated at $\mathbf{q} = \mathbf{q}_*$ is of the order of $v^2\tau$. Substituting the asymptotics obtained we find that the anisotropic decomposition law (9.2.35) should be applicable at more large times compared with the isotropic decomposition law (9.2.37) (compare with the validity analysis of asymptotics when $d = 1$ and $d > 1$ in Section 9.1).

Computational Example

An example of calculations based on Equation (9.2.37) was given in the previous section, where the decomposition in feldspars was studied. According to Putnis and McConnel,[25] this is an isostructural solid solution. The theory here is especially suited precisely for such solutions, since the development of spinodal structures in them reduces to the motion of cations in a relatively inert crystal framework, which remains continuous throughout the entire crystal. In the previous section the quantity z_o was set equal to 0.1. Averaging over the initial conditions changes the formulas. The starting data for feldspars were taken from Khisina.[10] In the previous section the reduction of the free energy of mixing to a symmetric form gave $b = 3.68$ and $a = -0.58$. Decomposition at $T = 773$ K was studied; at this temperature the unit cell is $\sim 10^{-9}$ m in size, whence we get $\rho \approx 10^{9d}\ m^{-d}$. The size of the spinodal structures at the end of the linear stage of the process is $\xi_M = (75 \pm 10) \cdot 10^{-10}$ m. The quantity λ_1 can be estimated using the formula

$$\frac{4\pi^2}{\xi_M^2} = -(a + 6b\phi_o^2)/2\lambda_1$$

whence $\lambda_1 \approx 4 \cdot 10^{-19}\ m^2$. If the solid solution is isotropic, then K can be determined using a similar formula. Let the spinodal decomposition proceed by means of a transition through the critical temperature T_c, i.e., $\phi_o = 0$. For an example, let $a_o = -a$, i.e., $T_c - T \approx T_o - T_c$. Performing the calculations using the relationships (9.2.41) and (9.2.42) we find that in the anisotropic case with $d = 1, 2$, and 3 the quantity y_* is equal to 0.71, 1.6, and 3.1, respectively. The growth of the structures can be described in the form (9.2.35) with the coefficient $y_*^{1/d}$ equal to 0.71, 1.25, and 1.48 with $d = 1, 2$, and 3, respectively. In the isotropic case with $d = 2$ and 3 the quantity y_* is equal to 0.9 and 1.6, respectively. The product in the argument of the logarithm in Equation (9.2.37) has the form $0.81\ \tau$ for $d = 2$ and $25.2\ \tau^3$ for $d = 3$.

Equations (9.2.35), (9.2.37), and (9.2.40) to (9.2.42) can serve as a basis for the calculation of the kinetics of the growth of structures in the process of spinodal decomposition of anisotropic and isotropic systems. If we transform in Equations (9.2.35) and (9.2.37) to dimensional variables and assume that the Onsager coefficient and the parameters of the thermodynamic potential are functions of the time-dependent temperature, we can study the effect of the rate of cooling on the kinetics of decomposition. This question is important in application to problems in metal science and geochemistry. Lastly, as we mentioned in Section 9.1, the exponents in the asymptotic formulas (9.2.35) and (9.2.37) are also the same when using the arbitrary form of function $F_o(\phi)$.

General Description of Diffusion-Controlled Phase Transitions and Connection With Results Obtained: Discussion

We shall discuss the generality of the results obtained in this section. Consider the most general form of the equation describing a diffusion-controlled phase transition in a system with a one-order parameter

$$\frac{\partial \phi}{\partial t} = \sum_{i,j=1}^{d} \frac{\partial}{\partial r_i}\left(\Lambda_{ij}\frac{\partial}{\partial r_j}\frac{\delta F}{\delta \phi}\right) \tag{9.2.43}$$

Here Λ_{ij} is an element of the symmetric matrix of the kinetic coefficients (for simplicity we assume below that Λ_{ij} are constants). We multiply Equation (9.2.43) by $\delta F/\delta\phi$ and integrate over the space. Using the identity

$$\int \frac{\delta F}{\delta \phi}\frac{\partial \phi}{\partial t}\, \mathbf{dr} \equiv \frac{dF}{dt}$$

we obtain

$$\frac{dF}{dt} = -\int \sum_{i,j=1}^{d} \Lambda_{ij}\left(\frac{\partial}{\partial r_i}\frac{\delta F}{\delta \phi}\right)\left(\frac{\partial}{\partial r_j}\frac{\delta F}{\delta \phi}\right) \mathbf{dr}$$

The only constraint of the theory is that the free energy must not increase with time,

$$dF/dt \leq 0$$

here the equality holds only on the stationary solutions $\mu = \text{const}$. For this reason the form $\Sigma\Lambda_{ij}q_iq_j$ should be positive-definite, which is equivalent to the possibility of rotating the coordinates so that Equation (9.2.43) assumes the form

$$\frac{\partial \phi}{\partial t} = \sum_{i=1}^{d} \Lambda_i \frac{\partial^2}{\partial r_i^2}\frac{\delta F}{\delta \phi}, \quad \Lambda_i > 0$$

Next, performing the transformation $r_i \to r_i\,(\Lambda/\Lambda_i)^{1/2}$, where Λ is the characteristic scale of the kinetic coefficient, we obtain Equation (9.1.1). As noted above, in the long-wavelength approximation the free energy functional has the general form (9.2.1). The coordinate transformations indicated above transform **K** once again into a symmetric matrix. Thus the description of a diffusion-controlled transition with a one-order parameter reduces to solving the problem (9.1.1) with the functional (9.2.1). Most importantly, in the dynamic description an anisotropic form cannot be reduced to an isotropic form, as can be done by rotating and then changing scales along the axes when analyzing the statics of critical phenomena based solely on the functional (9.2.1). This fact was apparently ignored in the theory of spinodal decomposition which was initially constructed precisely for applications to solid solutions. A general analysis of decomposition in an anisotropic system was not performed. The constructive development of a general theory, performed above, was made possible by the use of two quite simple, but fundamental, assumptions. The first assumption consists of the introduction of an order parameter, which means that the relaxation times of all other parameters are much shorter. These parameters actually change with their own equilibrium values, which depend on the order parameter, and thereby they "follow" the order parameter and are eliminated from the analysis. The second assumption consists of using the energy functional in the long-wavelength approximation and the diffusion equation for the anisotropic system in its general form.

Thus it is possible to propose, starting from first principles, a quite general scenario of phase separation in diffusion-controlled systems with one order parameter. Whether the decomposition is isotropic or anisotropic depends on the properties of the gradient-energy tensor, whose components can be determined from small-angle scattering data for a given substance. The exponent in the law of growth of the structures depends on the type of decomposition. The simplest form of the equation of decomposition in the anisotropic scenario is

$$\frac{\partial \phi}{\partial t} = \nabla^2\left(-\phi + \phi^3 - \sum_{i=1}^{d} \lambda_i \frac{\partial^2 \phi}{\partial r_i^2}\right)$$

which can be derived by transforming the dependent and independent variables after retaining in Equations (9.1.1) and (9.2.1) nonlinearities up to and including cubic terms.

Lastly, let us consider the question about the applicability of "mean field approximation" (9.1.18) by comparing the results of this section with the contemporary results in phase separation simulation. Generally, two main models, with the conserved and the not conserved order parameter, were used. For example, Mazenko et al.[26] studied the growth kinetics of the spin-flip and the spin-exchange kinetic Ising models using a combination of renormalization-group (RG) methods and Monte Carlo simulations.[26] In the case of spin-flip models (with the not conserved order parameter) they have obtained growth exponent 1/2, which is in agreement with the curvature-driven Lifshitz-Allen-Cahn dynamics.[27,28] In the case of spin-exchange models with the conserved order parameter (this case corresponds to the coalescence dynamics of Lifshitz and Slyozov[1]) they have found the structure growth law to be in the form $\xi \sim \ln t$. In a later paper of Mazenko et al.[29] the coupled technique of the RG method and numerical simulation leads to the results of the growth exponent 1/2 in the case of the not conserved order parameter and 1/4 in the case of the conserved order parameter.[29] Valls and Mazenko discussed the results of direct numerical calculations of structures growth in the model with the conserved order parameter.[30] They have reported a growth exponent of 1/4 and discussed the possibility of a later crossover from the 1/4 value to the classical 1/3 Lifshitz-Slyozov value. They mentioned such a crossover for the different phase separation model considered by Oono and Puri,[31] but informed about the absence of the crossover in their own calculations. The 1/4 value of growth exponent has been also observed by Mouritsen,[32,33] Langer et al.,[11] and Binder.[15]

On the other hand, the work of another group of authors also is concerned with the numerical simulation of systems with the conserved order parameter, and these authors have reported a growth exponent of 1/3. For example, Amar et al.[34] obtained such a value, and Huse,[35] and Rogers and co-workers have also obtained the same result.[36] They have concluded that the study carried out by Valls and Mazenko[30] was restricted to comparatively early (intermediate) stages of the separation process (see also Elder and Desai[13]).

Comparing the analytical results proposed in this section with these publications we can make conclusion about the applicability times for mean field approximation (9.2.18) in the phase separation theory. It can be seen that for the isotropic case (gradient term in form of $K(\nabla\phi)^2$) this approximation is applicable at the intermediate times of the structures growth process and probably is not applicable at a later stage. We would like to underline that the new analytical technique suggested for the mean field approximation does not use any additional assumptions except that it considers the asymptotical stage (after the linear stage of initial structures growth). According to the above discussion of validity of the asymptotics (9.2.35) and (9.2.37) we can expect that in the anisotropic case the mean field approximation (and, consequently, the exponent 1/2) is applicable at a later time. Quite probably, the phase separation of isotropic systems should demonstrate a consequent crossover from the 1/4 exponent to the 1/3 value in the structures growth law.

It is very interesting that for the common anisotropic case we have obtained the growth exponent of 1/2. As we mentioned, the same value of the growth exponent has been obtained for all the versions of phase separation theory with nonconserved (Lifshitz-Cahn-Allen) dynamics. Therefore it would be of interest to compare the transition from the curvature-driven (Lifshitz-Cahn-Allen) scenario to the coalescent (Lifshitz-Slyozov) scenario and the corresponding transition in the models (9.1.1) and (9.2.1) from the anisotropic case of general position (degeneracy of the minimum eigenvalue of **K** is 1) to the isotropic one (degeneracy of the only eigenvalue of **K** is d).

9.3. GRADIENT THEORY OF A THIN LIQUID INTERLAYER BETWEEN TWO SOLIDS

The thermodynamic properties of a thin liquid interlayer are different from those of the bulk liquid. Molecules inside the film possess an excess chemical potential that is manifested in an excess pressure. To describe this excess, the term 'disjoining pressure' was introduced. According to the Derjaguin-Landau-Verwey-Overbeek (DLVO) theory,[37-39] the total disjoining pressure in thin fluid film is generally the sum of Van der Waals, electrostatic, Born repulsion, and structural forces. The theoretical calculation of the first three contributions is reasonably well understood in terms of existing theories. The theory of structural surface forces is still being developed.

The introduction of structural forces has been dictated by the necessity to explain the observed deviation between experimentally measured values of disjoining pressure and theoretical predictions obtained from the DLVO theory.[40-43] The reported measurements of structural forces usually can be approximated by an exponential dependence on the film thickness L:

$$\Pi = B_1 \exp(-L/\xi) \tag{9.3.1}$$

Sometimes this simple dependence (9.3.1) is not applicable over the entire range of film thickness and a more complex empirical dependence may have to be assumed.

The origin of structural forces lies in the inhomogeneity of the order parameter (density, or concentration, or polarization) in the vicinity of the bounding surfaces. The principal goal of the theoretical analysis, therefore, is to determine the nature of this inhomogeneity and relate it to changes in the free energy of the interlayer region.

Marcelija and Radic considered the problem of two identical solid surfaces bounding a water film.[44] They wrote the free energy of the film in gradient form,

$$F = \int_0^L \left(a_1 \phi^2 + K \left(\frac{d\phi}{dx} \right)^2 \right) dx \tag{9.3.2}$$

where ϕ is the order parameter, later interpreted by Ninham as the local polarizability of film,[45] and a_1 and K are constants. By minimizing the free energy functional (9.3.2) and solving the corresponding linear boundary problem, Marcelija and Radic[44] obtained an expression for the interfacial energy which was calculated as the minimum value of F. By differentiating with respect to the film thickness they obtained an exponential dependence of the structural component of disjoining pressure. Their derivation provided the first theoretical basis for the experimentally observed dependence.

Cahn considered another problem for a system "flat surface + liquid half-space" in connection with wetting transition phenomena.[46] He solved the variational problem for the free energy functional

$$F = \gamma(\phi_1) + \int_0^\infty \left(W(\phi) + K \left(\frac{d\phi}{dx} \right)^2 \right) dx \tag{9.3.3}$$

where ϕ_1 is the value of the order parameter at the surface, $\phi_1 = \phi(0)$; γ is the contribution to the free energy from the molecular interaction of the solid surface and the liquid; $W(\phi)$ is the free energy density of the homogeneous liquid. The boundary condition is not defined arbitrarily but obtained directly from solving the variational problem (9.3.3):

$$x = 0 : \frac{d\gamma}{d\phi} - 2K \frac{d\phi}{dx} = 0$$

Cahn's approach is quite general and applicable to different physical systems with interacting solid boundaries. It is a valid approach, when the interactions between surface and fluid are sufficiently short-range so that their contribution to the free energy can be of "local" form (see also de Gennes[47]). The usefulness

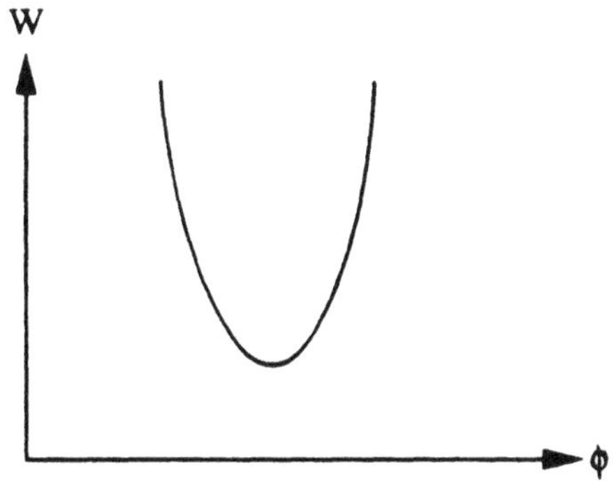

FIGURE 69. Common form of function W(ϕ) used.

of the model is enhanced by the fact that it can be studied analytically; particularly for the wetting transition problem, Cahn's model predicts all the main features of this phenomenon, which were later obtained quantitatively in numerical calculations of more complicated (nonlocal) models.[48]

In this section the above approaches[44,46] are combined and generalized. We consider the gradient theory with an arbitrary free energy density function $W(\phi)$ without inflection points, i.e., a phase transition is forbidden (Figure 69). The system "liquid interlayer + flat surface 1 + flat surface 2" will be discussed. The general solution of the corresponding variational problem and the discussion of possible distributions of order parameter in the interlayer are provided. The general thermodynamic relations, which are inherent in the gradient theory under consideration, are established. When the distance between surfaces (L) is sufficiently more than molecular interaction scale, we develop a new asymptotically exact technique for obtaining the dependences of several interlayer thermodynamic parameters on L. The exponential dependences of the structural component of disjoining pressure and the interfacial energy on L are obtained. The behavior of the preexponent factor is discussed; the possibility of sign reversal and nonmonotonic behavior of disjoining pressure is demonstrated. A physically plausible model of a multicomponent regular solution in contact with a surface is described. A possible scenario of hydrophobic effect within the framework of the model under consideration and the influence of surface roughness are discussed. The consideration of this and the next sections correspond to our work.[49]

Statement of the Problem and General Solution

Let us consider the following free energy functional

$$F = \gamma_1(\phi_2) + \gamma_2(\phi_2) + \int_0^L \left(W(\phi) + K\left(\frac{d\phi}{dx}\right)^2 - \mu\phi \right) dx \qquad (9.3.4)$$

Here $\gamma_m(\phi_m)$ is the energy contribution of the interaction between the m-th surface and liquid interlayer; L is the thickness of interlayer; ϕ_1 and ϕ_2 are defined at the two boundaries, $\phi_1 = \phi(0)$, $\phi_2 = \phi(L)$. The functional (9.3.4) describes the system of two parallel plates immersed in an large reservoir of liquid; μ is the chemical potential of the system. The variable ϕ, for instance, can be the concentration of a component in a binary incompressible liquid, or the density of a one-component compressible liquid.

The variation of the functional F is,

$$\delta F = \left.\frac{d\gamma_1}{d\phi}\right|_{x=0} \cdot \delta\phi_1 + \left.\frac{d\gamma_2}{d\phi}\right|_{x=L} \cdot \delta\phi_2 - 2K\left.\frac{d\phi}{dx}\right|_{x=0} \cdot \delta\phi_1$$

$$+ 2K\left.\frac{d\phi}{dx}\right|_{x=L} \cdot \delta\phi_2 + \int_0^L \left(\frac{dW}{d\phi} - 2K\frac{d^2\phi}{dx^2} - \mu\right) \delta\phi dx \quad (9.3.5)$$

At equilibrium, setting $\delta F = 0$, we obtain from Equation (9.3.5)

$$x = 0: \quad \frac{d\gamma_1}{d\phi} - 2K\frac{d\phi}{dx} = 0 \quad (9.3.6)$$

$$x = L: \quad \frac{d\gamma_2}{d\phi} + 2K\frac{d\phi}{dx} = 0 \quad (9.3.7)$$

$$\frac{dW}{d\phi} - 2K\frac{d^2\phi}{dx^2} - \mu = 0 \quad (9.3.8)$$

Integrating Equation (9.3.8) with respect to ϕ, we get

$$W(\phi) = K\left(\frac{d\phi}{dx}\right)^2 + \mu\phi + h \quad (9.3.9)$$

where h is the integration constant. Substituting Equation (9.3.9) into Equations (9.3.6) and (9.3.7) we obtain

$$\frac{d\gamma_1}{d\phi} = \pm 2[K(W(\phi) - \mu\phi - h)]^{1/2} \quad (9.3.10)$$

$$\frac{d\gamma_2}{d\phi} = \mp 2[K(W(\phi) - \mu\phi - h)]^{1/2} \quad (9.3.11)$$

The phase plane of Equation (9.3.9) is shown in Figure 70a. Figure 70b represents schematic dependences of W-$\mu\phi$ and W_ϕ-μ which are necessary to construct the phase plane in Figure 70a. Notice that the solution ϕ_* of equation $W_\phi = \mu$ corresponds to value of ϕ in the reservoir. The change of parameter h directly corresponds to a change in the interlayer thickness L. The separatrix in Figure 70a corresponds to infinite L.

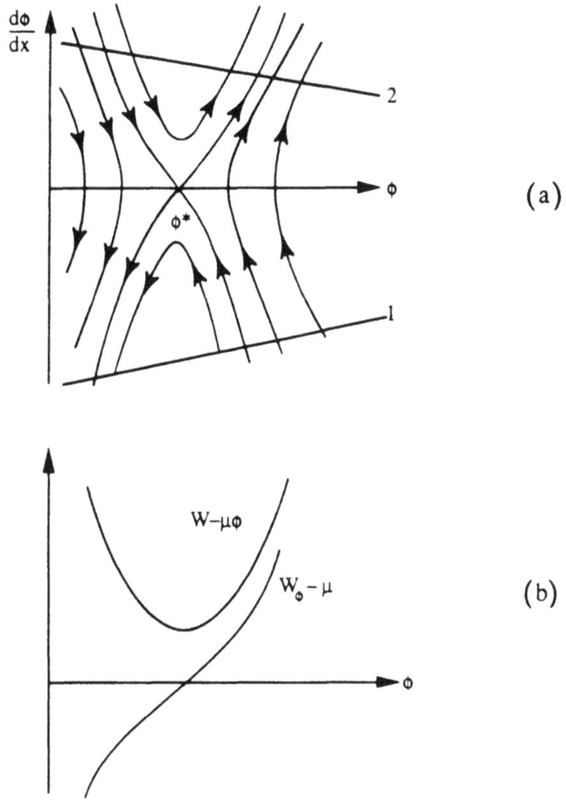

FIGURE 70. a: graphic representation of solution (9.3.6) and the boundary conditions imposed; b: dependences $W - \mu\phi$ and $W_\phi - \mu$ used for obtaining the graphic representation.

Line 1 in Figure 70a shows the function $(d\gamma_1/d\phi)/2K$, line 2 shows the function $-(d\gamma_1/d\phi)/2K$. All subsequent phase planes will be constructed in a similar manner. According to Equations (9.3.6) and (9.3.7) the appropriate solution must begin on line 1 and end on line 2. The arrows in Figure 70a show the direction of change in ϕ, which corresponds to increasing the spatial variable x, and give up the rule for selection of the appropriate phase curves.

The possible structure of the functions $\gamma_m(\phi)$ was discussed by de Gennes.[47] He pointed out that γ_m must have the form (for a one-component compressible liquid)

$$\gamma_m = \gamma_{o,m} - \gamma_{1,m}\,\phi + \gamma_{2,m}\,\phi^2 + \ldots \qquad (9.3.12)$$

The linear term should correspond to interactions between the liquid and surface, the quadratic term represents the reduction of the liquid/liquid interactions near the surface because of decreasing the number of interacting neighbors for the liquid molecules lying directly on the surface. We will calculate the functions

$\gamma_m(\phi)$ in the form of quadratic polynomials for multicomponent liquids later, using a simple lattice model. Notice here that the coefficient $\gamma_{2,m}$ is connected only with the properties of the liquid and must be the same for both functions γ_1 and γ_2. In the same time the coefficients $\gamma_{1,1}$ and $\gamma_{1,2}$ can be different for different solids. For the case of a single-component compressible liquid the linear term should be negative.[47] But for the case of binary incompressible liquids it can have any sign, depending on the energies of pair interaction, "solid-liquid" and "liquid-liquid". The quadratic term in Equation (9.3.12) is usually positive.

Let us consider qualitatively the possible types of liquid interlayer structure for the assumed quadratic $\gamma_m(\phi)$ dependences. This restriction on the form of γ_1 and γ_2 is not important; in fact, one should require the derivatives $d\gamma_1/d\phi$ and $d\gamma_2/d\phi$ to be monotonic in the range of γ variation.

The Case of Two Identical Surfaces

By identical surfaces we mean surfaces possessing identical functions $\gamma_m(\phi)$. Possible phase planes are shown in Figure 71. It should be noted that lines 1 and 2 are symmetric with respect to the ϕ axis.

The structure of the interlayer strongly depends on the locations of the saddle point $(\phi_*, 0)$ and the point $(\phi_{alter}, 0)$ of vanishing $d\gamma_m/d\phi$. In Figure 71a we show the case when there is no zeroes of $d\gamma_m/d\phi$ in the range of ϕ considered. The profile $\phi(x)$ decreases from the boundaries toward the middle of interlayer for all L. The thickness of the interlayer L decreases with increasing h, and L approaches infinity as h approaches to the value h_* which corresponds to the separatrix. Depending on the form of $W(\phi)$ near the boundaries of ϕ range, the maximum h value could be either infinite (for example, for the Peng-Robinson model[50]) or finite (for example, for the regular solution model[18]). In the first case the minimal possible equilibrium value of L is exactly zero; in the second case (shown in Figure 71a) this value is positive, implying that within the framework of the model it is impossible to make the distance between two plates less than a defined value. Note that the phase curves from the left of $(\phi_*,0)$ can not be chosen as appropriate solutions because they lead from line 2 to line 1.[46]

Similar solutions can be constructed for other locations of $d\gamma_m/d\phi$ as shown in Figures 71b through 71d. In Figures 71b and 71c $d\gamma_m/d\phi$ vanishes in the range of ϕ considered, and the minimum value of L is zero. Both minima (Figures 71a and 71b) and maxima (Figures 71c and 71d) in $\phi(x)$ are observed. The profiles are, however, always symmetric.

The Case of Two Different Surfaces

We will consider only those structures which do not disappear with a small change of the parameters.[51] Besides that, except in the case of two equivalent surfaces, we will not discuss the peculiarities of the form of $W(\phi)$ near the limiting values of ϕ. For instance, we considered cases "a" and "b" in Figure 71 as separate classes. We will not distinguish between them in the classification used below.

Figure 72 lists the possible cases. We can see that the number of possibilities is eight, when $\gamma_{2,1} = \gamma_{2,2}$ and $\gamma_{1,1}$ and $\gamma_{1,2}$ are arbitrary (Figures 72a to 72h). In the more general case, when $\gamma_{2,1}$ and $\gamma_{2,2}$ are equal but have arbitrary signs, two additional possibilities appear (Figures 72i to 72j).

Dependences h(*L*) shown in Figure 72 are obtained by the supposition that the interlayer thickness *L* increases as the distance between the saddle point (ϕ_*, 0) and the phase curve decreases.

We can see that in cases where the points of intersection of the separatrix and line 1 and the points of intersection of the separatrix and line 2 belong to different half-planes of the phase plane, the h(*L*) dependence is monotonic (Figures 72c and f) and equivalent to that presented in Figure 71.

In cases where the points of intersection of the separatrix and lines 1 and 2 lie in the the same half-plane, h(*L*) is nonmonotonic and contains two segments (Figure 72a). The decreasing segment of h(*L*) in Figure 72a corresponds to a typical trajectory X_1X_2 in the phase plane. The dependence $\phi(x)$ in this case has negative derivatives $d\phi/dx$ and $d^2\phi/dx^2$ inside the entire interlayer. The increasing segment of h(*L*) in Figure 70a corresponds to the typical trajectory $X_3X_1X_2$. We can conclude that the appearance of a minimum in h(*L*) corresponds to an appearance of an inflection point in the $\phi(x)$ profile.

We will show later that the quantity $P = -h$ is exactly the pressure acting on the surfaces. Therefore, subject on the mutual disposition of lines 1 and 2 in the phase plane, the dependence of the pressure or disjoining pressure on thickness can have maxima when the corresponding $\phi(x)$ profiles are monotonic and no maxima when the corresponding $\phi(x)$ profiles have extremum.

Lastly, we note that for example in Figure 72a h(*L*) can be smaller or larger than h_* as *L* changes, while in Figure 72g h(*L*) is always less than h_*.

Equilibrium Free Energy, Surface Energy, and Disjoining Pressure

The equilibrium value of free energy denoted below as F_{eq} can be calculated by inserting Equation (9.3.9) into Equation (9.3.4)

$$F_{eq} = \gamma_1(\phi_1) + \gamma_2(\phi_2) + 2\int_0^L K\left(\frac{d\phi}{dx}\right)^2 dx + L\,h$$

$$= \gamma_1(\phi_1) + \gamma_2(\phi_2) + 2\int [K(W - \mu\phi - h)]^{1/2}\, d\phi + Lh \qquad (9.3.13)$$

Note that the cases shown in Figure 72a, b, g, and j are qualitatively equivalent at large *L* when h(*L*) dependences in Figure 72 are purely monotonic and can be represented generally by one figure (Figure 73c). Corresponding expression for F_{eq} is obtained in this case from Equation (9.3.13):

$$F_{eq} = \gamma_1(\phi_1) + \gamma_2(\phi_2) + 2\int_{\phi_2}^{\phi_1} [K(W - \mu\phi - h)]^{1/2}\, d\phi + Lh \qquad (9.3.14)$$

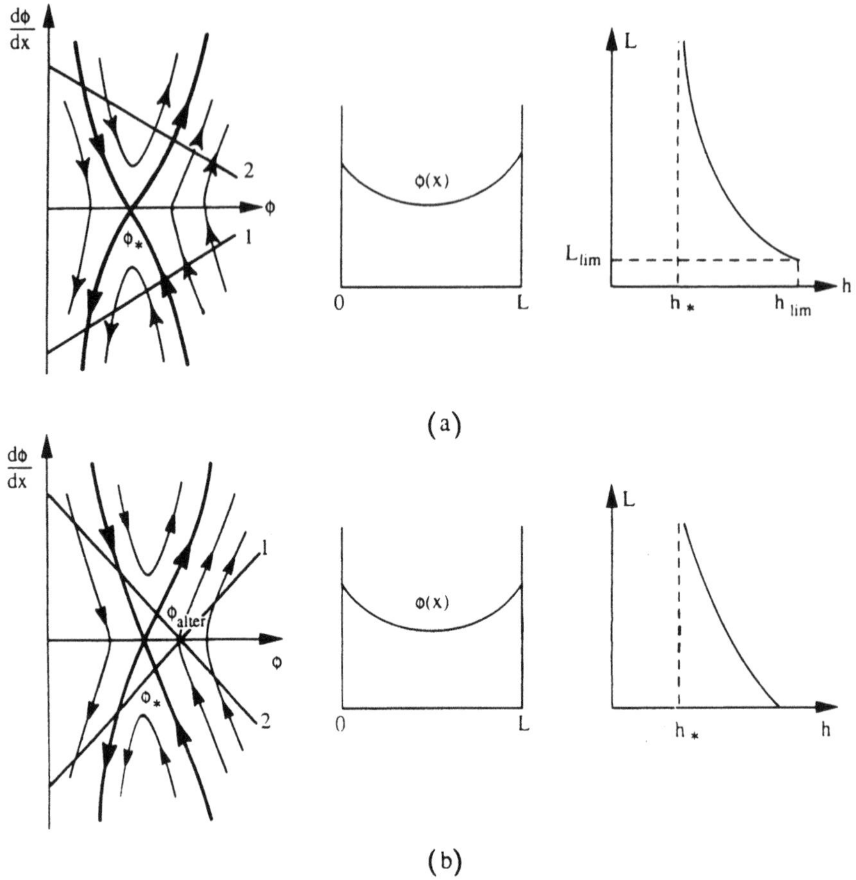

FIGURE 71. Possible liquid structures in the interlayer: case of identical dependences $\gamma_1(\phi)$ and $\gamma_2(\phi)$. Each possible situation (a through d) is presented via three drawings: first is the phase plane of the solution, second is the typical interlayer profile $\phi(x)$, and third is the relationship of the integration constant h and the interlayer thickness L.

and the length of the interlayer is

$$L = \int_{\phi_2}^{\phi_1} \left(\frac{K}{W - \mu\phi - h}\right)^{1/2} d\phi \tag{9.3.15}$$

where ϕ_1 and ϕ_2 are the appropriate roots of Equations (9.3.10) and (9.3.11).

Figures 72d, e, h, and i are also equivalent for large L and are represented by one figure (Figure 73d). The corresponding expressions for F_{eq} and L are

$$F_{eq} = \gamma_1(\phi_1) + \gamma_2(\phi_2) + 2\int_{\phi_1}^{\phi_2} [K(W - \mu\phi - h)]^{1/2}\, d\phi + Lh \tag{9.3.16}$$

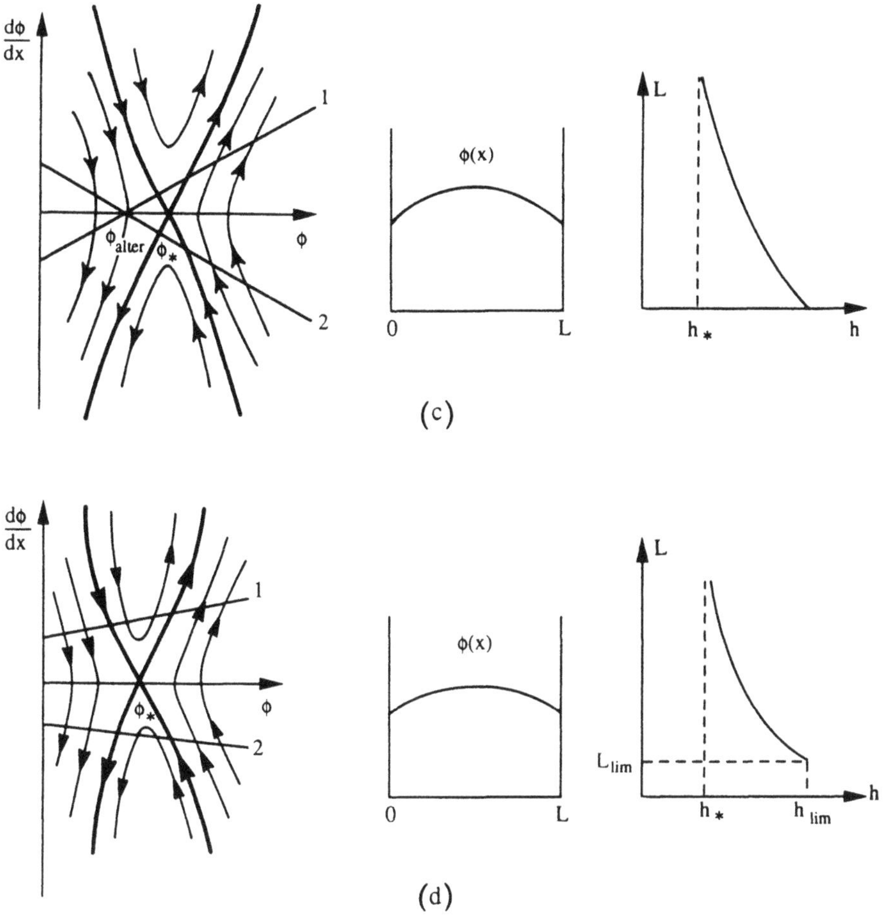

FIGURE 71c and d.

$$L = \int_{\phi_1}^{\phi_2} \left(\frac{K}{W - \mu\phi - h} \right)^{1/2} d\phi \qquad (9.3.17)$$

Figure 72c is separate and is represented in Figure 73a; F_{eq} and L are

$$F_{eq} = \gamma_1(\phi_1) + \gamma_2(\gamma_2) + 2 \sum_{m=1}^{2} \int_{\phi_o}^{\phi_m} [K(W - \mu\phi - h)]^{1/2} d\phi + Lh \qquad (9.3.18)$$

$$L = \sum_{m=1}^{2} \int_{\phi_o}^{\phi_m} \left(\frac{K}{W - \mu\phi - h} \right)^{1/2} d\phi \qquad (9.3.19)$$

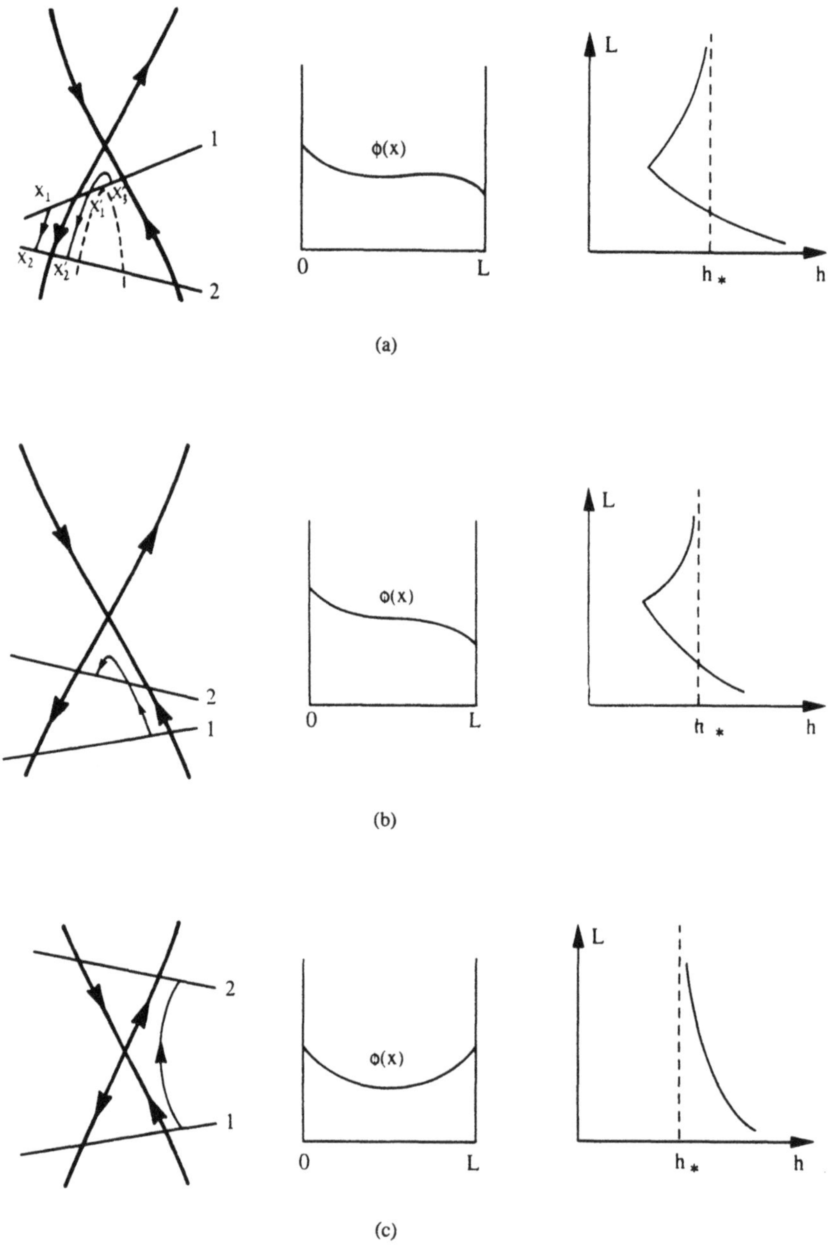

FIGURE 72. Possible liquid structures in the interlayer: case of two different dependences $\gamma_1(\phi)$ and $\gamma_2(\phi)$. The denotations and the form of representation of each possible situation (a through j) are equivalent to Figure 71.

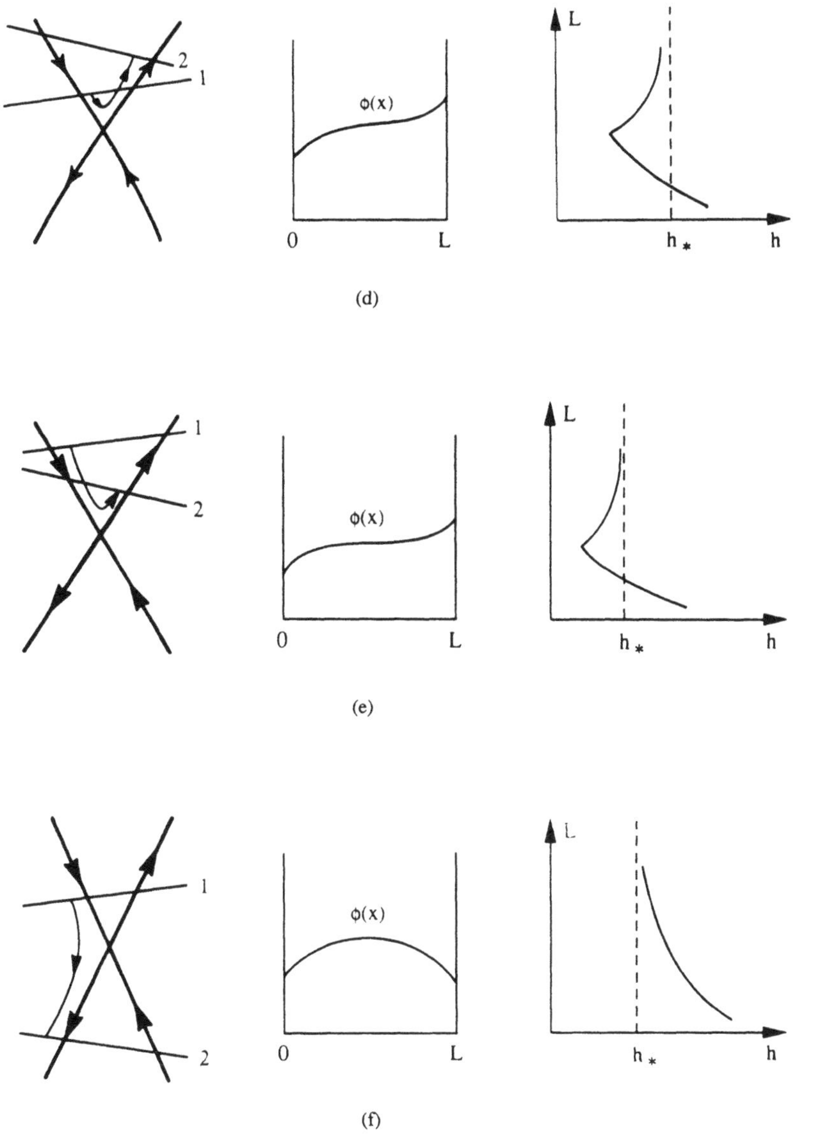

FIGURE 72 d, e, and f.

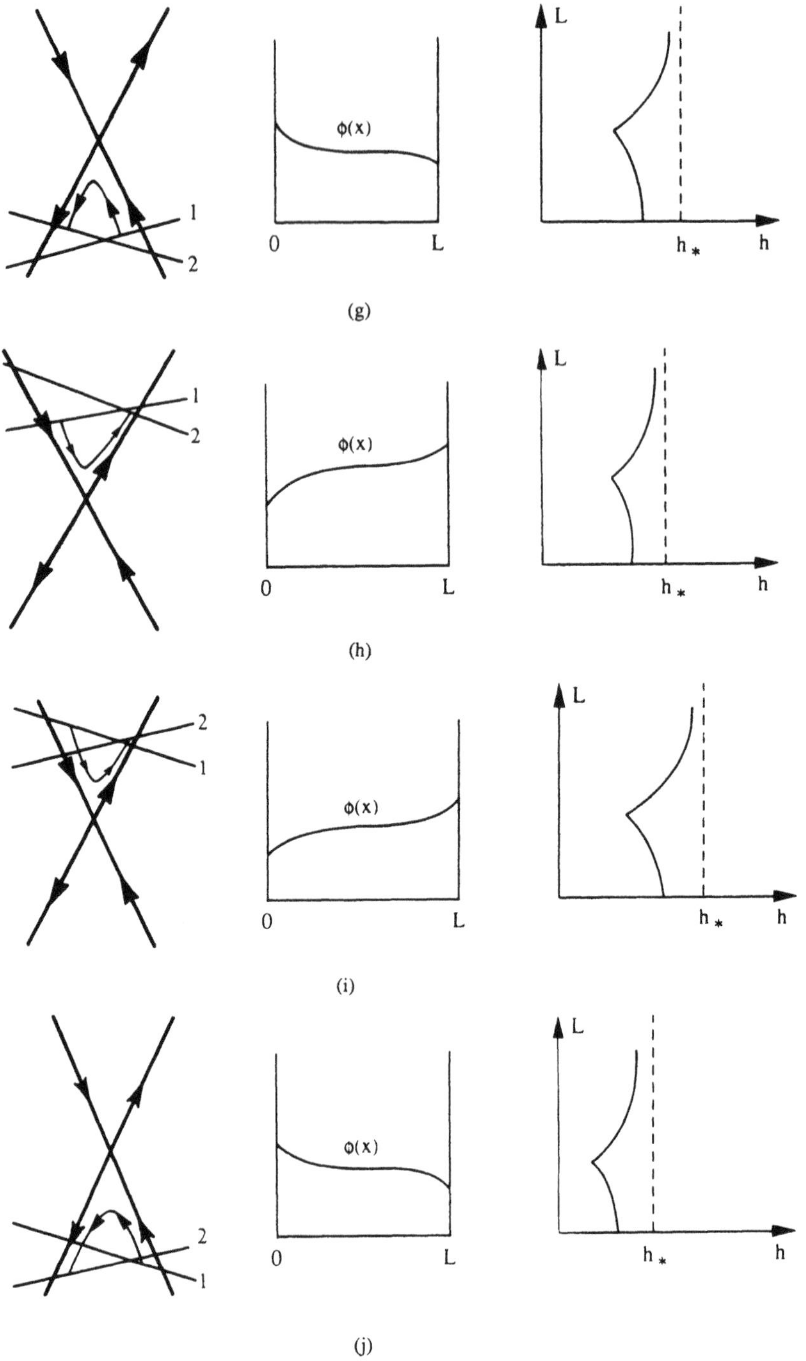

FIGURE 72 g, h, i, and j.

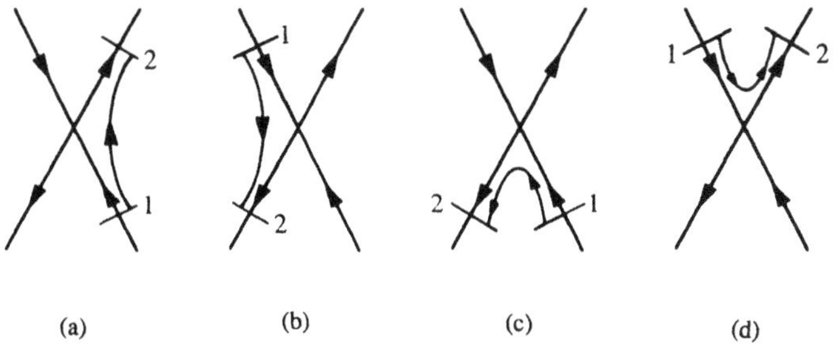

FIGURE 73. Possible qualitatively different phase planes of the problem under consideration at quite large values of thickness L.

where ϕ_o is defined as an appropriate root of the equation

$$W(\phi_o) - \mu\phi_o - h = 0 \qquad (9.3.20)$$

Lastly, Figure 72f is also separate and is represented in Figure 73b; F_{eq} and L are

$$F_{eq} = \gamma_1(\phi_1) + \gamma_2(\phi_2) + 2\sum_{m=1}^{2}\int_{\phi_m}^{\phi_o}[K(W - \mu\phi - h)]^{1/2}\,d\phi + Lh \qquad (9.3.21)$$

$$L = \sum_{m=1}^{2}\int_{\phi_m}^{\phi_o}\left(\frac{K}{W - \mu\phi - h}\right)^{1/2} d\phi \qquad (9.3.22)$$

and ϕ_o is also defined by Equation (9.3.20).

Though the free energy dependences obtained above differ from case to case, one can establish the universal character of $F_{eq}(L)$. Let us consider, as an example, the case in Figure 73a. The change in F_{eq} for a small change in h is

$$\Delta F_{eq} = \sum_{m=1}^{2}\frac{d\gamma_m}{d\phi_m}\Delta\phi_m + 2\Delta\left(\sum_{m=1}^{2}\int_{\phi_o}^{\phi_m}[K(W - \mu\phi - h)]^{1/2}\,d\phi\right) + \Delta(Lh)$$

$$= \sum_{m=1}^{2}\frac{d\gamma_m}{d\phi_m}\Delta\phi_m + 2\sum_{m=1}^{2}\{[K(W(\phi_m) - \mu\phi_m - h)]^{1/2}\Delta\phi_m - [K(W(\phi_o)$$

$$- \mu\phi_o - h)]^{1/2}\Delta\phi_o\} - \sum_{m=1}^{2}\Delta h\int_{\phi_o}^{\phi_m}\left(\frac{K}{W - \mu\phi - h}\right)^{1/2} d\phi + \Delta(hL) \qquad (9.3.23)$$

The quantity $[K(W(\phi_o) - \mu\phi_o - h)]^{1/2}$ vanishes by the definition of ϕ_o in Equation (9.3.20). Combining the terms and taking into account the boundary conditions in this case (see Equations (9.3.10) and (9.3.11))

$$\frac{d\gamma_m}{d\phi_n} + 2[K(W(\phi_m) - \mu\phi_m - h)]^{1/2} = 0\ ,\ m = 1,2$$

and from the definition of L in Equation (9.3.19) we finally obtain from Equation (9.3.23)

$$\Delta F_{eq} = -L\Delta h + \Delta(hL) = h\Delta L \tag{9.3.24}$$

The pressure acting on the surfaces is

$$P = -\frac{dF_{eq}}{dL} = -h \tag{9.3.25}$$

and thus the integration constant h given with the negative sign is identical to the pressure.

The change of surface energy σ can be defined as

$$\Delta\sigma = \Delta F_{eq} - h_*\Delta L = (h - h_*)\Delta L \tag{9.3.26}$$

where

$$h_* = -P_* = W(\phi_*) - \mu\phi_*\ , \quad W_\phi(\phi_*) = \mu \tag{9.3.27}$$

Formula (9.3.26) defines the change in the excess free energy. Equations (9.3.26) and (9.3.27) follow from the fact that the energy of a uniform system with the dimension L is h_*L.

The disjoining potential is defined as

$$G(L) = \int_L^\infty (P - P_*)dL = -\int_L^\infty (h - h_*)dL \tag{9.3.28}$$

and the surface energy $\sigma(L)$ is

$$\sigma(L) = \sigma_* + G(L) \tag{9.3.29}$$

where

$$\sigma_* = \gamma_1(\phi_1) + \gamma_2(\phi_2) + 2K\int_{-\infty}^{\infty}\left(\frac{d\phi}{dx}\right)^2 dx$$

$$= \gamma_1(\phi_{1*}) + \gamma_2(\phi_{2*}) + 2\int [K(W - \mu\phi - h_*)]^{1/2}\, d\phi \tag{9.3.30}$$

The limits of integration in the last integral of Equation (9.3.30) are defined according to the limits of integration in Equations (9.3.15), (9.3.17), (9.3.19), and (9.3.22), i.e., they depend on the type of case under consideration. Formula (9.3.29) is a generalization of the Cahn-Hilliard expression for surface energy.[18]

The disjoining pressure is defined as

$$\Pi = -\frac{dG}{dL} = P - P_* = h_* - h \qquad (9.3.31)$$

Consequently considering the cases in Figure 73, one can find that Equations (9.3.24) to (9.3.31) do not depend on a specific case. They are universal and derivable in general from the model (9.3.4).

Equations (9.3.14) to (9.3.22) together with Equations (9.3.24) to (9.3.31) provide the implicit relationships between the thermodynamic parameters of the system and L. Our next goal is to obtain the explicit dependences of G(L) and Π(L) for sufficiently large L.

Asymptotic Behavior of G(L) and Π(L)

We will develop a special asymptotic method of analysis of implicit relations between F_{eq} and L given by Equations (9.3.14) to (9.3.22).

Cases in Figure 73a and 73b

First we consider the case shown in Figure 73a, which corresponds to Equations (9.3.18) and (9.3.19). Let us expand the expression under the root in Equation (9.3.19) in a Taylor series near the point ϕ_o. This is justified because $L \to \infty$ when ϕ_o approaches ϕ_*, and the main contribution to the integral (9.3.19) is confined to the small vicinity of ϕ_o. Therefore we can leave only three first terms of the series. The first term vanishes by the definition of ϕ_o; as a result we have

$$\sum_{m=1}^{2} \int_{\phi_o}^{\phi_m} \frac{d\phi K^{1/2}}{(W - \mu\phi - h)^{1/2}} = \left(\frac{2K}{W_{\phi\phi}(\phi_o)}\right)^{1/2} \left[\sum_{m=1}^{2} \int_0^{\phi_m - \phi_o} \frac{d\phi}{(b\phi + \phi^2)^{1/2}} + Z\right] \qquad (9.3.32)$$

We will return later to the calculation of the last term (Z) on the right-hand side of Equation (9.3.32). Simplifying the above integrals

$$\int_0^{\phi_m - \phi_o} \frac{d\phi}{\left[\left(\phi + \frac{b}{2}\right)^2 - \frac{b^2}{4}\right]^{1/2}} = \ln\left[\phi + \frac{b}{2} + (\phi^2 + b\phi)^{1/2}\right] \Big|_0^{\phi_m - \phi_o}$$

$$= \ln\left[\phi_m - \phi_o + \frac{b}{2} + ((\phi_m - \phi_o)^2 + b(\phi_m - \phi_o))^{1/2}\right] - \ln\left(\frac{b}{2}\right) \qquad (9.3.33)$$

$$b = \frac{2(W_\phi(\phi_o) - \mu)}{W_{\phi\phi}(\phi_o)} = 2(\phi_o - \phi_*) + O((\phi_o - \phi_*)^2) \qquad (9.3.34)$$

The leading term in the representation is proportional to ln(b/2). It tends to infinity when $\phi_o \to \phi_*$ as long as other terms remain finite. Therefore we obtain from Equations (9.3.33) and (9.3.34) the representation of L in the form,

$$L = \sum_{m=1}^{2} \left(\frac{2K}{W_{\phi\phi}(\phi_o)}\right)^{1/2} \left\{ \ln\left[\phi_m - \phi_o + \frac{b}{2} + ((\phi_m - \phi_o)^2 + b(\phi_m - \phi_o))^{1/2}\right] - \ln\left(\frac{b}{2}\right) + \frac{Z}{2}\right\} = \sum_{m=1}^{2} \left(\frac{2K}{W_{\phi\phi}(\phi_o)}\right)^{1/2}$$

$$\left[\ln(2(\phi_m - \phi_o)) - \ln\frac{b}{2} + \frac{Z}{2}\right] + O(\phi_o - \phi_*) \tag{9.3.35}$$

Clearly, the condition $L \to \infty$ corresponds to $\phi_o \to \phi_*$. In this asymptotic limit we can write

$$L = \left(\frac{2K}{W_{\phi\phi}(\phi_o)}\right)^{1/2} \cdot [\ln(4(\phi_{1*} - \phi_*)(\phi_{2*} - \phi_*) - 2\ln(\phi_o - \phi_*) + Z] + O(\phi_o - \phi_*) \tag{9.3.36}$$

where the ϕ_{m*} values correspond to the separatrix. Equations (9.3.35) and (9.3.36) are exact up to $O(\phi_o - \phi_*)$. Thus, the singularity of the integral for L is separated.

Let us consider now the term Z:

$$Z = \left(\frac{W_{\phi\phi}(\phi_o)}{2K}\right)^{1/2} \sum_{m=1}^{2} \left[\int_{\phi_o}^{\phi_m} \left(\frac{K}{W - \mu\phi - h}\right)^{1/2} d\phi - \int_{\phi_o}^{\phi_m} \left(\frac{K}{(W_\phi(\phi_o) - \mu)(\phi - \phi_o) + \frac{1}{2} W_{\phi\phi}(\phi_o)(\phi - \phi_o)^2}\right)^{1/2} d\phi\right]$$

It can be easily shown that as $\phi_o \to \phi_*$ and $\phi_m \to \phi_{m*}$, Z converges to a finite value. Finally we can write the following asymptotic expression for L:

$$L = \left(\frac{2K}{W_{\phi\phi}(\phi_*)}\right)^{1/2} [-2\ln(\phi_o - \phi_*) + \ln(4(\phi_{1*} - \phi_*)(\phi_{2*} - \phi_*)) + Z] + O(\phi_o - \phi_*) \tag{9.3.37}$$

where

$$Z = \sum_{m=1}^{2} \int_{\phi_*}^{\phi_{m*}} \frac{\left[\frac{1}{2} W_{\phi\phi}(\phi_*)(\phi - \phi_*)^2 - W(\phi) + \mu\phi + h_*\right] d\phi}{(W - \mu\phi - h_*)^{1/2} \cdot |\phi - \phi_*| \cdot \left[(W - \mu\phi - h_*)^{1/2} + |\phi - \phi_*| \left(\frac{1}{2} W_{\phi\phi}(\phi_*)\right)^{1/2}\right]} \tag{9.3.38}$$

Using Equation (9.3.37) we obtain

$$\phi_o - \phi_* = B^{1/2} \exp\left[- \left(\frac{W_{\phi\phi}(\phi_*)}{8K}\right)^{1/2} L \right] \tag{9.3.39}$$

where

$$B = 4(\phi_{1*} - \phi_*)(\phi_{2*} - \phi_*)\exp Z \tag{9.3.40}$$

According to Equation (9.3.25) the pressure P increases with increasing L. Near the value $h = h_*$ one can approximate h as

$$h = W(\phi_o) - \mu\phi_o = h_* + \frac{1}{2} W_{\phi\phi}(\phi_*)(\phi_o - \phi_*)^2 + O((\phi_o - \phi_*)^3) \tag{9.3.41}$$

Substituting Equation (9.3.39) into Equation (9.3.41) we obtain

$$P(L) = P_* - \frac{1}{2} W_{\phi\phi}(\phi_*) B \exp(- L/\xi) \tag{9.3.42}$$

where

$$\xi = \left(\frac{2K}{W_{\phi\phi}(\phi_*)}\right)^{1/2} \tag{9.3.43}$$

is the correlation radius in bulk liquid (see Landau and Lifshitz[19]). The disjoining pressure and disjoining potential are given by

$$\Pi = - \frac{1}{2} W_{\phi\phi}(\phi_*) B \exp(- L/\xi) \tag{9.3.44}$$

$$G = - \frac{1}{2} W_{\phi\phi}(\phi_*) B \xi \exp(- L/\xi) \tag{9.3.45}$$

One can see that Π and G are negative. Similar behavior of G was observed in the numerical calculations.[52]

It is straightforward to repeat the calculation for the case in Figure 73b. The asymptotic equation for L is obtained in the form

$$L = \left(\frac{2K}{W_{\phi\phi}(\phi_*)}\right)^{1/2} [- 2\ln(\phi_* - \phi_o) + \ln(4(\phi_{1*} - \phi_*)(\phi_{2*} - \phi_*)) + Z] + O(\phi_* - \phi_o) \tag{9.3.46}$$

where

$$Z = \sum_{m=1}^{2} \int_{\phi_{m*}}^{\phi_*} \frac{\left[\frac{1}{2} W_{\phi\phi}(\phi_*)(\phi - \phi_*)^2 - W(\phi) + \mu\phi + h_*\right] d\phi}{(W - \mu\phi - h)^{1/2} \cdot |\phi - \phi_*| \cdot \left[(W - \mu\phi - h_*)^{1/2} + |\phi - \phi_*|\left(\frac{1}{2} W_{\phi\phi}(\phi_*)\right)^{1/2}\right]} \tag{9.3.47}$$

The expression (9.3.46) is transformed as

$$\phi_* - \phi_o = B^{1/2} \cdot \exp\left[-\left(\frac{W_{\phi\phi}(\phi_*)}{8K}\right)^{1/2} \cdot L\right] \tag{9.3.48}$$

where

$$B = 4(\phi_{1*} - \phi_*)(\phi_{2*} - \phi_*) \exp Z \tag{9.3.49}$$

After substituting Equations (9.3.48) and (9.3.49) into Equations (9.3.25) to (9.3.31) we again obtain Equations (9.3.42) to (9.3.45).

Cases in Figures 73c and 73d

Cases in Figures 73c and 73d differ principally from Figures 73a and 73b, as will be shown below. We will start from the case in Figure 73c described by Equations (9.3.14) and (9.3.15). Let us consider the expression for L in the following form

$$L = \int_{\phi_2}^{\phi_1} \left(\frac{K}{W - \mu\phi - h}\right)^{1/2} d\phi = \int_{\phi_2}^{\phi_*} \left(\frac{K}{W - \mu\phi - h}\right)^{1/2} d\phi + \int_{\phi_*}^{\phi_1} \left(\frac{K}{W - \mu\phi - h}\right)^{1/2} d\phi \tag{9.3.50}$$

Here ϕ_* corresponds to an ϕ value at which $d\phi/dx$ is a maximum. Rewrite Equation (9.3.50) in the form

$$\int_{\phi_2}^{\phi_*} \left(\frac{K}{W - \mu\phi - h}\right)^{1/2} d\phi + \int_{\phi_*}^{\phi_1} \left(\frac{K}{W - \mu\phi - h}\right)^{1/2} d\phi = \left(\frac{2K}{W_{\phi\phi}(\phi_*)}\right)^{1/2} \left[\int_0^{\phi_* - \phi_2} \frac{d\phi}{(b + \phi^2)^{1/2}} + \int_0^{\phi_1 - \phi_*} \frac{d\phi}{(b + \phi^2)^{1/2}} + Z\right] + O(b) \tag{9.3.51}$$

where

$$b = \frac{2(W(\phi_*) - \mu\phi_* - h)}{W_{\phi\phi}(\phi_*)} = \frac{2(h_* - h)}{W_{\phi\phi}(\phi_*)}, \quad h_* = W(\phi_*) - \mu\phi_* \tag{9.3.52}$$

One can see that the basic integrals in the right-hand side of Equation (9.3.51) and the parameter b (Equation [9.3.52]) have a different form than the cases in Figures 73a and 73b considered above. This is because Equation (9.3.51) is obtained by expanding as a Taylor series the expression in the denominator of the integrand in Equation (9.3.50). Clearly at $\phi = \phi_*$ the second term of the series vanishes identically, and the divergence of the integral (9.3.50) results in the first term of the series tending to zero. Then,

$$L = \left(\frac{2K}{W_{\phi\phi}(\phi_*)}\right)^{1/2} \cdot [\ln(4(\phi_* - \phi_{2*})(\phi_{1*} - \phi_*)) - \ln b + Z] + O(b) \qquad (9.3.53)$$

where

$$Z = \left(\frac{W_{\phi\phi}(\phi_*)}{2K}\right)^{1/2} \cdot \left[\int_{\phi_2}^{\phi_1} \left(\frac{K}{W - \mu\phi - h}\right)^{1/2} d\phi - \int_{\phi_2}^{\phi_1} \left(\frac{K}{(W(\phi_*) - \mu\phi_* - h) + \frac{1}{2} W_{\phi\phi}(\phi_*)(\phi - \phi_*)^2}\right)^{1/2} d\phi\right]$$

or

$$Z = \int_{\phi_{2*}}^{\phi_{1*}} \frac{\left[\frac{1}{2} W_{\phi\phi}(\phi_*)(\phi - \phi_*)^2 - W(\phi) + \mu\phi + h_*\right] d\phi}{(W - \mu\phi - h_*)^{1/2} \cdot |\phi - \phi_*| \left[(W - \mu\phi - h_*)^{1/2} + \left(\frac{1}{2} W_{\phi\phi}(\phi_*)\right)^{1/2} \cdot |\phi - \phi_*|\right]} + O(b)$$

(9.3.54)

From Equations (9.3.53) and (9.3.54) we obtain

$$b = B \exp(-L/\xi), \; B = 4(\phi_* - \phi_{2*})(\phi_{1*} - \phi_*) \exp Z \qquad (9.3.55)$$

Combining Equations (9.3.31), (9.3.52), and (9.3.55) we obtain the expression of disjoining pressure for large L. In this case Π is positive

$$\Pi = P - P_* = h_* - h = B \frac{W_{\phi\phi}(\phi_*)}{2} \exp(-L/\xi) \qquad (9.3.56)$$

The disjoining potential is positive as well,

$$G = \frac{1}{2} W_{\phi\phi}(\phi_*) \, B\xi \exp(-L/\xi) \qquad (9.3.57)$$

Surface energy is defined by Equations (9.3.29), (9.3.30), and (9.3.57).

One could regard this case to be a generalized Marcelija-Radic model.[44]

The case in Figure 73d is basically the same as Figure 73c. The expression for L is obtained in the form

$$L = \left(\frac{2K}{W_{\phi\phi}(\phi_*)}\right)^{1/2} \cdot [\ln(4(\phi_{2*} - \phi_*)(\phi_* - \phi_{1*})) - \ln b + Z] + O(b) \qquad (9.3.58)$$

$$Z = \left(\frac{W_{\phi\phi}(\phi_*)}{2K}\right)^{1/2} \cdot \left[\int_{\phi_1}^{\phi_2} \left(\frac{K}{W - \mu\phi - h}\right)^{1/2} d\phi - \int_{\phi_1}^{\phi_2} \left(\frac{K}{(W(\phi_*) - \mu\phi_* - h) + \frac{1}{2} W_{\phi\phi}(\phi_*)(\phi - \phi_*)^2}\right)^{1/2} d\phi\right] = \qquad (9.3.59)$$

$$\int_{\phi_{1*}}^{\phi_{2*}} \frac{\left[\frac{1}{2} W_{\phi\phi}(\phi_*)(\phi - \phi_*)^2 - W(\phi) - \mu\phi + h_*\right] d\phi}{(W - \mu\phi - h_*)^{1/2} \cdot |\phi - \phi_*| \left[(W - \mu\phi - h_*)^{1/2} + \left(\frac{1}{2} W_{\phi\phi}(\phi_*)\right)^{1/2} \cdot |\phi - \phi_*|\right]} + O(b)$$

Expressions (9.3.55) to (9.3.57) for b, Π, and G are the same in this case.

Thus, it is shown that in the limit of large L the expressions for Π and G have a leading term in the form $\exp(-L/\xi)$. The signs of Π and G are defined by the signs of $d\gamma_m/d\phi|_{\phi=\phi_{m*}}$. If

$$\text{sign}\left.\frac{d\gamma_1}{d\phi}\right|_{\phi=\phi_{1*}} = \text{sign}\left.\frac{d\gamma_2}{d\phi}\right|_{\phi = \phi_{2*}}$$

then Π and G are negative; in the case of opposite signs, these quantities are positive.

We can rewrite the relations obtained, in the following common form:

$$\Pi = \frac{W_{\phi\phi}(\phi_*)B}{2} \exp\left(-\frac{L}{\xi}\right)\left(1 + O\left(\exp\left(-\frac{L}{2\xi}\right)\right)\right) \qquad (9.3.60)$$

$$G = \frac{W_{\phi\phi}(\phi_*)B\xi}{2} \exp\left(-\frac{L}{\xi}\right)\left(1 + O\left(\exp\left(-\frac{L}{2\xi}\right)\right)\right) \qquad (9.3.61)$$

$$P = P_* + \frac{1}{2} W_{\phi\phi}(\phi_*) B \exp\left(-\frac{L}{\xi}\right)\left(1 + O\left(\exp\left(-\frac{L}{2\xi}\right)\right)\right) \qquad (9.3.62)$$

where

$$B = 4(\phi_* - \phi_{1*})(\phi_{2*} - \phi_*) \exp Z \qquad (9.3.63)$$

and Z is defined by Equations (9.3.38), (9.3.47), (9.3.54), or (9.3.59), depending on the case under consideration.

Qualitative Behavior of the Preexponent Term

We can describe the properties of preexponent B_1 in Equation (9.3.1) within the framework of our theory. From Equation (9.3.60), it has the form

$$B_1 = 2W_{\phi\phi}(\phi_*)(\phi_* - \phi_{1*})(\phi_{2*} - \phi_*) \exp Z \qquad (9.3.64)$$

B_1 can change signs as ϕ_* is changed. Below we explore the variation of $B_1(\phi_*)$ under different conditions.

First let us consider the case of two surfaces with identical dependences $\gamma_m(\phi)$. Notice that $B_1 = O$ only when $d\gamma_m/d\phi = 0$ in the ϕ-domain of interest. This is shown in Figure 74; when μ (or ϕ_*) changes, there is a specific value of μ corresponding to $\phi_* = \phi_{m*}$. At this point the function $B_1(\phi_*)$ goes to zero. As discussed above, the relative locations of the saddle point and ϕ_{alter} (the root of equation $d\gamma_m/d\phi = 0$) defines the type of $\phi(x)$ distribution inside the film. When ϕ_* is zero to the right of $d\gamma_m/d\phi$, the $\phi(x)$ profile is convex; when ϕ_* is to the left of it, the $\phi(x)$ profile is concave (see Figure 71). One can describe the ϕ-value corresponding to $d\gamma_m/d\phi = 0$, as a point at which the effective surface properties change from attractive to repulsive as ϕ_* changes. The function $B_1(\phi_*) \leq 0$, and at $\phi_* = \phi_{alter}$ the derivative $dB_1/d\phi_* = 0$.

In general, the functions γ_1 and γ_2 are different. Possible variants of the mutual disposition of the curves $d\gamma_m/d\phi$ and the separatrix and the corresponding $B_1(\phi_*)$ dependences are presented in Figure 75.

As above, we will consider the case when at least one of the functions $d\gamma_m/d\phi$ tends to zero within the region of definition for ϕ.

The first case is shown in Figure 75a and corresponds to both possibilities $d\gamma_1/d\phi = 0$ and $d\gamma_2/d\phi = 0$. The function $B_1(\phi_*)$ vanishes two times. It is positive between the zeroes and negative in the remaining domain of ϕ_*. This is similar to some reported experimental observations.[53,54] If the solid-liquid properties provide the possibility of $d\gamma_m/d\phi = 0$, the theory predicts the existence of a narrow "window" of ϕ_* where $B_1(\phi_*) > 0$. We will consider this point in detail later in connection with the hydrophobic effect.

The second and third cases are shown in Figures 75b and 75c. They correspond to the vanishing of only one function $d\gamma_m/d\phi$ ($d\gamma_2/d\phi = 0$; Figure 75b, and $d\gamma_1/d\phi = 0$; Figure 75c). Correspondingly, the preexponent $B_1(\phi_*)$ can either increase or decrease with increasing ϕ_*.

We did not consider the influence of Z on B_1 behavior. Z characterizes the deviation of $W(\phi)$ from an exact quadratic form, i.e., the presence of higher-order terms in the Taylor series for $W(\phi)$. Such changes cannot provide sign changes in $B_1(\phi_*)$, noted previously, but can lead to quantitative changes in $B_1(\phi_*)$ in structural force calculations.

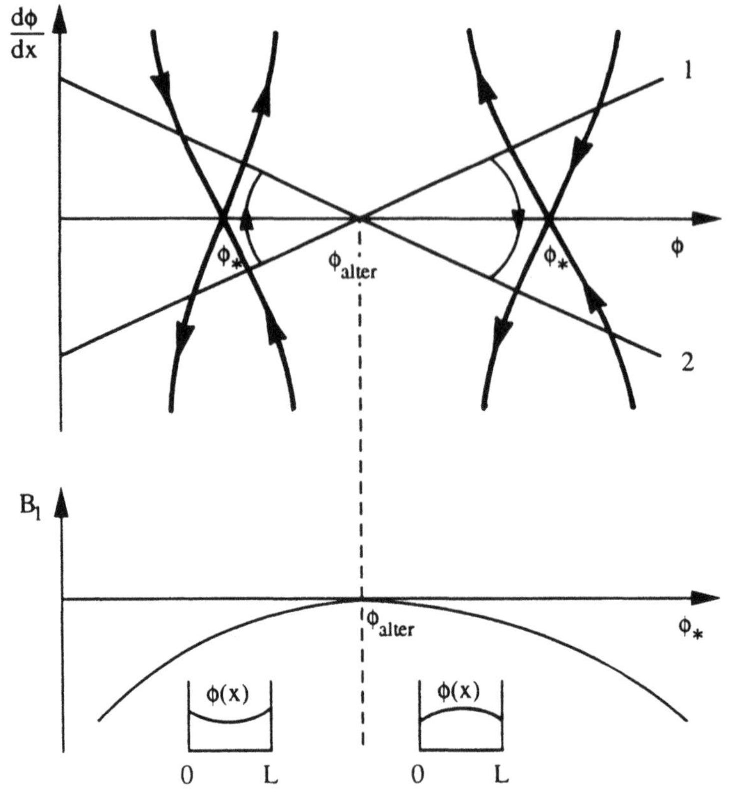

FIGURE 74. Case of two identical dependences $d\gamma_1/d\phi$ and $d\gamma_2/d\phi$ having a zero within ϕ domain. Above: possible dispositions of separatrix with regard to the boundary conditions imposed (lines 1 and 2) by changing the chemical potential. Below: corresponding dependence of the preexponent in the asymptotical $\Pi(L)$ expression on the bulk value of the order parameter, ϕ_*.

The above analysis shows the possibility of positive, negative, and zero structural force contributions (Π) when we take into account the nonlinear terms in expanding (9.3.12). Let us outline again that the slopes of the lines 1 and 2 must be opposite because the second coefficient in the series (9.3.12) describes the "liquid-liquid" interaction and is invariant for both γ_1 and γ_2. The property of opposite slopes for lines 1 and 2 does not depend on the sign of the second coefficient.

Model of a Multicomponent Regular Solution in Contact With a Surface

Below, an example of the free energy functional (9.3.4) is obtained from a lattice model of a multicomponent regular solution.

We will consider the model of a $(l + 1)$-component mixture on the d-dimensional cubic lattice with a lattice characteristic length, a. The Ising Hamiltonian has the following form

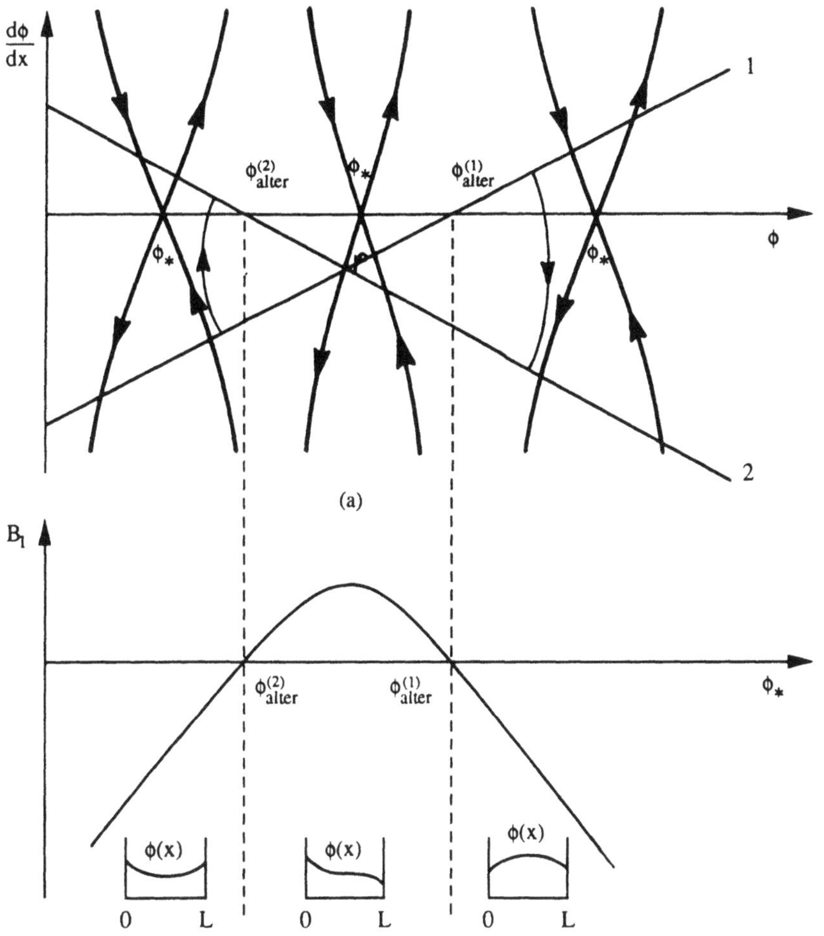

FIGURE 75. Case of two different dependences $d\gamma_1/d\phi$ and $d\gamma_2/d\phi$, from which at least one has a zero within ϕ domain. The denotations and the form of representation of each possible situation are equivalent to Figure 74.

$$H = \sum_{\eta \neq \nu} \omega_{\eta\nu} \epsilon_\eta \epsilon_\nu \tag{9.3.65}$$

where $\omega_{\eta\nu}$ are the pair interaction energy constants between sites η and ν, ϵ_η is an index showing the presence of monomer of either sort in a position η of the lattice. Assume that the interactions take place only between two nearest neighbors. Let the probability of a monomer of type i being present at location **r** be $\psi_i(\mathbf{r})$, i.e., the volumetric fraction of component i. Equation (9.3.65) can be rewritten as follows,

$$U = \langle H \rangle = \frac{1}{2} \sum_{\eta} \sum_{\kappa=1}^{d} \sum_{i,j=1}^{l} [\psi_i(\mathbf{r}_\eta)\psi_j(\mathbf{r}_\eta + \mathbf{a}_\kappa)\omega_{ij} + \psi_i(\mathbf{r})\psi_j(\mathbf{r}_\eta - \mathbf{a}_\kappa)\omega_{ij}] \tag{9.3.66}$$

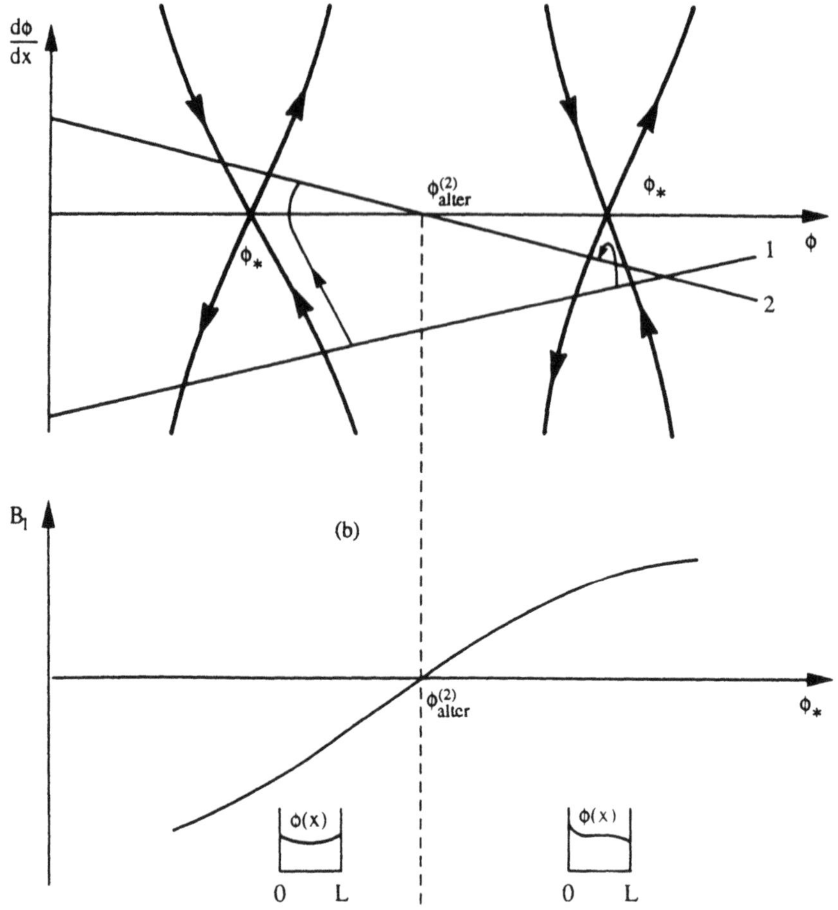

FIGURE 75b.

where ω_{ij} is the pair interaction energy between monomers of the sort i and j; a_κ is a vector of length a in the direction of the κ-th axis of the coordinate system; r_η is the vector location for the position η of the lattice. In the long-wavelength approximation, expanding Equation (9.3.66) into a series and retaining only the terms which contain derivatives not higher than the second order we obtain the form,

$$U = \sum_{\eta} \sum_{\kappa=1}^{l} \sum_{i,j=0}^{l} \left(\omega_{ij}\psi_i\psi_j + \frac{1}{2}\,\psi_{ij}a^2\psi_i\,\frac{d^2\psi_j}{dr_\kappa^2} \right)$$

$$= \sum_{\eta} \sum_{i,j=0}^{l} \left(d\omega_{ij}\psi_i\psi_j - \frac{a^2}{2} \sum_{\kappa=1}^{d} \omega_{ij}\,\frac{d\psi_i}{dr_\kappa}\,\frac{d\psi_j}{dr_\kappa} \right)$$

$$= \sum_{\eta} \sum_{i,j=0}^{l} \left(d\omega_{ij}\psi_i\psi_j - \frac{a^2}{2}\,\omega_{ij}(\nabla\psi_i, \nabla\psi_j) \right) \qquad (9.3.67)$$

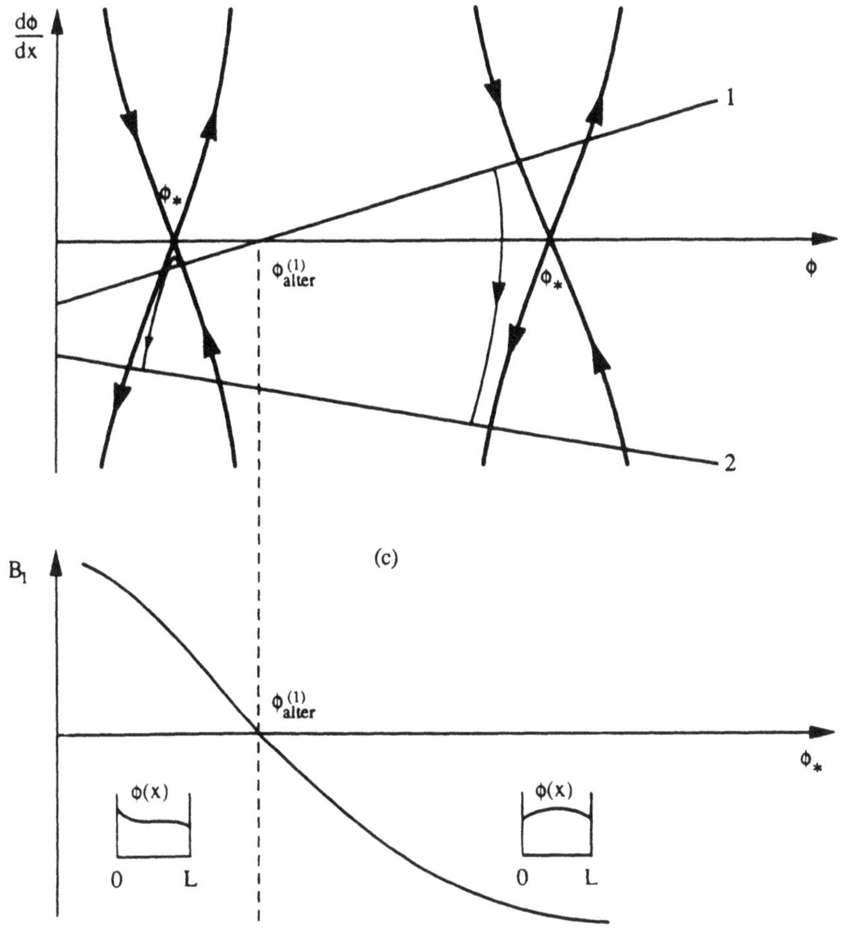

FIGURE 75c.

Notice that

$$\psi_o + \psi_1 + \ldots + \psi_l = 1 \quad (9.3.68)$$

Substituting Equation (9.3.68) into Equation (9.3.67) and excluding variable ψ_o we obtain

$$U = \sum_{\eta} \sum_{i,j=0}^{l} \left[- dC_{ij}\psi_i\psi_j + \frac{a^2}{2} C_{ij}(\nabla\psi_i, \nabla\psi_j) \right] + \text{Linear terms} \quad (9.3.69)$$

$$C_{ij} = \omega_{oj} + \omega_{oi} - \omega_{o0} - \omega_{ij} \quad (9.3.70)$$

Let us discuss the case of a flat interactive surface bounding the liquid. Suppose that the position of this boundary coincides with the plane x = 0 and the region $x \leq 0$ is occupied by the solid. Also suppose that there are no ϕ gradients in the directions transferse to the x-axis i.e., a one-dimensional problem is considered. To calculate the difference between the interaction energies of the liquid half-space in contact with the interactive and noninteractive surface, we replace a "liquid" monomer in the position x = 0 with a "solid" monomer. We obtain, using Equation (9.3.66), the following expression for interaction energy per unit of surface area,

$$
\begin{aligned}
U_{ss} &= \sum_{n=1}^{\infty} \sum_{i,j=0}^{l} \omega_{ij}\, \psi_i(na)\psi_j((n+1)a) + \sum_{i=0}^{l} \psi_i(a)\omega_{is} \\
&= \frac{1}{2} \sum_{n=1}^{\infty} \sum_{i,j=0}^{l} \omega_{ij}[\psi_i(na)\psi_j((n+1)a) + \psi_i(na)\psi_j((n-1)a)] \\
&- \frac{1}{2} \sum_{i,j=0}^{l} \omega_{ij}\psi_i(a)\psi_j(0) + \sum_{i=0}^{l} \psi_i(a)\omega_{is}
\end{aligned}
\qquad (9.3.71)
$$

where ω_{is} is the effective energy of interaction between the surface and a monomer of the i-th species. Note that the "solid properties" of the new monomer in the zero-th position of lattice are reflected in the linearity of the last summation of Equation (9.3.71). The energy contribution from the solid-liquid interaction is linear because the "solid" monomer occupies a position with probability 1, while the "liquid" monomer has an occupation probability of ψ_i.

The expression (9.3.71) can be transformed as follows

$$
\begin{aligned}
U_{ss} &= \sum_{n=1}^{\infty} \sum_{i,j=0}^{l} \left[\omega_{ij}\psi_i(na)\psi_j(na) + \frac{\omega_{ij}}{2} a^2\psi_i(na) \frac{d^2\psi_j}{dx^2}(na) \right] \\
&- \frac{1}{2} \sum_{i,j=0}^{l} \omega_{ij}\psi_i(a)\psi_j(o) + \sum_{i=0}^{l} \omega_{is}\psi_i(a) = \\
&\sum_{n=1}^{\infty} \sum_{i,j=0}^{l} \left[\omega_{ij}\psi_i(na)\psi_j(na) - \frac{\omega_{ij}a^2}{2} \frac{d\psi_i}{dx}(na) \frac{d\psi_j}{dx}(na) \right] \\
&- \sum_{i,j=0}^{l} \frac{\omega_{ij}}{2} a\psi_i(a) \frac{d\psi_i}{dx}(a) + \sum_{i=0}^{l} \psi_i(a)\omega_{is} - \\
&\frac{1}{2} \sum_{i,j=0}^{l} \left[\omega_{ij}\psi_i(a)\psi_j(a) - \omega_{ij}\psi_i(a)a \frac{d\psi_j}{dx}(a) \right]
\end{aligned}
$$

Here we have taken into account that

$$
\psi_i(a)\psi_j(0) = \psi_i(a)\psi_j(a) - a\psi_i(a) \frac{d\psi_j}{dx}(a) + O(a^2)
$$

As a result, we obtain

$$U_{ss} = \sum_{n}^{\infty} \sum_{i,j=0}^{l} \left(\omega_{ij}\psi_i\psi_j - \frac{\omega_{ij}}{2} \frac{d\psi_i}{dx} \frac{d\psi_j}{dx} \right)$$

$$+ \sum_{i=0}^{l} \psi_i(a)\omega_{is} - \frac{1}{2} \sum_{i,j=0}^{l} \omega_{ij}\psi_i(a)\psi_j(a) \qquad (9.3.72)$$

The above approach will be applicable in cases of an arbitrary form of solid surface, when the properties of the surface (curvature and interaction energies) change continuously so that the characteristic length scale of the changes is sufficiently larger than a. Combining Equations (9.3.67) and (9.3.72), we have

$$U = \sum_{i,j=0}^{l} \int_V \rho_V \left[d \cdot \omega_{ij}\psi_i\psi_j - \frac{\omega_{ij}a^2}{2} (\nabla\psi_i, \nabla\psi_j) \right] dV$$

$$+ \int_\Gamma \rho_\Gamma \left[\sum_{i=0}^{l} \omega_{is}\psi_i - \frac{1}{2} \sum_{i,j=0}^{l} \omega_{ij}\psi_i\psi_j \right] d\Gamma \qquad (9.3.73)$$

where ω_{ij} are constants, and ω_{ij} generally depend on the material properties; $\rho_v = 1/a^3$ and $\rho_\Gamma = 1/a^2$ are bulk and surface densities of monomers respectively.

The global free energy of the system is

$$\frac{F}{k_BT} = \int_V \rho_V \left[\sum_{i=0}^{l} \frac{\psi_i}{N_i} \ln \psi_i + \sum_{i,j=0}^{l} \left(d\omega'_{ij}\psi_i\psi_j - \frac{\omega'_{ij}a^2}{2} (\nabla\psi_i, \nabla\psi_j) \right) \right.$$

$$\left. - \sum_{i=0}^{l} \mu_i\psi_i \right] dV + \int_\Gamma \rho_\Gamma \left[\sum_{i=0}^{l} \omega'_{is}\psi_i - \frac{1}{2} \sum_{i,j=0}^{l} \omega'_{ij}\psi_i\psi_j \right] d\Gamma,$$

$$\omega'_{ij} = \frac{\omega_{ij}}{k_BT}, \quad \omega'_{is} = \frac{\omega_{is}}{k_BT} \qquad (9.3.74)$$

The first summation in the right-hand side of Equation (9.3.74) is the entropy density. N_i represents the degree of polymerization of the i-th component.[55] The quantity μ_i is the chemical potential of species i. Excluding the variable ψ_o and using Equations (9.3.69) and (9.3.70) we have an equivalent expression for F in terms of $\psi_1, \ldots, \psi_l$

$$\frac{F}{k_BT} = \int_V \rho_V \left\{ \frac{\left(1 - \sum_{i=1}^{l} \psi_i\right)}{N_o} \ln \left(1 - \sum_{i=1}^{l} \psi_i\right) + \sum_{i=1}^{l} \frac{\psi_i \ln \psi_i}{N_i} \right.$$

$$\left. + \sum_{i,j=1}^{l} \left[- dC'_{ij}\psi_i\psi_j + \frac{a^2}{2} C'_{ij} (\nabla\psi_i, \nabla\psi_j) \right] - \sum_{i=1}^{l} \mu_i\psi_i \right\} dV \qquad (9.3.75)$$

$$+ \int_\Gamma \rho_\Gamma \left[\sum_{i=1}^{l} (\omega'_{is} - \omega'_{os} - \omega'_{io} + \omega'_{oo})\, \psi_i + \frac{1}{2} \sum_{i,j=1}^{l} C'_{ij}\psi_i\psi_j \right] d\Gamma$$

Here $C'_{ij} = C_{ij}/k_BT$, and each chemical potential is renormalized according to the transformations carried out.

In the particular case of a binary mixture between two flat plates we can simplify Equation (9.3.75):

$$\frac{F}{k_BT} = \frac{1}{a}\int_0^L \left[\frac{\phi \ln \phi}{N_1} + \frac{(1-\phi)\ln(1-\phi)}{N_2} + \chi\phi(1-\phi) + \frac{\chi a^2}{6}\left(\frac{d\phi}{dx}\right)^2 - \mu\phi\right] dx + \left[-\beta^{(1)}\phi + \frac{\chi}{6}\phi^2\right]\Bigg|_{x=0} + \left[-\beta^{(2)}\phi + \frac{\chi}{6}\phi^2\right]\Bigg|_{x=L} \tag{9.3.76}$$

Here F is the free energy per unit of surface area and ϕ is the fraction of the first component,

$$\beta^{(m)} = \omega'_{2s} + \omega'_{12} - \omega'_{1s} - \omega'_{22}\,,\ m = 1,2 \tag{9.3.77}$$

$$\chi = 3(2\omega'_{12} - \omega'_{11} - \omega'_{22}) \tag{9.3.78}$$

Equation (9.3.78) defines the effective interaction constant for the liquid, which is usually positive.[55,56] The parameter $\beta^{(m)}$ depends on the interaction constants "liquid-liquid" and "liquid-solid" and in general can be of arbitrary sign; if the plates are not identical, the value of $\beta^{(m)}$ differs from one surface to the other.

In the case of a one-component compressible liquid, ϕ represents the dimensionless density. Since "holes" provide no contribution to the interaction energy, relationships (9.3.77) and (9.3.78) can be transformed as follows,

$$\beta^{(in)} = -\omega'_{1s}\,,\quad m = 1,2 \tag{9.3.79}$$

$$\chi = -3\omega'_{11} \tag{9.3.80}$$

The lattice model presented above provides a specific expression for the free energy functional, the general form of which is provided by Equation (9.3.4).

Let us estimate the value of the order parameter corresponding to changing the sign of structural force. Within the framework of the lattice model presented, we obtain from Equation (9.3.76)

$$\phi^{(m)}_{alter} = \frac{3\beta^{(m)}}{\chi}\,,\quad m = 1,2 \tag{9.3.81}$$

where $\phi^{(m)}_{alter}$ is the value of ϕ at which the structural force can change sign, the index m = 1,2 designates the m-th boundary. Substituting Equations (9.3.77) and (9.3.78) into Equation (9.3.81) we can rewrite Equation (9.3.81) in the form

$$\phi_{alter}^{(m)} = \frac{\omega_{2s}^{(m)} + \omega_{12} - \omega_{1s}^{(m)} - \omega_{22}}{2\omega_{12} - \omega_{11} - \omega_{22}}, \quad m = 1,2 \qquad (9.3.82)$$

If $\phi_{alter}^{(m)}$ does not belong to interval (0, 1) for all m (which is equivalent to the $d\gamma_m/d\phi$ sign conservation), the structural force sign cannot change within (0, 1). From Equation (9.3.76) we obtain the conditions of constant sign of structural force

$$\left(-\beta^{(m)} + \frac{\chi}{3}\right)\beta^{(m)} < 0, \quad m = 1,2 \qquad (9.3.83)$$

or after substituting Equations (9.3.77) and (9.3.78) into Equation (9.3.83)

$$(\omega_{1s}^{(m)} - \omega_{2s}^{(m)} + \omega_{12} - \omega_{11})(\omega_{1s}^{(m)} - \omega_{2s}^{(m)} + \omega_{22} - \omega_{12}) > 0$$

$$m = 1,2 \qquad (9.3.84)$$

If the conditions (9.3.83) do not hold, the structural force changes its sign.

In the particular case of a one-component compressible liquid, Equations (9.3.82) and (9.3.84) can be rewritten using Equation (9.3.79) and (9.3.80)

$$\phi_{alter}^{(m)} = \frac{\omega_{12}^{(m)}}{\omega_{11}} \qquad (9.3.85)$$

$$(\omega_{1s}^{(m)} - \omega_{11})\,\omega_{1s}^{(m)} > 0, \quad m = 1,2 \qquad (9.3.86)$$

In case of dominant Van-der-Waals attraction between solid and liquid ($\omega_{is}^{(m)} < 0$) we can rewrite Equation (9.3.86) in the form

$$\omega_{1s}^{(m)} < \omega_{11}, \quad m = 1,2 \qquad (9.3.87)$$

Since in this case both $\omega_{is}^{(m)}$ and ω_{11} are negative, Equation (9.3.87) gives

$$|\omega_{1s}^{(m)}| > |\omega_{11}|, \quad m = 1,2 \qquad (9.3.88)$$

i.e., when the attraction between the liquid-solid molecules of both surfaces is stronger than the attraction between the liquid-liquid molecules, the structural force conserves its sign. If the liquid-solid attraction is weaker than the liquid-liquid attraction, for at least one surface, one should expect a change in the sign of the structural force when changing the bulk liquid density.

The relationships obtained above can be simplified when the interactions between two molecules of the system amount mainly to the Van der Waals interaction which must be proportional to the product of the electron polarizabilities α_i and α_j of both molecules of kinds i and j

$$\omega_{ij} = -k_{VW}\alpha_i\alpha_j \tag{9.3.89}$$

where k_{vw} is a positive coefficient.[57] For example, after substituting Equation (9.3.89), Equation (9.3.82) becomes

$$\phi^{(m)}_{alter} = \frac{\alpha_s^{(m)} - \alpha_2}{\alpha_1 - \alpha_2}, \quad m = 1,2 \tag{9.3.90}$$

where $\alpha_s^{(m)}$ is the polarizability of a solid molecule belonging to the m-th surface, and α_j is the polarizability of a liquid molecule of kind j. From Equation (9.3.90) it follows that one can expect the structural force to change its sign when changing the bulk liquid composition when,

$$0 < \frac{\alpha_s^{(m)} - \alpha_2}{\alpha_1 - \alpha_2} < 1 \tag{9.3.91}$$

for at least one m, which is equivalent to $\alpha_s^{(m)}$ lying within the interval (α_1, α_2).

Alteration of Structural Force: Influence of Surface Roughness

Let us consider the structural force for the case of two surfaces whose materials are equivalent. As we showed above, the structural force for the system of two ideally smooth surfaces with a liquid interlayer does not change sign, even though it has a zero (Figure 74). The calculated force between two surfaces that are rough can be obtained from a generalization of the model presented earlier.

To understand the main idea let us consider a two-dimensional case (Figure 76). The ideal surface is plane, and all the "solid" monomers taking part in the interactions with liquid belong to a continuous straight line. Let us perturb the shape of the solid surface (dashed line). Within the framework of a lattice model this means that the "solid" monomers interacting with liquid will take new positions so that the dotted line is approximated by a piecewise continuous line. The positions of the "solid" surface monomers are denoted in Figure 76 by circles. The positions occupied by "solid" monomers which belong to the interior of the solid are shown by dots. The positions occupied by liquid are shown by crosses.

Assume that the characteristic scale of the inhomogeneity of the perturbed solid surface is essentially more than a. The additional solid-liquid interaction energy per monomer between the perturbed and unperturbed cases is the energy contribution from the additional surface area created (Figure 76). Averaging the

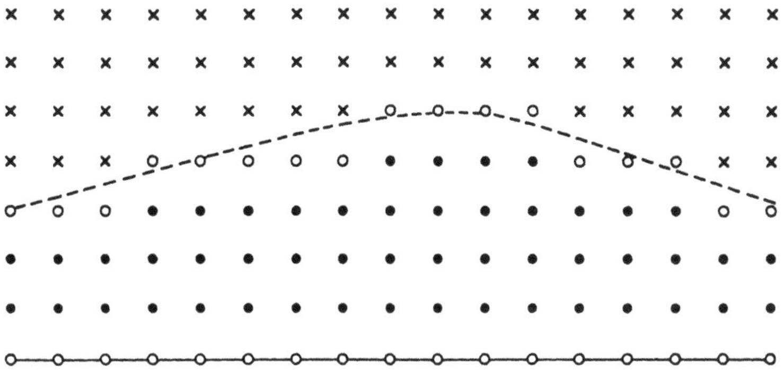

FIGURE 76. Local structure of surface roughness within the framework of a lattice model. Crosses correspond to the sites occupied by liquid; dots correspond to the sites occupied by solid monomers which do not take part in interactions with liquid; circles designate the solid monomers interacting with liquid (unperturbed and perturbed positions of the boundary are shown).

additional contribution over the number of interacting solid surface monomers we obtain, that ω_{is} value is replaced in the perturbed case to

$$\tilde{\omega}_{is} = \omega_{is}\left(1 + \left|\frac{\partial u}{\partial y}\right|\right) \tag{9.3.92}$$

where u is the deviation of the surface from the unperturbed level. A correction similar to Equation (9.3.92) appears in the theory for contact angles on rough surfaces.[47]

In a three-dimensional problem, u will be a function of two space coordinates on the unperturbed surface. Generalizing the two-dimensional consideration, we can show that the correction to ω_{is} will be found proportional to the modulus of grad u

$$\tilde{\omega}_{is} = \omega_{is}\,(1 + |\nabla u(y,z)|) \tag{9.3.93}$$

We would like to establish a region in ϕ over which the structural force is positive. Assuming that the distance between the surfaces is larger than the magnitude of the surface inhomogeneity, one can neglect the local changes of the distance between them in the above theory. From Equation (9.3.82) we find

$$(\Delta\phi_{alter})^2 = \left(\frac{\omega_{2s}^{(1)} - \omega_{2s}^{(2)} - \omega_{1s}^{(1)} + \omega_{1s}^{(2)}}{2\omega_{12} - \omega_{11} - \omega_{22}}\right)^2 \tag{9.3.94}$$

Substituting Equation (9.3.93) into Equation (9.3.94) and averaging over the spatial coordinates, y and z, we obtain in the case of two surfaces of the same material ($\omega_{is}^{(1)} = \omega_{is}^{(2)} = \omega_{is}$)

$$\langle(\Delta\phi_{alter})^2\rangle = \left(\frac{\omega_{2s} - \omega_{1s}}{2\omega_{12} - \omega_{11} - \omega_{22}}\right)^2 \langle(|\nabla u_1| - |\nabla u_2|)^2\rangle \quad (9.3.95)$$

where

$$\langle(|\nabla u_1| - |\nabla u_2|)^2\rangle = \frac{\int (|\nabla u_1| - |\nabla u_2|)^2 \, dy\,dz}{\int dy\,dz} \quad (9.3.96)$$

The mean value of ϕ_{alter} is

$$\langle\phi_{alter}\rangle = \phi_{alter} + \frac{\omega_{2s} - \omega_{1s}}{2\omega_{12} - \omega_{11} - \omega_{22}} \langle|\nabla u|\rangle$$

This differs from the ideal ϕ_{alter} value for a perfectly smooth surface.

If both surfaces have random heterogeneities so as to satisfy ergodicity, we can replace the space average in Equation (9.3.95) by a statistical average. For $|\nabla u_m|$ assumed to be random noise we obtain

$$\langle(|\nabla u_1| - |\nabla u_2|)^2\rangle = 2\langle|\nabla u|^2\rangle - 2\langle|\nabla u|\rangle^2 = 2\sigma^2_{|\nabla u|} \quad (9.3.97)$$

where $\sigma_{|\nabla u|}$ is the mean square deviation of $|\nabla u|$($|\nabla u_1|$ and $|\nabla u_2|$ are assumed to have the same random properties).

Combining Equations (9.3.95), (9.3.97), and (9.3.82) we have

$$\frac{\langle(\Delta\phi_{alter})^2\rangle}{\phi^2_{alter}} = 2\sigma^2_{|\nabla u|}\left(\frac{\omega_{2s} - \omega_{1s}}{\omega_{2s} - \omega_{1s} + \omega_{12} - \omega_{22}}\right)^2 \quad (9.3.98)$$

In the case of a one-component compressible liquid Equation (9.3.98) simplifies to,

$$\frac{\langle(\Delta\phi_{alter})^2\rangle}{\phi^2_{alter}} = 2\sigma^2_{|\nabla u|}$$

As an example, let us consider the regular surface inhomogeneity in form of the one-dimensional billow set with period T_{sur} and magnitude A_{sur}

$$u = A_{sur} \sin\left(\frac{2\pi x}{T_{sur}}\right)$$

Equation (9.3.96) gives

$$\sigma^2_{|\nabla u|} = \frac{A^2_{sur}}{T^2_{sur}} 2(\pi^2 - 8)$$

i.e., the ratio of "window" scale to the ϕ_{alter} value is of the same order that the ratio of the magnitude is to the period of surface fluctuations.

Thus, within the framework of the theory, the rough surfaces give rise to either monotonic or nonmonotonic ϕ profiles, depending on the local structure of both surfaces. One can calculate the disjoining pressure between two rough surfaces using the one-dimensional solution (9.3.60) with correction (9.3.93) and averaging it over y and z. Correspondingly, the integral disjoining pressure should be defined by its local, in every (y, z), values; the latter can be at random either positive or negative.

Concluding Remarks

Thus, we have used the gradient theory to provide a complete description of the interlayer structure of a fluid between two dissimilar surfaces. The dependences of disjoining pressure, disjoining potential, and the order parameter profiles in the interlayer region are derived as functions of interlayer thickness and fluid composition.

A new analytical method is suggested, which enables us to obtain the asymptotically exact dependences of $\Pi(L)$ and $G(L)$ for any arbitrary form of $W(\phi)$. It is shown that Π and G demonstrate an exponential dependence on film thickness (L), when L sufficiently exceeds the molecular scale.

Two possible types of interlayer structures are observed. For the first type, the order parameter distribution within the interlayer is monotonic. For the second type, it has an extremum in the interlayer. For the structures of first type, P and Π first increase, then decrease with decreasing L. For the structures of second type, P and Π decrease monotonically with decreasing L. For the structures of first type, Π and G are positive at large L; for the structures of second type, they are negative.

The decay length for $\Pi(L)$ is the correlation radius ξ of the bulk liquid. In general $\Pi(L)$ is represented by series of exponents with the decay lengths $2\xi/n$, $n = 2, 3, \ldots$.

The preexponent term B_1 in $\Pi(L)$ depends on the deviation of $W(\phi)$ from a quadratic form and on the interaction parameters of both surfaces. Depending on the type of structure, Π can be either positive or negative, and can also change sign as the order parameter is changed.

A simple illustration of the theory is presented through a lattice model for a binary liquid mixture.

We did not consider the subject of how the results of this section should change in the case of nonlocal solid-fluid interactions. The generalized model, with the nonlocal solid-fluid potential, has the form

$$F = \gamma_1(\phi_1) + \gamma_2(\phi_2) + \int_0^L \left(W(\phi) + K\left(\frac{d\phi}{dx}\right)^2 - \mu\phi + \omega_1(x)\phi + \omega_2(L - x)\phi \right) dx$$

where $\omega_m(x)$ depends on the distance between the m-th surface and a given point of the interlayer, $\gamma_1(\phi) = \gamma_2(\phi)$ is the energy contribution because of the reduction of the energy of liquid-liquid interactions near each solid surface. In fact, in the consideration of this section we supposed that the functions $\omega_1(x)$ and $\omega_2(x)$ have sharp extremum at $x = 0$. Another limiting case corresponds to replacement of the real potential by a step function, whose dimension of the localization area is essentially larger than the L range of interest. Clearly, in this case the functions $\omega_m\phi$ do not depend explicitly on a spatial coordinate, i.e., they have the same form as $\mu\phi$. As a result, the disjoining pressure cannot be positive, even for two different solid surfaces, i.e., only the possibilities shown in Figures 73a and 73b can materialize. We can see that the change in the radius of solid-fluid interactions do change the interlayer structure and disjoining pressure. A similar question is discussed by de Gennes,[47] in connection with the consideration of second-order phase transitions in the wetting transition theory.

Recently, an extension of the approach described above was considered in order to take into account the van der Waals tail in solid-fluid interaction.† The main idea is to approximate the nonlocal interaction, $\omega_m(x)$, as a sum of a short-range (δ-function-like) contribution and a large-range contribution, which depends on L but does not depend on x. For a single-component compressible fluid, an asymptotic analysis shows that, when the van der Waals tail is nonzero, the structural part of the disjoining pressure has the following form at large L

$$\Pi_{st} = \frac{1}{2} W_{\phi\phi}(\phi_*) B\exp\left[-\frac{L}{\xi}\right] - L\frac{dD}{dL}\Delta\phi_{mean}$$

where

$$D(L) = D_1(L) + D_2(L), \Delta\phi_{mean} = \frac{1}{L}\int_0^L (\phi - \phi_*)\, dx$$

and $D_m(L) = \omega_m(L) \sim 1/L^3$ at large L.

One can see that for $D(L) = 0$, we recover Equation (9.3.60). Since the shift in the mean value of the order parameter in the film from its bulk value, $\Delta\phi_{mean}$, is proportional to 1/L (see Section 9.4 below), the Van der Waals contribution to the structural force varies as $1/L^4$. Thus, in a system with long-range solid-fluid interactions the power-law behavior should dominate at large separation distances; but at shorter (still much larger than the molecular diameter) separation distances, the exponential and power contributions become competitive. Depending on the strength of the power-law effect, the exponential behavior may or may not show up. Particularly, in numerical calculations[52] the exponential

† Mitlin, V. S. and Sanchez, I. C., Presented in the Thermodynamics and Phase Equilibria II Symposium at the Spring 1993 AIChE National Meeting (Houston).

behavior of the structural force was observed at film thicknesses of the order of 10 molecular diameters and the $1/L^4$ behavior dominated at L of the order of 20 molecular diameters.

9.4. GRADIENT THEORY AND PETROLEUM ENGINEERING

Based on the consideration of the previous section, let us discuss some points of the porous medium influence in the estimation of the petroleum reservoir resources and in well testing.

Deviation of Liquid Properties in a Narrow From Bulk Properties: One-Phase Case

Let us obtain the asymptotical expression of the order parameter ϕ averaged over the narrow thickness, for the case shown in Figure 73a. The interlayer thickness is

$$L = \sum_{m=1}^{2} \int_{\phi_o}^{\phi_m} \left(\frac{K}{W - \mu\phi - h} \right)^{1/2} d\phi \tag{9.4.1}$$

and the total amount of ϕ per unit of area is

$$\tilde{Q} = \sum_{m=1}^{2} \int_{\phi_o}^{\phi_m} \phi \left(\frac{K}{W - \mu\phi - h} \right)^{1/2} d\phi \tag{9.4.2}$$

From Equations (9.4.1) and (9.4.2) we find the mean ϕ value and its deviation from the bulk value

$$\phi_{mean} = \frac{\tilde{Q}}{L}, \quad \Delta\phi_{mean} = \phi_{mean} - \phi_* \tag{9.4.3}$$

Let us consider Equation (9.4.2) in the same way as Equation (9.3.19) in the previous section. Expanding the expression under the root in Equation (9.4.2) in a Taylor series near the value ϕ_o and retaining the three first terms of the series we have

$$\tilde{Q} = \left(\frac{2K}{W_{\phi\phi}(\phi_o)} \right)^{1/2} \cdot \left[\sum_{m=1}^{2} \int_{0}^{\phi_m - \phi_o} \frac{(\phi + \phi_o)\, d\phi}{(b\phi + \phi^2)^{1/2}} + Y \right] \tag{9.4.4}$$

where b was defined by Equation (9.3.34), and Y will be discussed further.

After calculation of the integrals in the right side of Equation (9.4.4) we obtain

$$\frac{\tilde{Q}}{L} = \frac{1}{L}\left(\frac{2K}{W_{\phi\phi}(\phi_o)}\right)^{1/2} \cdot \left\{\sum_{m=1}^{2} [(\phi_m - \phi_o)^2 + b(\phi_m - \phi_o)]^{1/2}\right.$$

$$+ Y + \left(\phi_o - \frac{b}{2}\right) \sum_{m=1}^{2} \left[\ln\left(\phi_m - \phi_o + \frac{b}{2} + ((\phi_m - \phi_o)^2 + b(\phi_m - \phi_o))^{1/2}\right)\right.$$

$$\left.\left. - \ln\left(\frac{b}{2}\right)\right]\right\} = \frac{1}{L}\left(\frac{2K}{W_{\phi\phi}(\phi_o)}\right)^{1/2} \cdot \left\{\sum_{m=1}^{2} [(\phi_m - \phi_o)^2 + b(\phi_m - \phi_o)]^{1/2}\right.$$

$$\left. + Y - \left(\phi_o - \frac{b}{2}\right) Z + \left(\phi_o - \frac{b}{2}\right) L\right\} \tag{9.4.5}$$

The final expression in Equation (9.4.5) is carried out by taking into account Equation (9.3.35) for L.

Now, for quite large L, as in the previous section, one can replace ϕ_o, ϕ_1, and ϕ_2 to ϕ_*, ϕ_{1*}, and ϕ_{2*} correspondingly. Taking into account that

$$\phi_o - \frac{b}{2} = \phi_* + O((\phi_o - \phi_*)^2)$$

we obtain from Equation (9.4.5)

$$\Delta\phi_{mean} = \frac{1}{L}\left(\frac{2K}{W_{\phi\phi}(\phi_*)}\right)^{1/2}\left[\sum_{m=1}^{2} |\phi_{m*} - \phi_*| + Y - \phi_* Z\right] + O\left(\frac{\phi_o - \phi_*}{L}\right) \tag{9.4.6}$$

where

$$Y - \phi_* Z = \sum_{m=1}^{2} \int_{\phi_*}^{\phi_{m*}} \frac{\left[\frac{1}{2} W_{\phi\phi}(\phi_*)(\phi - \phi_*)^2 - W(\phi) + \mu\phi + h_*\right] d\phi}{(W - \mu\phi - h_*)^{1/2}\left[(W - \mu\phi - h_*)^{1/2} + |\phi - \phi_*| \left(\frac{1}{2} W_{\phi\phi}(\phi_*)\right)^{1/2}\right]} \tag{9.4.7}$$

According to Equation (9.3.39) the precision of expression (9.4.6) is

$$O\left(\frac{\phi_o - \phi_*}{L}\right) = O\left(\frac{\exp(-L/2\xi)}{L}\right)$$

Considering by analogy the case shown in Figure 71b, we obtain

$$\Delta\phi_{mean} = \frac{1}{L}\left(\frac{2K}{W_{\phi\phi}(\phi_*)}\right)^{1/2}\left[-\sum_{m=1}^{2}|\phi_{m*} - \phi_*| + Y - \phi_* Z\right] + O\left(\frac{\exp(-L/2\xi)}{L}\right) \quad (9.4.8)$$

where Y − ϕ*Z is also defined by Equation (9.4.7).

Combining Equations (9.4.6), (9.4.8), and (9.3.43) one can write generally

$$\Delta\phi_{mean} = \frac{\xi}{L}(\phi_{1*} + \phi_{2*} - 2\phi_* + Y - \phi_* Z) + O\left(\frac{\exp(-L/2\xi)}{L}\right) \quad (9.4.9)$$

Notice now that the situation under consideration is directly connected with the problem of the estimation of the reservoir fluid density (composition) changing due to its contact with the surfaces of the grains forming the porous collector. If a given collector is characterized by a pore size distribution $\rho_{fr}(L)$ with a sharp maximum at the pore size value L_{fr}, one can average Equation (9.4.9) with the weight

$$\langle\Delta\phi_{mean}\rangle = \int \rho_{fr}(L)\Delta\phi_{mean}\, dL \quad (9.4.10)$$

Since Equation (9.4.9) is an asymptotical expression, Equation (9.4.10) is justified if the main change of $\rho_{fr}(L)$ is concentrated in a small vicinity of its sharp maximum. Particularly in the case of a sharp Gaussian distribution, Equation (9.4.10) transforms in the general approximation of the Laplace method up to Equation (9.4.9) by replacing L with L_{fr}; the next terms of the asymptotics also depend on the width of ρ_{fr}.

Deviation of Liquid Properties in a Narrow From Bulk Properties: Two-Phase Case

Both in this and in the previous section we still did not discuss the situation when a phase transition was possible (wetting transition phenomena).[46] To admit this possibility, one should consider the function W(ϕ) having inflection points, as in Figure 66. The original work of Cahn[46] will be discussed further in connection with the contact angles behavior in porous collectors. Here we consider a generalization of Cahn's analysis for the phase transition of a binary system in a narrow. The most interesting case (for petroleum engineering) is the case of two equivalent solid surfaces. We also will restrict ourselves considering the situation when one has the exact coexistence of two phases in the bulk liquid. All possible qualitatively different cases are considered below; we will not pay attention to the possible specific features of $W(\phi)$ behavior near the boundary of ϕ definition.

The first general case is presented graphically in Figure 77a. One can see that the possible interlayer structures correspond (for a given h) to phase trajectories AA′, BB′, CC′, and DD′. But not all of them are appropriate. Actually, the trajectory AA′ corresponds to decreasing the spatial variable x by going from point A to point A′ (see the previous section), and it cannot be selected as appropriate. Then, the curves BB′ and CC′ both fit regarding the selection rule. However, the curve CC′ does not fit in sense of stability; let us demonstrate that.

Actually, the energy of a equilibrium state is given by the following expression (see Equation (9.3.28))

$$G = \int_L^{\infty} (h_* - h)\, dL$$

For the phase curve containing the points B, B′, C, and C′, the corresponding h value is larger than h_*. The interlayer thickness L is larger for the structure CC′ than for BB′. This means that the structure CC′ possesses larger value of equilibrium energy than BB′, though both the structures have the same value of pressure $P = -h$. Thus, the transition of the system from state BB′ to state CC′ due to removing the plates leads to an increase of the equilibrium energy. Consequently, the state CC′ is unstable.

As a result we have two appropriate solutions, BB′ and DD′. For the case of a one-component compressible liquid, the first of them corresponds to the gaseous phase, the second corresponds to the liquid phase. Notice that the solutions correspond to the same h but to different L. This means that the liquid and gaseous solutions with the same L should correspond to the different phase pressures P.

The two-phase situation seems to be essentially more complicated compared with the one-phase one. But in fact this is only a quantitative complication, and one can apply the asymptotical method of the previous section to calculate G_{w-}, G_g, P_w, P_g, $\Delta\phi_{mean,w}$, and $\Delta\phi_{mean,g}$ (the index "w" relates to the liquid phase, index g relates to the gaseous one). The reason for the simplification is that the different solutions (liquid, gas) of the variational problem can be considered independently. As a result we obtain

$$P_w(L) = P_* - \frac{1}{2} W_{\phi\phi}(\phi_{*,w}) B_w \exp(-L/\xi_w) \tag{9.4.11}$$

$$P_g(L) = P_* - \frac{1}{2} W_{\phi\phi}(\phi_{*,g}) B_g \exp(-L/\xi_g) \tag{9.4.12}$$

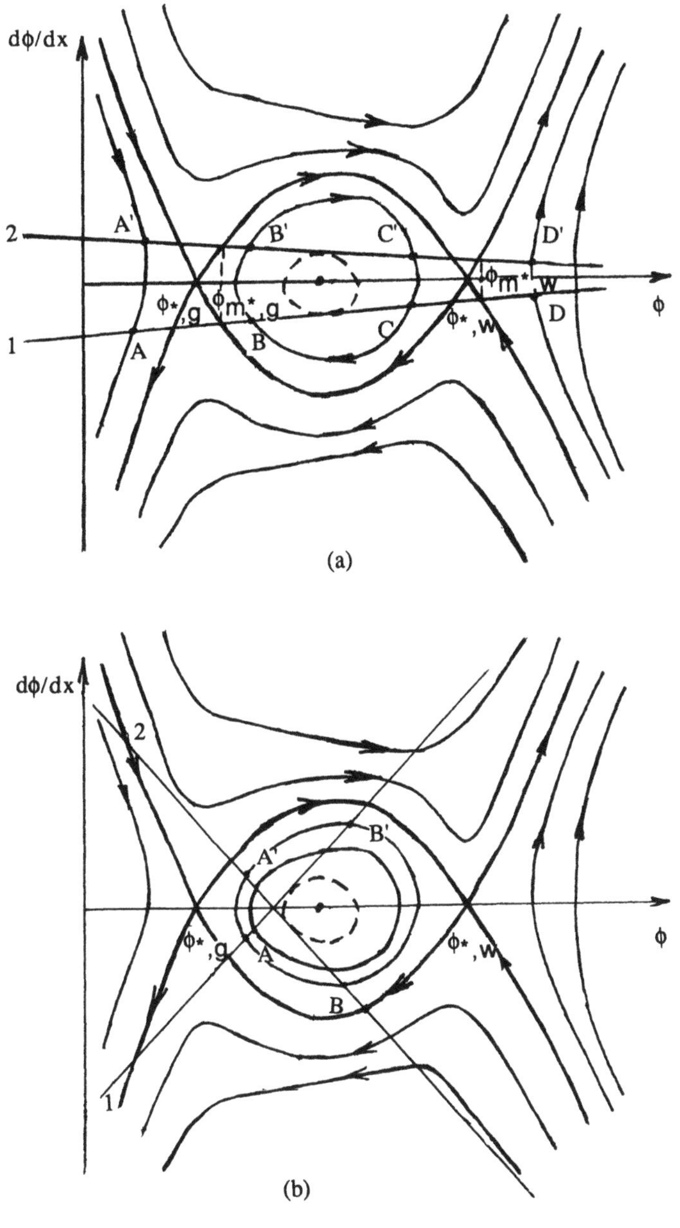

FIGURE 77. Phase transition of a binary liquid system in a narrow; a, b, c: possible qualitatively different types of phase planes.

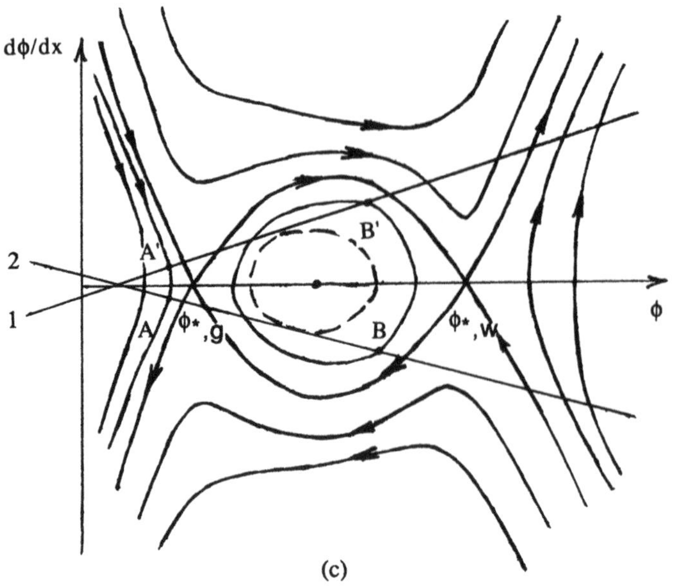

FIGURE 77c.

where $\xi_{w,g}$ and $B_{w,g}$ are given by Equations (9.3.43) and (9.3.40), so that one should replace ϕ_* and ϕ_{m*} in Equation (9.3.40) to $\phi_{*,w}$ and $\phi_{m*,w}$ for Equation (9.4.11), and to $\phi_{*,g}$ and $\phi_{m*,g}$ for Equation (9.4.12), (see denotations in Figure 77a). The equations of the previous section are written for the common case of different $\gamma_m(\phi)$. In the case under consideration we have just $\phi_{1*,w} = \phi_{2*,w}$ and $\phi_{1*,g} = \phi_{2*,g}$.

We can also estimate the deviation of ϕ_{mean} from bulk value for both phases

$$\Delta\phi_{mean,w} = \frac{\xi_w}{L}[\phi_{1*,w} + \phi_{2*,w} - 2\phi_{*,w} + Y - Z\cdot\phi_{*,w}] \quad (9.4.13)$$

$$\Delta\phi_{mean,g} = \frac{\xi_g}{L}[\phi_{1*,g} + \phi_{2*,g} - 2\phi_{*,g} + Y - Z\cdot\phi_{*,g}] \quad (9.4.14)$$

and the expression (9.4.7) for $Y - \phi_* Z$ transforms by replacing ϕ_* and ϕ_{m*} either to $\phi_{*,w}$ and $\phi_{m*,w}$ (in the case of the liquid phase), and to $\phi_{*,g}$ and $\phi_{m*,g}$ (in the case of the gaseous phase). The expressions (9.4.13) and (9.4.14) derive the change of the phase diagram for a two-phase system in a narrow with thickness L.

We can see from Figure 77a that for quite large L we have two phases. When L decreases, there is a special L_{sp} value, at which the gaseous phase disappears (that corresponds to touching the curves 1 and 2 and the trajectory shown by the dashed line in Figure 77a). At $L < L_{sp}$ one has a one-phase (liquid) state.

One can also average Equations (9.4.11) to (9.4.14) over the pore size distribution. For a Gaussian form of the distribution with a narrow peak at the pore size L_{fr}, the quantity L in Equations (9.4.11) to (9.4.14) only has to be replaced to L_{fr}.

The second case corresponds to the disposition of $d\gamma/d\phi$ zero between $\phi_{*,w}$ and $\phi_{*,g}$ (Figure 77b). In this case the appropriate solutions correspond to the phase trajectories AA′ and BB′. Again, at large L we have two phases. When L decreases, there is a special L_{sp} value, at which one phase disappears. But this can be either a liquid or gaseous phase. The liquid phase disappears, if $d\gamma/d\phi$ zero lies to the left of the vortex point of the phase plane. If it lies to the right of the vortex point, the gaseous phase disappears by decreasing L.

The third case is shown in Figure 77c. In this case the appropriate solutions correspond to the phase trajectories AA′ and BB′. At large L we have two phases. When L decreases, there is a special L_{sp} value, at which the "liquid" phase disappears. At $L < L_{sp}$ we have a one-phase (gaseous) state.

It is necessary to note that the value L_{sp} for the cases shown in Figures 77a and 77c can be infinite. That would be true if line 1 has no points of intersection with the separatrix between $\phi_{*,w}$ and $\phi_{*,g}$. One can see that at certain relationships between the energies of interaction liquid-liquid and liquid-solid the reservoir system can be in a one-phase state inside the pore channels with quite an arbitrary size, though the corresponding bulk system is surely in a two-phase state.

The profiles of the phase structures in the case shown in Figure 77a are both convex downward. The profile of liquid structure in the case shown in Figure 77b is convex upward, the corresponding profile of the gaseous structure is convex downward. The profiles of the phase structures in the case shown in Figure 77c are both convex upward.

In the case shown in Figure 77a the quantity $\Delta\phi_{mean}$ is positive for both phases. In the case shown in Figure 77b the quantity $\Delta\phi_{mean}$ is positive for the gaseous phase and negative for the liquid one. In the case shown in Figure 77c the quantity $\Delta\phi_{mean}$ is negative for both phases. Thus, the phase diagram can change in different ways depending on the interaction constants of solid and fluid.

Analogously, the consideration of the two-phase situation can be carried out in the general case of two different solid surfaces. But in any event the phase pressures should obey the exponential law of the kind in Equations (9.4.11) and (9.4.12), and the deviations of the mean ϕ phase values in a narrow from their bulk values should obey the 1/L law of the kind in Equations (9.4.13) and (9.4.14).[49,58]

Mobility of Phases of a Reservoir Mixture Near the Critical Point

Here we will show how the gradient theory can be used for estimation of the phase mobilities of a reservoir system in the near-critical region.

Let us consider first some main concepts. If a flat surface of a solid phase "s" contacts two fluid phases (for example, liquid phase "w" and gaseous

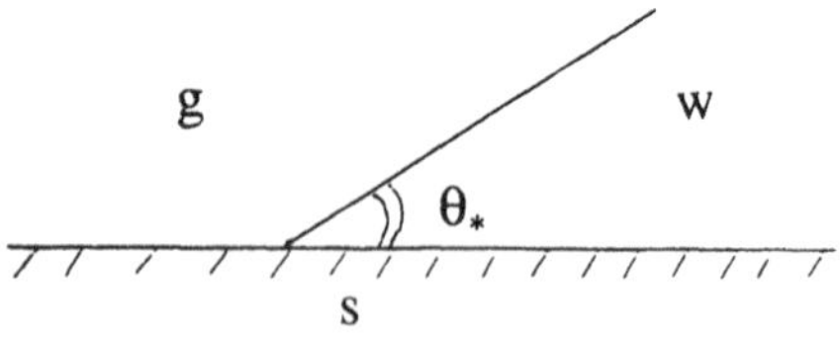

FIGURE 78. Region of contact of two fluid phases and a solid surface.

phase "g"), the contact angle θ_* (the angle between the interface surface of fluid phases and the solid surface) is derived from the Young equation

$$\sigma_{sg} - \sigma_{sw} - \sigma_{gw} \cos\theta_* = 0 \tag{9.4.15}$$

Here σ_{mn} is the free energy of the interface surface of phases m and n (Figure 78). Clearly, the solution of Equation (9.4.15) does not always exist. Actually, when the inequality

$$\sigma_{gw} < |\sigma_{sg} - \sigma_{sw}| \tag{9.4.16}$$

holds, the solution of Equation (9.4.15) is absent. This means that in the case of holding the condition (9.4.16) one of the fluid phases wets the solid completely, and the second fluid phase has no contact with the solid. If Equation (9.4.16) does not hold, one has a finite contact angle so that one can visually observe the drops of a more wettable phase contacting the solid surface and surrounded by a less wettable phase which also contacts the solid.

The following consideration is based on the theory of phase transitions in a fluid contacting a solid (wetting transition). In 1973, Heady and Cahn reported a new surface phenomenon.[59] They discovered experimentally that the contact angle near the critical point of a binary two-phase fluid system vanished before the critical point was reached, so that a thin film of a more dense fluid phase formed above a less dense fluid phase (density inversion). The consequent theory of Cahn has explained this effect.[46]

The description of a wetting transition in the system "fluid half-space-solid" is based on the minimization of the free energy functional in form (9.3.3). The quantity F is the excess free energy which appears as a result of contact between fluid and solid. The function $\gamma(\phi_1)$ in Equation (9.3.3) is the interaction energy of fluid and solid. Based on Cahn's theory,[46] we will discuss below the case of an absence of $d\gamma/d\phi$ zeroes within the region of ϕ definition.

Let us discuss the situation of exact coexistence of phases 1 and 2 so that the function $W(\phi)$ in Equation (9.3.3) is always positive except for vanishing at two points of minimum. After minimization of Equation (9.3.3) we obtain

$$W(\phi) = K\left(\frac{d\phi}{dx}\right)^2 \tag{9.4.17}$$

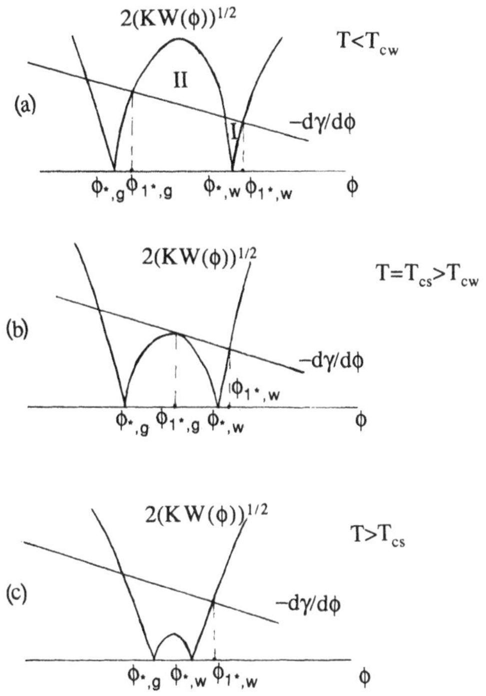

FIGURE 79. Cahn's graphical construction in the wetting transition theory. (Adapted from J. W. Cahn, *J. Chem. Phys.*, 66, 3667, 1977).

with the boundary condition at the point $x = 0$

$$\frac{d\gamma}{d\phi_1} = -2(KW(\phi_1))^{1/2} = 2K\frac{d\phi}{dx}\bigg|_{x=0} \tag{9.4.18}$$

Graphic representation of solution (9.4.18) is given in Figure 79. The situation taking place at the temperature T, essentially less than the critical temperature T_{cr}, is presented in Figure 79a. Equation (9.4.18) has four roots. The smallest one does not satisfy the selection rule discussed in Section 9.3. One of the remaining three roots (intermediate one) corresponds to the solution (9.4.17) which provides the free energy maximum and is therefore unstable. The two appropriate roots of Equation (9.4.18) are designated as $\phi_{1*,g}$ and $\phi_{1*,w}$ in Figure 79a. The first corresponding solution decreases from $\phi_{1*,w}$ at $x = 0$ to $\phi_{*,w}$ at $x \to \infty$. The second one decreases from $\phi_{1*,g}$ at $x = 0$ to $\phi_{*,g}$ at $x \to \infty$. The equilibrium energies of these two states are

$$\sigma_{sg} = \gamma(\phi_{1*,g}) + 2\int_{\phi_{*,g}}^{\phi_{1*,g}} (KW(\phi))^{1/2}\, d\phi,$$

$$\sigma_{sw} = \gamma(\phi_{1*,w}) + 2\int_{\phi_{*,w}}^{\phi_{1*,w}} (KW(\phi))^{1/2}\, d\phi \tag{9.4.19}$$

and the interfacial energy[18] of the liquid-gas system

$$\sigma_{gw} = 2\int_{\phi_{*,w}}^{\phi_{*,g}} (KW(\phi))^{1/2}\, d\phi \tag{9.4.20}$$

Let us define the spreading coefficient,[47]

$$\text{Spr} = \sigma_{sg} - \sigma_{sw} - \sigma_{gw}$$

The negativity of Spr exactly corresponds to a finite value of the contact angle, the opposite situation corresponds to complete wetting. Compare this with Equations (9.4.15) and (9.4.16).

One can show that Spr equals to the difference between areas I and II in Figure 79a.[46] Clearly, at $T \ll T_{cr}$ the quantity $\text{Spr} < 0$, which is equivalent to the partial wetting. At a certain temperature T_{cw} the areas I and II become equal, and the contact angle vanishes.

The temperature region $T_{cw} < T < T_{cs}$ corresponds to the coexistence of two film fluid phases, the phase "w" completely wets the solid, and the contact angle is zero. This region was discussed in details by Teletzke et al.[48] for a more general mathematical model of the wetting transition. The temperature T_{cs} corresponds to touching the curves $-d\gamma/d\phi$ and $(2KW(\phi))^{1/2}$ and can be considered as the critical temperature of fluid in the presence of a solid (Figure 79b). As was shown,[48] T_{cw} depends, first, on the ratio of the interaction energy fluid-solid to the interaction energy fluid-fluid, and secondly on the ratio of the characteristic radius of molecular interaction fluid-fluid to the characteristic radius of molecular interaction fluid-solid. When one of the dimensionless parameters is sufficiently small, T_{cw} is close to T_{cr}.

In the temperature region $T > T_{cs}$ only one solution of Equation (9.4.18) exists (Figure 79c). Notice that the one-phase state is reached at a lower temperature than T_{cr}.

Thus, there are three main regimes of wetting the solid by a two-phase fluid. The partial wetting corresponds to the temperatures $T < T_{cw}$, the complete wetting and the film phases coexistence correspond to the interval $T_{cw} < T < T_{cs}$, and the one-phase state corresponds to $T > T_{cs}$ ($T_{cs} < T_{cr}$).

More general models, taking into account the nonlocal interactions between solid and liquid, have been investigated.[48,60] The results of these papers confirm the issue about the existence of a special region disposed at the two-phase region of the phase plane, where a complete wetting takes place. This enables us to look from another point of view at the process of compositional reservoir flow in collectors with a low porosity.

As is known, a macroscopic motion of a phase of a multiphase fluid is possible, if the drops of the phase contact each another and form a connective cluster in the porous space. We considered in section 5.3 the phase mobilities behavior near the critical point, provided that the porous medium influence was small enough. From the consideration of this section one can see that, first, the critical point of the reservoir mixture itself can be shifted essentially due to the influence of a solid surface (T_{cr} is replaced to T_{cs}). Secondly, at $T_{cw} < T < T_{cs}$ the reservoir system does not need to reach any non-zero residual saturation value regarding a phase, for the phase would become mobile. This region of parameters corresponds to the coexistence of film fluid phases, and the residual saturations there should be equal to zero. Actually, a phase can move in its own continuous film; in fact there is no lower limit of the pore volume which should be occupied by the film in order to provide the motion of the phase. We then arrive at the following interesting deduction: for the reservoir systems with strong pore surface influence one should expect that by approaching the critical point the residual saturations tend to zero (which corresponds to $T = T_{cw}$ in the above consideration), but the remaining properties of phases still can be quite far each from another. Then, by further approaching the critical point the form of phase permeabilities changes so that they should be proportional to phase saturations.

Such a scenario enables us to suggest a reasonable explanation of the temporary productivity increase of a gas-condensate well when the reservoir pressure becomes less than the pressure of the condensation beginning. This strange phenomenon is well known by reservoir engineers, and it is not in accordance with a possible description within the framework of the usual concept of phase permeabilities. Actually, after getting the system into the two-phase region of parameters one should expect the appearance of drops of the new (liquid) phase. They are disposed separately to each another as long as the reservoir conditions are close to the point of phase transition; therefore, the corresponding phase should be immobile. As a result, one could expect a productivity decrease at a well just because of decreasing the amount of moving fluid in the flow at the same pressure gradient. However, if the parameters of the reservoir system correspond to the film phases coexistence, then forming the single drops of the new phase is a thermodynamically disadvantageous process. It is more advantageous to form the fluid film structure inside the system of pore channels. The motion in the film phases should provide an essential increase in the well productivity compared to the reservoir system behavior expected in the framework of the usual phase permeabilities concept (Figure 80).

One can estimate how the finiteness of the two-phase fluid system changes the region of parameters corresponding to a complete wetting. Let ϕ mean the density, i.e., one has the coexistence of liquid and gas. As was mentioned above, the transition from partial to complete wetting corresponds to vanishing the spreading coefficient, Spr, which is negative for partial wetting. Using Young's Equation (9.4.15) and Spr definition, and applying Equations (9.3.28) and (9.3.29) to both phases, one can derive the excess (L-depending part) of Spr,

$$\Delta \mathrm{Spr} = \frac{1}{2}\int_{L}^{\infty} (P_g - P_w)\, dL \qquad (9.4.21)$$

or the excess (L-depending part) of $\cos\theta_*$

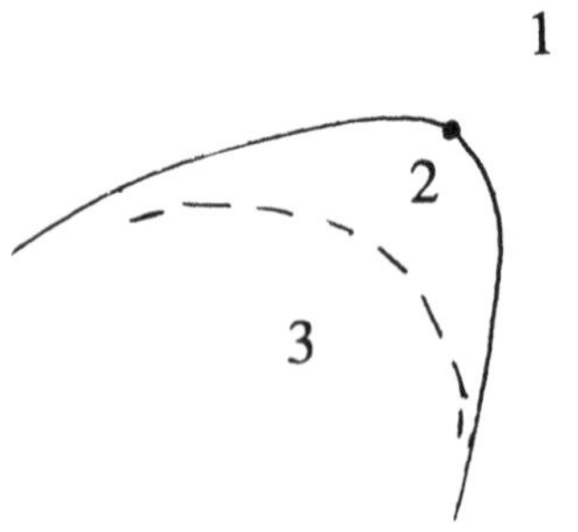

FIGURE 80. Qualitative picture of thermodynamic behavior of fluid in porous media based on Cahn's theory. Continuous line is boundary 1 of a one-phase state at the phase diagram of a reservoir fluid (influence of porous medium is taken into account). Region 2 corresponds to coexistence of film phases. The dashed line is the boundary of the region of partial wetting, region 3.

$$\Delta(\cos\theta_*) = \frac{1}{2\sigma_{gw}} \int_L^\infty (P_g - P_w)\, dL$$

Substituting Equations (9.4.11) and (9.4.12) into Equation (9.4.25), taking into account the positivity of the product $W_{\phi\phi}(\phi_{*,g(w)})B_{g(w)}$ in Equations (9.4.11) and (9.4.12), and using the obvious inequality,

$$\xi_w > \xi_g$$

(the correlation of molecules in the liquid phase is stronger than in the gaseous) one can see that ΔSpr is positive. Thus, a decrease of the separation distance between two equivalent solid plates can lead to the wetting transition with the formation of film phases, even if one had partial wetting at $L = \infty$, and the region of the film phases coexistence should increase for a confined system (such as fluid between two solid grains of a porous medium), compared to an unbounded fluid system contacting a solid.

An approximate estimation of the phase permeabilities shape for such a film flow region of parameters can be suggested based on the exact solution of the steady-state flow problem in a cylindrical channel, for the piece-constant shape of fluid parameters along the radial coordinate. To determine the region of the phase diagram where the residual saturations vanish, it is necessary to carry out measurements of contact angles for systems "a multiphase fluid — a solid" near the critical point of the fluid.

The system "saturated porous medium + free fluid in a well" is unique because the reservoir fluid can move through the porous medium inside film phases or just can be there in the one-phase state, though the same fluid in the well acquires the usual multiphase properties. The difference of the properties of the fluid in a porous medium and in bulk conditions is connected with the reservoir solid-fluid interactions which seem to be the cause of many unusual phenomena appearing at the early stage of development of oil-gas-condensate reservoirs.

REFERENCES

1. **Lifshitz, I. M. and Slyozov, V. V.,** *Zh. Eksp. Teor. Fiz.*, 35, 479, 1958; *Sov. Phys. JETP,* 8, 331, 1959.
2. **Volmer, M.,** *Kinetik der Phasenbilding,* Steinkopf, Dresden, 1939.
3. **Cahn, J. W.,** *Acta Metall.*, 9(9), 795, 1961.
4. **Mitlin, V. S.,** *Zh. Eksp. Teor. Fiz.*, 95, 1826, 1989; *Sov. Phys. JETP,* 68, 1056, 1989.
5. **Scripov, V. P. and Scripov, A. V.,** *Usp. Fiz. Nauk,* 128, 193, 1979; *Sov. Phys. Usp.*, 22, 389, 1979.
6. **Mitlin, V. S., Manevich, L. I., and Erukhimovich, I. Ya.,** *Zh. Eksp. Teor. Fiz.*, 88, 495, 1985; *Sov. Phys. JETP,* 61, 290, 1985.
7. **Langer, J. S.,** *Ann. Phys.*, 65, 53, 1971.
8. **Fedoryuk, M. V.,** *Asymptotics: Integrals and Series,* Nauka, Moscow, 1987 (in Russian).
9. **Huston, E. L., Cahn, J. W., and Hilliard, J. E.,** *Acta Metall.*, 14, 1053, 1966.
10. **Khisina, N. R.,** *Subsolidus Transformations of Solid Solutions of Rock-Forming Minerals,* Nauka, Moscow, 1987 (in Russian).
11. **Langer, J. S., Bar-on, M., and Miller, H. D.,** *Phys. Rev. A,* 11, 1417, 1975.
12. **Ustinovshchicov, Yu. I.,** *Precipitation of a Second Phase in Solid Solutions,* Nauka, Moscow, 1988 (in Russian).
13. **Elder, K. R. and Desai, R. C.,** *Phys. Rev. B,* 40, 243, 1989.
14. **Marro, J., Bortz, A. B., Kalos, M. H., et al.,** *Phys. Rev. B,* 12, 2000, 1975.
15. **Binder, K.,** *Phys. Rev. B,* 15, 4425, 1977.
16. **Vasil'ev, V. A., Romanovskii, Yu. M., and Yakhno, V. G.,** *Autowave Processes,* Nauka, Moscow, 1987 (in Russian).
17. **Cahn, J. W.,** *Acta Metall.*, 10, 179, 1962.
18. **Cahn, J. W. and Hilliard, J. E.,** *J. Chem. Phys.*, 28, 258, 1958.
19. **Landau, L. D. and Lifshitz, E. M.,** *Statistical Physics. Part 1,* 3rd ed.; Pergamon Press, New York, 1980.
20. **Ma, Shang-Keng,** *Modern Theory of Critical Phenomena,* W. A. Benjamin, Readind, MA, 1976.
21. **Mitlin, V. S.,** *Fiz. Zemli,* 1, 46, 1990; *Physics of the Solid Earth,* 26(1), 31, 1990.
22. **Haken, G.,** *Advanced Synergetics. Instability Hierarchies of Self-Organizing Systems and Devices,* Springer-Verlag, Berlin, 1983.
23. **Mitlin, V. S.,** *Zh. Eksp. Teor. Fiz.*, 98(2), 4554, 1990; *Sov. Phys. JETP,* 71(2), 308, 1990.
24. **Gelfand, I. M.,** *Lectures on Linear Algebra,* 2nd ed., Interscience, New York, 1961.
25. **Putnis, A. and McConnel, J.,** *Main Features of Mineral Behavior,* Russ. ed., Mir, Moscow, 1983.
26. **Mazenko, G. F., Valls, O. T., and Zhang, F. C.,** *Phys. Rev. B,* 31, 4453, 1985.
27. **Lifshitz, I. M.,** *Zh. Eksp. Teor. Fiz.*, 42, 1354, 1961; *Sov. Phys. JETP,* 15, 939, 1962.
28. **Allen, M. and Cahn, J. W.,** *Acta Metall.*, 27, 1085, 1979.
29. **Mazenko, G. F., Valls, O. T., and Zannetti, M.,** *Phys. Rev. B,* 38, 520, 1988.
30. **Valls, O. T. anb Mazenko, G. F.,** *Phys. Rev. B,* 38, 1165, 1988.
31. **Oono, Y. and Puri, S.,** *Phys. Rev. A,* 38, 434, 1988.
32. **Mouritsen, O. G.,** *Phys. Rev. B,* 28, 3150, 1983.
33. **Mouritsen, O. G.,** *Computer Studies of Phase Transitions and Critical Phenomena,* Springer-Verlag, Berlin, 1984.
34. **Amar, J. G., Sullivan, F. E., and Mountain, R. D.,** *Phys. Rev. B,* 37, 196, 1988.
35. **Huse, D. A.,** *Phys. Rev. B,* 34, 7845, 1986.
36. **Rogers, T. M., Elder, K. R., and Desai, R. C.,** *Phys. Rev. B,* 37, 9638, 1988.
37. **Derjaguin, B. V. and Landau, L. D.,** *Acta Phys. Chim. USSR,* 14, 633, 1941.
38. **Derjaguin, B. V. and Landau, L. D.,** *Zh. Eksp. Teor. Fiz.*, 11, 802, 1941.
39. **Verwey, E. J. W. and Overbeek, J. Th. G.,** *Theory of the Stability of Lyophobic Colloids,* Elsevier, Amsterdam, 1948.

40. **Derjaguin, B. V. and Churaev, N. V.,** *J. Colloid Interface Sci.*, 49, 249, 1974.
41. **Peshel, G. and Belaschek, P.,** *Progr. Colloid Polym. Sci.*, 60, 108, 1976.
42. **Israelachvili, J. N. and Adams, G. E.,** *J. Chem. Soc. Faraday Trans. I*, 74, 975, 1978.
43. **Israelachvili, J. N. and Pashley, R.,** *Nature*, 300, 341, 1982.
44. **Marcelija, S. and Radic, N.,** *Chem. Phys. Lett.*, 42, 129, 1976.
45. **Ninham, B. W.,** *J. Chem. Phys.*, 84, 1423, 1980.
46. **Cahn, J. W.,** *J. Chem. Phys.*, 66, 3667, 1977.
47. **de Gennes, P. G.** *Rev. Mod. Phys.*, 57, 827, 1985.
48. **Teletzke, G. F., Scriven, L. E., and Davis, H. T.,** *J. Colloid. Interface Sci.*, 87, 550, 1982.
49. **Mitlin, V. S.,** *Int. J. Eng. Sci.*, in press; and **Sharma, M. M.,** *J. Colloid Intert. Sci.*, in press.
50. **Peng, D. Y. and Robinson, D. B.,** *Ind. Eng. Chem., Fundam.*, 15, 59, 1976.
51. **Arnol'd, V. I.,** *Catastrophe Theory*, Springer-Verlag, New York, 1984.
52. **Chan, D. Y. C., Mitchell, D. J., Ninham, B. W., et al.,** *J. Chem. Soc. Faraday Trans., 2*, 75, 556, 1979.
53. **Tanford, C.,** *The Hydrophobic Effect*, John Wiley & Sons, New York, 1973.
54. **Menezis, J. L., Yan, J., and Sharma, M. M.,** *Coll. Surf.*, 38, 365, 1989.
55. **Flory, P.,** *Principles of Polymer Chemistry*, Cornell University Press, Ithaca, NY, 1971.
56. **de Gennes, P.-G.,** *Scaling Concepts in Polymer Physics*, Cornell University Press, Ithaca, NY, 1979.
57. **Hiemenz, P. C.,** *Principles of Colloid and Surface Chemistry*, 2nd ed., Marcel Dekker, New York, 1986.
58. **Lipovski, R. and Gompper, G.,** *Phys. Rev. B*, 29, 5213, 1984.
59. **Heady, R. B. and Cahn, J. W.,** *J. Chem. Phys.*, 58, 896, 1973.
60. **Ebner, C. and Saam, W. F.,** *Phys. Rev. Lett.*, 38, 1486, 1977.

10

Conclusion: Actual Problems of the Theory of Multicomponent Reservoir Flow

In this final chapter the author would like to enumerate some essential questions of TMRF which require further work.

Role of Different Variables in Mathematical Description

Progress in the analytical studies of TMRF is connected with development in two general directions.

The first direction is to study the analytical solutions of the problems of reservoir mixture inflow to a well or a system of wells. Particularly, Khristianovich[1] considered the stationary solutions in the theory of reservoir flow of gasified liquid and showed that the flow structure is determined by the pressure only, and all the remaining variables can be expressed as some functions of pressure[1] (the generalization for the common case of multicomponent reservoir flow is trivial[2]). Gurevich and Nikolaevskii[3] showed that by the simulation of such processes as contact condensation, differential condensation, and the exploitation of a reservoir, one can trace back the three-component systems (and generally also the multicomponent ones) to a certain conditional binary one.[2,3] This means once again that the process is determined only by pressure. The nonstationary problem about the exploitation of a circular seam with an impermeable boundary has been considered.[4] The problem is reduced to solving a system of ordinary differential equations with pressure as the independent variable. This simplification is connected with the fact that the reservoir composition can be expressed through the pressure at the impermeable boundary of symmetric formation; see also the generalization for the nonsymmetric formation — the conditions of compatibility outlined in Chapter 3. Panfilov studied the self-similar problems of the nonstationary inflow of a multicomponent reservoir mixture to a well.[5] Using a natural small parameter and applying the asymptotical technique, he also showed that the process is determined only by pressure, and one can use the conditional quasi-binary systems. In Chapter 7 we considered the relaxation of fluctuations to the spatially homogeneous solution and showed that the pressure was the leading variable.

The second direction is connected with obtaining analytical solutions of the problems of reservoir mixture displacement by different active agents, and one can cite the remarkable progress in the investigations of this sort that has been achieved in the last few years. The construction of analytical solutions of the self-similar and some nonself-similar displacement problems in the theory of multicomponent reservoir flow also have been carried out.[6-9] However, these analytical achievements became possible because of the use of the equation of state in a degenerate form. Namely, the limitations of these results are connected with the supposition about the reservoir mixture incompressibility or about the conservation of the summa of partial component volumes. By such suppositions the total flow is invariant, and one can exclude one equation and obtain the reduced system of a hyperbolic type for the reservoir composition, which then can be studied using the methods of the theory of quasi-linear equations. Clearly, the pressure is excluded from the consideration. Though it is not quite clear how to combine the results[1-5] and[6-9] (particularly, how to combine the assumption of conservation of the summa of partial component volumes with the natural physical description within the framework of the equations of state of real multicomponent fluids), there are no doubts that the component fractions are leading variables by the description of many displacement processes (at least in those cases where the characteristic scale of pressure change is quite small). At the same time, requirement of the non-zero compressibility of reservoir mixture is necessary,[1-5] and solving the problems of inflow to a well leads to a contradiction if we are to use the main supposition of some investigations.[6-9] Some displacement problems with the variable total flow have been studied analytically.[10,11] However, in general, the point of the role of different variables in TMRF, particularly the role of the pressure in the mathematical models of displacement processes, is still open. It seems to be the point of hierarchy of the fast and slow variables in TMRF.

Critical Dynamics of Multicomponent Fluids in Porous Media

This subject has taken on concrete meaning in recent years. It is connected with the beginning of the development of deep oil-gas-condensate fields, the reservoir mixtures of which are in near-critical or over-critical conditions. As is known, the standard description of the properties of physical systems (liquids, gases, and magnetics) becomes incorrect near the critical point, and special methods of analysis are necessary.[12] The equations of motion of reservoir fluids in the near-critical conditions also should be specially obtained and studied.

Influence of Porous Media on Phase Equilibria and Reservoir Flow

We considered an approach to this problem in Chapter 9. This is the fundamental question, especially for the collectors with low porosity. The influence of porous media can lead to the shear of phase transition parameters compared to the fluid without porous media.[13,14] In "bad" situations one can obtain results which are wrong not only quantitatively but qualitatively (for example, for estimation of reservoir resources[15]). It is possible that the "bad" situations occur quite often, however we still are not able (or do not want) to analyze them at the contemporary stage of development of TMRF.

For collectors with low porosity and a complex reservoir system it is considered to be extremely important to accomplish the following program:

- Solve the one-dimensional plane-radial problem about the distribution of components in a capillary, depending on the bulk composition and pressure, temperature, and radius (some corresponding statements of the problem have been published[16,17]);
- Determine the change of phase diagram due to porous media influence by averaging the results over the pore size distribution for a given collector;
- Determine the region of parameters corresponding to the coexistence of film phases, and develop the TMRF for these conditions.

The first step in accomplishing the solution of such a problem should be by replacing the standard procedure of thermodynamic calculations, which uses a multicomponent equation of state, with the procedure of solving the equilibrium problem for a multicomponent fluid (with a given equation of state) contacting the solid wall with the interaction potential inherent to the material of a given porous collector. Fulfillment of such a program requires numerical solving of quite cumbersome systems of nonlinear ordinary differential equations, whose solution is often ambiguous. Nevertheless, this seems to be unavoidable because currently the significant part of the hydrocarbon resources is concentrated in deep, low-porous collectors.

REFERENCES

1. **Khristianovich, S. A.,** *Prikl. Mat. Mekh.*, 5(2), 277, 1941 (in Russian).
2. **Nikolaevskii, V. N.,** *Mechanics of Porous and Fractured Media,* Singapore World Pub., 1990.
3. **Gurevich, G. R. and Nikolaevskii, V. N.,** Taking into account the change of gas composition by the interpretation of results of gas-condensate well testing, in *News in Theory of Development and Exploitation of Gas and Gas-Condensate Formations,* Moscow, VNIIGAS, 1966, 28 (in Russian).
4. **Rozenberg, M. D., Shovkrinskii, G. Yu., Kurbanov, A. K., et al.,** Calculations of development of formations of light hydrocarbon mixtures, in *Scientific and Technical Collected Articles of Oil Production,* Moscow, VNIINEFT, 1966, 26, 80 (in Russian).
5. **Panfilov, M. B.,** *Izv. Akad. Nauk SSSR, Mekh. Zhid. Gaza,* 4, 94, 1985; *Fluid Dyn. (USSR),* 20(4), 574, 1985.
6. **Barenblatt, G. I., Entov, V. M., and Ryzhik, V. M.,** *Motion of Fluids Through Natural Rocks, Klumer Academic Publ.*, Dordrecht, 1990.
7. **Entov, V. M.,** *Usp. Mekh.*, 4(3), 41, 1981.

8. **Bedrikovetskii, P. G.,** *Dokl. Akad. Nauk SSSR,* 262(1), 49, 1982; *Sov. Phys. Dokl.,* 27(1), 14, 1982.
9. **Zazovskii, A. F.,** *Izv. Akad. Nauk SSSR, Mekh. Zhid. Gaza,* 5, 116, 1985; *Fluid Dyn. (USSR),* 20(5), 765, 1985.
10. **Ziskin, E. A. and Ryzhik, V. M.,** *Izv. Akad. Nauk SSSR, Mech. Zhid. Gaza,* 4, 66, 1981; *Fluid Dyn. (USSR),* 16(4), 541, 1981.
11. **Zazovskii, A. F.,** *Izv. Akad. Nauk SSSR, Mekh. Zhid. Gaza,* 5, 116, 1985; *Fluid Dyn. (USSR),* 20(3), 433, 1985.
12. **Ma, S.,** *Modern Theory of Critical Phenomena,* W. A. Benjamin, New York, 1976.
13. **Goh, M. C., Goldburg, W. I., and Knobler, C. M.,** *Phys. Rev. Lett.,* 58, 1008, 1987.
14. **Huang Yanzhand and Chzhu-Vai-Yao,** in Int. Conf. Development Gas-Condensate Fields, Sect. 6. Fundamental and Research Scientific Investigation, Rep., Krasnodar, 1990, 50.
15. **Tankayev, R. U.,** in Int. Conf. Development Gas-Condensate Fields, Sect. 6. Fundamental and Research Scientific Investigation, Rep., Krasnodar, 1990, 50.
16. **Teletzke, G. F., Scriven, L. E., and Davis, H. T.,** *J. Colloid Interface Sci.,* 87(2), 550, 1982.
17. **Tarazona, P.,** *Phys. Rev. A,* 31, 2672, 1985.

INDEX

G

H

I

J

K

L

M

N

O

P

Q

R

S

T

V

W

Y